7

Chemistry and Physics of
One-Dimensional Metals

NATO ADVANCED STUDY INSTITUTES SERIES

A series of edited volumes comprising multifaceted studies of contemporary scientific issues by some of the best scientific minds in the world, assembled in cooperation with NATO Scientific Affairs Division.

Series B: Physics

RECENT VOLUMES IN THIS SERIES

Volume 15 — Nuclear and Particle Physics at Intermediate Energies
edited by J. B. Warren

Volume 16 — Electronic Structure and Reactivity of Metal Surfaces
edited by E. G. Derouane and A. A. Lucas

Volume 17 — Linear and Nonlinear Electron Transport in Solids
edited by J. T. Devreese and V. E. van Doren

Volume 18 — Photoionization and Other Probes of Many-Electron Interactions
edited by F. J. Wuilleumier

Volume 19 — Defects and Their Structure in Nonmetallic Solids
edited by B. Henderson and A. E. Hughes

Volume 20 — Physics of Structurally Disordered Solids
edited by Shashanka S. Mitra

Volume 21 — Superconductor Applications: SQUIDs and Machines
edited by Brian B. Schwartz and Simon Foner

Volume 22 — Nuclear Magnetic Resonance in Solids
edited by Lieven Van Gerven

Volume 23 — Photon Correlation Spectroscopy and Velocimetry
edited by E. R. Pike and H. Z. Cummins

Volume 24 — Electrons in Finite and Infinite Structures
edited by P. Phariseau and L. Scheire

Volume 25 — Chemistry and Physics of One-Dimensional Metals
edited by Heimo J. Keller

The series is published by an international board of publishers in conjunction with NATO Scientific Affairs Division

A	Life Sciences	Plenum Publishing Corporation
B	Physics	New York and London
C	Mathematical and Physical Sciences	D. Reidel Publishing Company Dordrecht and Boston
D	Behavioral and Social Sciences	Sijthoff International Publishing Company Leiden
E	Applied Sciences	Noordhoff International Publishing Leiden

Chemistry and Physics of One-Dimensional Metals

Edited by
Heimo J. Keller

Anorganisch-Chemisches Institut der Universität
Heidelberg, Germany

PLENUM PRESS • NEW YORK AND LONDON
Published in cooperation with NATO Scientific Affairs Division

Library of Congress Cataloging in Publication Data

Nato Advanced Study Institute on Chemistry and Physics of One-Dimensional Metals,
   Bolzano (City), 1976.
   Chemistry and physics of one-dimensional metals.

   (NATO Advanced study institutes series: Series B, Physics; v. 25)
   "Published in cooperation with NATO Scientific Affairs Division."
   Includes index.
   1. One-dimensional conductors—Congresses. 2. Organometallic compounds—Con-
gresses. I. Keller, Heimo J., 1935-   II. North Atlantic Treaty Organization.
Division of Scientific Affairs. III. Title. IV. Series.
QC176.8.E4N33 1976   530.4   77-5135
ISBN 0-306-35725-9

Lectures presented at the NATO Advanced Study Institute on Chemistry and Physics
of One-Dimensional Metals held in Bolzano, Italy, August 17–29, 1976

PREFACE

   Theoretical and experimental work on solids with low-dimensional cooperative phenomena has mushroomed in the last few years, and
it seems to be quite fashionable to contribute to this field,
especially to the problem of one-dimensional metals. The huge amount
of recent investigations on quasi one-dimensional metals could be
divided into two parts although there is much overlap between these
regimes, namely investigations on organic materials such as TTF-
TCNQ and its derivatives or the TTT halides, and secondly research
on inorganic solids with the highly conducting polymer $(SN)_x$ and
transition metal complexes like KCP, platinum oxalates or the mixed
valence dicarbonylhaloiridates. There is, of course, much overlap
between these fields and in some cases theories have been tested on
both types of compounds simultaneously. In fact, however, most of
the scientific groups in this area could be associated roughly with
one of these categories and, in addition, a separation between
theoreticians and experimentalists in each of these groups leads to
a further splitting of interests. Furthermore, the small number of
compounds studied thoroughly so far suggests an "information gap"
between chemists and physicists which hampers the testing of new
materials. More cooperation and understanding between scientists
working on quasi 1d conductors should appreciably stimulate further
development. With a better interdisciplinary understanding, new
ideas could possibly help chemists in synthesizing tailor-made
solids. This would in return give experimentalists new phenomena to
examine and finally would stimulate new theoretical work.

   It was the purpose of the NATO ASI held in Bolzano, Italy, to
stimulate this cooperation and to develop a common "language", which
means that the theoretical background of solid state physics of
extremely anisotropic solids would be taught from the beginning to
chemists and experimental physicists. Furthermore, the experimental
physicists gave a detailed account of what they had done on "metallic" 1d solids and what type of experiments could be done in the
future, together with a mention of technical application of these
ideas. Finally, the preparative chemists gave an up-to-date summary
of the progress made in this field regarding the preparation of

tailor-made molecules and indicated what kind of compounds could
be prepared in the near future.

In several evening and weekend sessions some participants
presented summaries of their recent work and these and other new
results were discussed. A draft of these discussions could not be
added in printed form because of the limitations set by the total
page number of this volume, but to give at least an idea of the
problems touched upon during these sessions, a list of the main
contributors together with the title of the conribution discussed
is given as an appendix. The reader might contact these authors
directly if interested in special recent results.

I hope that the participants have profited from the meeting
and, furthermore, that at least some of the readers of the following
papers are stimulated to high-dimensional cooperative efforts on
low-dimensional conductive solids.

Primarily I have to thank NATO who made this project possible
through generous financial support. Especially I would like to mention gratefully the excellent cooperation with Dr. T. Kester of the
NATO Scientific Affairs Division, whose personal efforts helped in
the preparation and organization of the meeting. The Advanced Study
Institute could not have taken place without the efforts of Mrs.
G. Egerland who did all the correspondence, the typing of particu-
lars and most of the book-keeping. In addition the organizational
work of the Co-directors, Dr. D. Nöthe and Dr. H. Endres is thank-
fully acknowledged. The generosity of IBM Germany and National
Science Foundation travel grants awarded to three US-participants
contributed to the completeness of the school.

                                                    Heimo J. Keller
Heidelberg
November 1976

CONTENTS

The Basic Physics of One-Dimensional Metals   1
    V. J. Emery

The Organic Metallic State: Some Chemical Aspects   25
    D. Cowan, P. Shu, C. Hu, W. Krug,
    T. Carruthers, T. Poehler, and A. Bloch

The Organic Metallic State: Some Physical Aspects
            and Chemical Trends   47
    A. N. Bloch, T. F. Carruthers, T. O. Poehler,
    and D. O. Cowan

Charge Density Wave Phenomena in TTF-TCNQ and
            Related Organic Conductors   87
    A. J. Heeger

The Role of Coulomb Interactions in TTF-TCNQ   137
    J. B. Torrance

Electronic Properties of the Superconducting
            Polymer (SN)$_x$ 167
    R. L. Greene and G. B. Street

Collective States in Single and Mixed Valence
            Metal Chain Compounds   197
    P. Day

The Structure of Linear Chain Transition Metal
            Compounds with 1d Metallic Properties   225
    K. Krogmann

The Chemistry of Anisotropic Organic Materials   233
    F. Wudl

Superconductivity and Superconducting Fluctuations   257
    W. A. Little

Instabilities in One-Dimensional Metals   267
     H. Gutfreund

Physical Considerations and Model Calculations
          for One-Dimensional Superconductivity   279
     H. Gutfreund and W. A. Little

Organic Linear Polymers with Conjugated Double Bonds . . .   297
     G. Wegner

X-Ray and Neutron Scattering Investigations of
          the Charge Density Waves in TTF-TCNQ   315
     R. Comès

Electronic Properties of Organic Conductors:
          Pressure Effects   341
     D. Jérome and M. Weger

Charge Density Waves in Layered Compounds   369
     F. J. DiSalvo

Comparison of Columnar Organic and Inorganic
          Solids   391
     Z. G. Soos and H. J. Keller

Appendix   413

Subject Index   415

THE BASIC PHYSICS OF ONE-DIMENSIONAL METALS*

V. J. Emery

Department of Physics, Brookhaven National Laboratory

Upton, NY 11973

I. INTRODUCTION

The study of nearly one-dimensional metals has reached an interesting stage of development. In the past few years there have been so many advances in theory and experiment that we are just about at a point where it is possible to understand much of what is going on in particular materials or at least to plan a rational program to find out. Of course there are many problems to solve and different points of view to reconcile, but this makes it an all the more appropriate time to hold a NATO Advanced Study Institute at which many of these questions can be described and explored. These introductory lectures will give a qualitative description of the physical concepts upon which our present understanding of such systems is based. They will be largely non-mathematical, leaving the details to be found in the original papers or in other lectures at this Institute.

Nearly one-dimensional metals are characterised by their very anistropic properties, exemplified by the electrical conductivity which takes place almost entirely along one particular direction. This behaviour is a consequence of their structure - the molecules or ions are typically arranged in chains or stacks along which the electrons move rather easily, whereas motion in a perpendicular direction is more difficult because it involves a "hopping" over relatively large intermolecular distances. The Physicist's interest in this field stems largely from the opportunity to study an almost one-dimensional electron gas, which has many characteristic and unusual properties. It has been spurred by the prospect of finding high temperature superconductivity[1] or at least an unusually high conductivity produced by some collective mechanism, and much

1

of the discussion at this Institute will be concerned with this
idea.

The structures of particular materials will be described in
detail in succeeding lectures and, for the present, it will be
sufficient to use a relatively simple model in terms of which the
physical properties may be discussed. It is assumed that each
molecular site has only one spatial state for a conduction electron.
The exclusion principle then allows at most two electrons (one of
each spin) to occupy a single site. This assumption simplifies the
argument and is justified when the spacing of electronic levels on
a particular molecule is larger than the other energies in the
problem, so excited states may be ignored. To draw a clearer distinction between the various physical effects, it will be assumed
at first that the molecules form a rigid lattice, and discussion of
the very important role of phonons will be deferred until later.
Then the system is specified by the following parameters:

$t_\parallel$ - amplitude for hopping from site to site along the
chain

$t_\perp$ - amplitude for hopping from chain to chain

U - interaction energy between electrons on the same
molecular site

V_{ij} - interaction between an electron on site i and an
electron on site j.

In a nearly one-dimensional material, $t_\parallel \gg t_\perp$. It may also be
that V_{ij} is largest for two sites within the same chain, but that
is not necessary. The magnitudes of these quantities will be discussed material by material in the later lectures but, for values
to keep in mind, it is estimated that in TTF-TCNQ, $t_\parallel$ is about
0.1 eV and $(t_\perp/t_\parallel) \approx 0.05$. There is more uncertainty about U.
Coulomb interactions make $U > 0$ but they are diminished by delocal-
isation of electrons within a molecule and opposed by an effective
electron-electron interaction due to polarisation of the molecule
that could conceivably be large enough to make U negative. At
present there is no universal agreement about the outcome of this
competition in many of the interesting materials, and several of
the lectures will be concerned with what the current experiments
have to tell us about it. The most significant connection between
the chains is provided by V_{ij}. It is mainly a Coulomb interaction,
although there is also a contribution from electron-phonon coupling.

Much of the physics of almost one-dimensional metals is concerned with understanding the conditions for various kinds of phase
transitions which lead to collective states governing the low temperature behaviour. This is of central importance in the search
for high temperature superconductivity which, for example, could be

thwarted by the intervention of a metal-insulator transition. The
most frequently discussed transitions are to states showing singlet
superconductivity (SS), triplet superconductivity (TS), static
charge-density waves (CDW) or static spin-density waves (SDW).
Superconductivity requires the formation of bound pairs of electrons
which are (roughly speaking) bosons and form a superfluid. The
pairs may form with singlet spin and even angular momentum as in a
metal[2] or with a triplet spin and odd angular momentum[3] as in liquid
He^3 (the latter is a superfluid but not a superconductor because
He^3 atoms are not charged). In a static CDW state, the electron
density is not uniform but has a periodic modulation which, in the
simplest situation, has the form

$$n(\underset{\sim}{r}) \quad = \quad n_0 + n_1 \cos(\underset{\sim}{q} \cdot \underset{\sim}{r} + \phi) \qquad\qquad (1.1)$$

at position $\underset{\sim}{r}$. Such states have been studied extensively in layered
compounds as will be described by Dr. diSalvo. In an SDW state it
is the spin density which varies from point to point to give a loose
kind of antiferromagnetism. All of these states have been observed
in more isotropic materials, but there may be others, as yet undis-
covered, which are peculiar to almost one-dimensional metals.

The low temperature phases are characterised by the existence
of a new kind of order which extends throughout the system. They
are described by one or more order parameters - in Eq. (1.1) by
n_1 and ϕ, the amplitude and phase of the CDW. The magnitude of n_1
is determined by minimizing the free energy, but unless there is a
specific coupling to something external (such as the lattice), ϕ is
free to vary and the state is degenerate. This has a number of
important consequences which will become clear in the subsequent
sections. Above the transition temperature, particularly in a
nearly one-dimensional metal, the order can extend over quite long
but finite distances. Within such regions the system has the appearance of the ordered phase. As the temperature is lowered, the range
of the order increases and it becomes infinite at the transition
temperature T_c.

In order to go into more detail, it is necessary first to
describe the properties of a purely one-dimensional electron gas.
This subject forms the main part of the lectures and is the content
of Sec. 2. A real system has interchain coupling, impurities and
a compressible lattice. Their effects will be described in Secs.
3 and 4.

2. ONE-DIMENSIONAL SYSTEMS

The first property of purely one-dimensional systems with
short-range forces is that fluctuations prevent them from undergoing phase transitions at all. To see how this comes about,
imagine that the system breaks into two segments with the phase of

the order parameter taking one value to the left of a point P and
another to the right. There are two contributions to the change
δF in the free energy - the surface free energy σ due to the mismatch at P and an entropy ℓnM which arises because P can be at
any of the M sites:

$$\delta F \;=\; \sigma - T\,\ell n\,M \qquad\qquad (2.1)$$

Since the contact between segments occurs at a point, σ is finite
for large M and the gain from the entropy far outweighs the cost
in surface free energy so δF $<$ 0. Then it is favourable to break
the system into further segments and to continue to do this until
there is no remnant of the original long range order. This argument does not work in two (or more) dimensions because, if the
region of fluctuation contains a macroscopic number λM of particles,
the surface energy σ is proportional to $M^{1/2}$ (or to $M^{1-d^{-1}}$ in d-
dimensions) and is bigger than ℓnM. The existence of a fluctuation then depends upon the sign of the surface term. The discussion is also modified if there are long-range forces which give
interactions between the segments. But none of this concerns us
here; the point is that fluctuations must be taken seriously and
it is not useful to do a rough calculation of the properties of a
one-dimensional system. This is true even at T = 0 where the entropy
contribution to Eq. (2.1) vanishes because, in general, the transition is removed by quantum fluctuations which, so far, have been
ignored.

There is a second effect which severely limits the applicability of simple theories which are often useful for three-dimensional systems. As the temperature is lowered, long range correlations corresponding to the various kinds of transition begin to
build up and they cannot avoid interfering with each other because
the available phase space is so small in one-dimension. It is then
not useful to do a theory of one kind of transition and to make
sure it is suppressed by including fluctuations - the situation
virtually demands an exact solution, or at least a controlled
approximation which can perhaps be obtained in various limiting
situations, where there are small parameters. Fortunately, this
is not totally impossible, for the mathematics of the many-body
problem simplifies sufficiently in one-dimension, and there are,
by now, a number of exact results upon which one can draw to get a
good qualitative picture of what is going on. The simplest case
to visualise is the large $|U|$ limit[4,5] and this will be described
in some detail. For weak and intermediate coupling there are
field theory models which will be considered rather more briefly.

In order to fix our ideas, the Hamiltonian will be written
down, although the discussion will make little use of its explicit
mathematical form. In terms of the parameters defined in Sec. 1,

$$H = H_0 + H_1 + H_2 \qquad (2.2)$$

where

$$H_0 = U \sum_i n_{i\uparrow} n_{i\downarrow} \qquad (2.3)$$

$$H_1 = -t_{\parallel} \sum_{i,\sigma} a_{i\sigma}^+ a_{i+1,\sigma} + \text{Hermitian conjugate} \qquad (2.4)$$

$$H_2 = \sum_{\substack{i,j \\ \sigma,\sigma'}} V_{ij} n_{i\sigma} n_{j\sigma'} \qquad (2.5)$$

Here $n_{i\sigma}$ is the number of electrons of spin σ ($\uparrow$ or $\downarrow$) at site i and $a_{i\sigma}^+ a_{i+1,\sigma}$ is a combination of fermion creation and annihilation operators which transfers an electron of spin σ from site i+1 to site i. This Hamiltonian defines what is often called an extended Hubbard model. Before considering strong coupling, it is useful to list some properties of the non-interacting system, $H_0 = 0 = H_2$.

2.1 Non-Interacting System

It is imagined that there are M molecular sites in a length L, beyond which the system is extended periodically. The lattice spacing s = L/M. For the non-interacting system, H_1 may be diagonalised by Fourier transformation $a_{m\sigma}^+ = M^{-\frac{1}{2}} \sum_k e^{ikms} a_{k\sigma}^+$ where the wave vectors k have values $2\pi n/L$, with n an integer. The energy spectrum is $2t_{\parallel} \cos ks$ (illustrated in Fig. 1) and the ground state is a "Fermi sea" with two electrons ($\uparrow$ and $\downarrow$ spin) in each state of wave vector k less than the Fermi wavevector $k_F = 2\pi N_0/L$. The Fermi "surface" consists of two points $k = \pm k_F$ and the value of k_F is related to the number N_e of electrons:

$$N_e = \sum_{n=-N_0}^{N_0} 2$$

$$= 4 N_0$$

$$= 2k_F L/\pi \qquad (2.6)$$

which may be expressed in terms of the reciprocal lattice vector

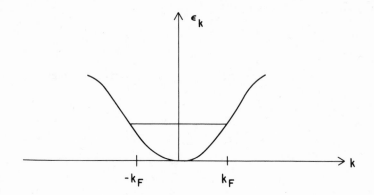

Fig. 1   Energy spectrum of non-interacting electrons
          on a lattice.

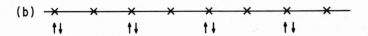

Fig. 2   Some static configurations of electrons for the
          strong coupling limit with $U < 0$. The arrows
          show the electron spin.

$G \equiv 2\pi/s$ as

$$N_e/M = 4k_F/G \qquad (2.7)$$

This relation is used to calculate N_e once k_F and G are obtained from x-ray and neutron scattering experiments. An important special case is a half-filled band, $N_e = M$, for which $G = 4k_F$.

2.2 Strong Coupling

The states of the system will now be described for the large $|U|$ limit, in which H_1 and H_2 may be treated as perturbations. It is not necessarily assumed that this applies to any particular material, although Dr. Torrance will present arguments to suggest that U is large and positive in TTF-TCNQ. For the present purpose, large $|U|$ is a limit in which the properties of one-dimensional conductors can easily be visualised. They depend upon the sign of U.

$U < 0$ When $H_1 = 0 = H_2$, the electrons occupy molecular sites in pairs to take advantage of the onsite attraction. This is illustrated in Fig. 2a where the crosses represent molecules and the arrows refer to up or down spin electrons. The ground state is very degenerate because the energy does not depend upon which sites are occupied.

CDW states occur in an extreme form when there is an intersite repulsion V_{ij} but still no hopping. To minimise the energy, the pairs are equally spaced, as shown in Fig. 2b for a half-filled band. The charge density varies periodically from 1 to 0 in a distance L/N_e so, using Eq. (2.6), the wave vector is just $2k_F$. This is a possible ground state because the system is classical when $H_1 = 0$. More realistically, when $H_1 \neq 0$, the picture is not so static and there is a much smaller modulation of the charge density. Also, if N_e/M is irrational, the electrons cannot be distributed so neatly among the sites. However, the wave vector $2k_F$ persists for, in another view, the CDW is formed by hybridisation of one-electron states near opposite Fermi points. This occurs self-consistently because the hybridisation is allowed by umklapp scattering off the periodic potential generated by the CDW itself.

Singlet superconductivity can arise when hopping (H_1) is included. The electron pairs are bosons, bound in a singlet state, and it is possible that they become superfluid (and hence superconducting since they are charged) at low enough temperatures. Actually this does not happen in a purely one-dimensional system since quantum mechanical fluctuations[4,5] prevent superconductivity even at zero temperature. When hopping between chains is allowed (see Sec. 3) a transition does take place at a temperature T_c determined partly by the value of $t_\perp$. Usually T_c is not the same

as the temperature $|U|/k_B$ which characterises pair formation, a
distinction which persists for weak coupling. In this respect
nearly one-dimensional conductors differ from isotropic three-
dimensional metals where, in weak coupling, the pairs form and Bose
condense at the same temperature.[2]

Triplet superconductivity will not occur because the electrons
are bound into singlet pairs before long range triplet correlations
can build up.

The excited states are of two kinds. Charge density wave
excitations require the movement of pairs from site to site and
they can be phonons or plasmons according to whether the neutral-
ising background (ions or holes) moves in phase or out of phase
with the electrons. However in a spin-wave excitation, a spin is
turned over and a pair must be broken since two electrons of the
same spin cannot occupy the same site. This costs an energy $|U|$
which appears as a gap in the spin wave spectrum. A static version
of such an excitation is shown in Fig. 2c. In the same way, the
Pauli susceptibility χ is proportional to $e^{-|U|/k_B T}$ at low temp-
eratures since only thermally broken pairs can respond to a weak
magnetic field. The spin-wave gap and exponential susceptibility
are general features of one-dimensional systems with attractive
interactions and they occur for weak coupling also. Later it will
be shown that, in the field theory models, they are related to the
excitation of soliton-antisoliton pairs.

$U > 0$ In this case the electrons avoid the strong repulsive on-
site interaction and the ground state has no doubly occupied sites.
(The discussion will assume $N_e \leqq M$ but it is easily adapted to
$N_e > M$). When $H_1 = 0 = H_2$, there is degeneracy because neither the
spin directions nor the locations of the unoccupied sites have any
effect on the energy.

CDW states occur again in an extreme form when there is an
intersite repulsion V_{ij} but no hopping. In contrast to $U < 0$,
however, single electrons rather than pairs are equally spaced so
the period of the CDW is halved and its wave vector is $4k_F$. This
is shown in Fig. 3a for a quarter-filled band which is very close
to the situation in TTF-TCNQ. (No particular significance is
attached to the spin orientations at this stage.) Once again,
hopping makes the CDW weaker and less static. It also mixes in
doubly occupied sites and restores the Fermi sea, which may lead to
an additional $2k_F$ periodicity by the mechanism described for $U < 0$.
This will be of importance in discussing the x-ray and neutron
scattering experiments on TTF-TCNQ.

The SDW instability is most clearly visualised for a half-
filled band, which has exactly one electron per site so that only

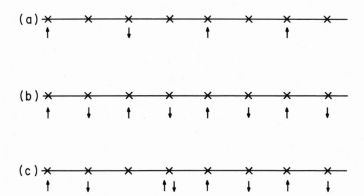

Fig. 3   Some static configurations of electrons for the
           strong coupling limit with $U > 0$. The arrows show
           the electron spin.

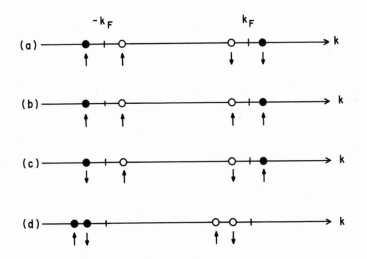

Fig. 4   Scattering processes for the field-theory model.
           The arrows show the electron spin.

the spin degrees of freedom have to be considered. Virtual hopping
produces an effective antiferromagnetic exchange interaction[5] and
the ground state has a modulation of the spin density, which is
illustrated in Fig. 3b. (This picture is correct if only the z-
components of the spins are coupled). The period is $2L/N_e$ and the
wave vector is $2k_F$. For weaker coupling and a different number of
electrons, the state is more dynamic, but the wave vector is still
$2k_F$. (The mechanism is the same as for the CDW except that the
umklapp scattering across the Fermi surface is accompanied by spin
flip).

Superconductivity will occur when $N_e \neq M$ and V_{ij} is attractive. The electrons form Cooper pairs[2] with either singlet[2] or
triplet[3] spin. It requires a coupling to the lattice or to collective modes of the electrons themselves[1] to overcome the intersite Coulomb interactions and make V_{ij} attractive. As for $U < 0$,
a connection between the chains is required for the phase transitions to take place.

The excited states are rather different in character from
those described for $U < 0$. When $N_e \neq M$, there is no difficulty in
rearranging spins and charges to produce an excited state but, for
a half-filled band, the CDW excitations have an energy gap, for
they require the double occupancy of a site as shown in Fig. 3c.
The spin waves are known exactly for this case[5,6] and have no gap.

The mathematical basis of this discussion can be found in
references 4 and 5. For weaker coupling, it is expected that there
will be the same qualitative behaviour and the same general conditions for the particular phase transitions to take place. However
various physical quantities, such as the energy gaps, will be
different functions of the basic parameters of the Hamiltonian.
Weak coupling leads naturally to the field theory models and to a
vivid picture of the excited states which will now be described.
The discussion of phase transitions is more mathematical and may be
found in the original papers.

2.3 Field Theory Models and Solitons

For weak coupling and low temperatures, the most important
single-electron states are those in the neighbourhood of the Fermi
points $\pm k_F$, for the faraway states are not significantly occupied
either by scattering or by thermal excitation. Then it is possible
to expand the energy spectrum (e.g., that shown in Fig. 1) to obtain

$$\epsilon_k \approx \epsilon_{k_F} + v_F(k - k_F) \tag{2.8}$$

where $v_F = (\partial \epsilon_k / \partial k)_{k_F}$ is the Fermi velocity. Frequently this
expansion is introduced towards the end of a calculation when it

is necessary to evaluate integrals. But, in one-dimensional systems, it produces an enormous technical simplification to use Eq. (2.8) at the outset, for the resultant linear spectrum is just that of a massless relativistic particle (with v_F replacing the velocity of light) and, in the limit $s \to 0$, the model becomes identical to a relativistic quantum field theory in one space dimension, which in certain circumstances, can be solved. In this approach, the interactions are viewed as scattering electrons from one plane wave state to another and some of the important processes are shown in Fig. 4, where the open circles represent holes in the states from which electrons have been removed and the closed circles are states occupied by the electrons excited out of the Fermi sea. In this model, the properties depend upon the different coupling constants g_i assigned to the processes shown in Fig. 4. In real systems, not all sets of values g_i are physically attainable, but a knowledge of the conditions in which, for example, superconductive ordering dominates over all others offers some guidance for the synthesis of new materials.

Technically, the field theory models have been tackled either by the use of boson representations[7-10] or by the renormalisation group method,[11] each of which has a different region of validity. The details may be found in the original papers and only a qualitative discussion will be given here.

The first essential feature is that, as soon as there are interactions, the electrons disappear from the problem. More precisely, it takes energies of the order of the bandwidth to detect them and the low-lying states are entirely collective modes, charge- and spin-density waves. This is analogous to the confinement of quarks in elementary particle physics and is a reason why particle theorists are interested in these models.[12]

The SDW and CDW excitations do not couple to each other and, in the absence of the interactions of Figs. 4c and 4d, they are harmonic and the Hamiltonian may be diagonalised exactly. Introduction of backward scattering[9] (Fig. 4c) gives a nonlinear interaction between SDW's and, for a half-filled band, umklapp scattering (Fig. 4d) has a similar effect[10] on the CDW's. The charge- or spin-density ϕ satisfies an equation of the form[9,10,13]

$$\frac{\partial^2 \phi}{\partial x^2} - \frac{1}{v_F^2} \frac{\partial^2 \phi}{\partial t^2} - \frac{\alpha}{\beta} \sin\beta\phi = 0 \qquad (2.9)$$

where α and β are functions of the g_i. Equation (2.9) is known as the Sine-Gordon equation and has been studied extensively in classical mechanics[14] where the solutions are <u>solitons</u>[14]

$$\phi = 4/\beta \text{ arc tan exp } \sqrt{\alpha} x \qquad (2.10)$$

A moving soliton is obtained from Eq. (2.10) by Lorentz transformation

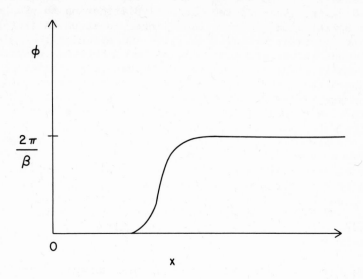

Fig. 5   A classical soliton.

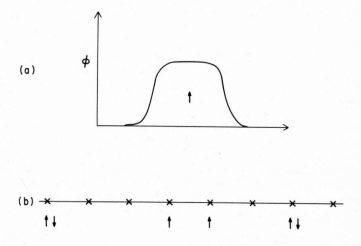

Fig. 6   (a) Classical spin-wave soliton-antisoliton pair.
(b) The comparable static configuration in the
strong-coupling lattice model for $U < 0$.

As shown in Fig. 5, ϕ changes from zero to $2\pi/\beta$ in a distance of order $1/\sqrt{\alpha}$. The energy[14] of a soliton is $8\sqrt{\alpha}/\beta^2$. If the sign of ϕ is reversed in Eq. (2.10) it is still a solution of Eq. (2.9) and is known as an <u>antisoliton</u>. The characteristic properties of classical solitons are that they do not dissipate and that they regain their original shape after scattering from each other. In this sense they are models of elementary particles.

In quantum mechanics, free solitons exist[9,13] only for $\beta^2 = 4\pi$ and, for larger values of β, they can form bound soliton-antisoliton pairs.[15] With this picture it is now possible to make contact with the strong coupling states discussed in Sec. 2.2. It is simplest to describe the case $H_2 = 0$ for which all of the coupling constants g_i can be expressed in terms of one quantity U.

$U < 0$ In this case, the solitons appear in the SDW excitations. The classical picture is shown in Fig. 6a, which shows the excess spin between the soliton and antisoliton. To make clear the analogy with strong coupling, Fig. 2c is repeated as Fig. 6b. The existence of a finite soliton energy implies that there is a gap in the SDW spectrum as in the strong-coupling limit.

$U > 0$ Here the solitons and the associated energy gap appear in the CDW spectrum for a half-filled band. The two pictures of a CDW excitation are shown in Fig. 7 where there are two soliton-antisoliton pairs because the net charge is unchanged by the excitation.

2.4 Interplay of Fermions and Bosons

One of the most intriguing results of the field-theory approach is that solitons obey Fermi statistics,[9,13] despite the fact that they are composed of interacting CDW's or SDW's which are bosons. A crude feeling for this can be obtained from Figs. 6 and 7 where it can be seen that putting two soliton-antisoliton pairs in the same place would correspond to having two electrons of the same spin on a given site. It is an example of the way in which interactions can bring about a change in the statistics of the elementary excitations of a one-dimensional system:

(a) With no interactions the excitations are electrons and holes which are fermions. The couplings in Figs. 4a and 4c bind them into CDW and SDW excitations which are bosons.

(b) Processes 4c and 4d give a nonlinear coupling between SDW's and CDW's respectively and, for appropriate sign of the coupling constants, produce solitons which are fermions.

This is a very important aspect of the quantum mechanics of one-dimensional systems. Yet another example is a point hard-core

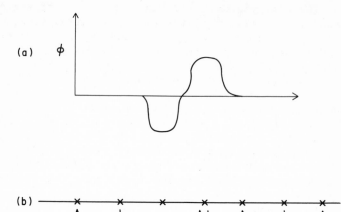

Fig. 7   (a) Two classical charge-density soliton-antisoliton
          pairs.   (b) The comparable static configuration in
          the strong-coupling lattice model for $U > 0$.

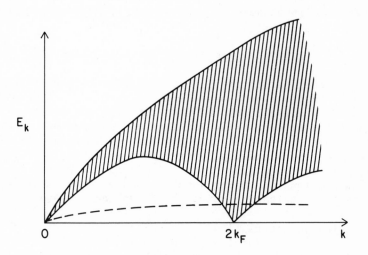

Fig. 8   Electron-hole spectrum for a non-interacting
          electron gas.   The dashed line shows a phonon
          branch.

force which turns bosons into fermions.[16,4,5]

2.5 Renormalization Group

Mathematically the very characteristic one-dimensional effects described above occur because in every order of perturbation theory in the interaction there are powers of $ln(T/t_{\parallel})$, which diverge as $T \to 0$. The exact solutions which have been obtained[8-10] show that the perturbation series may be summed to give powers $(T/t_{\parallel})^{\alpha}$, where α depends on the coupling constants. The sign of α then determines whether there is a divergent $(\alpha < 0)$ or convergent $(\alpha > 0)$ result as $T \to 0$. But exact solutions[8-10] cannot always be found and so far they have not been found for the SDW excitations when $U > 0$ and for the CDW's when $U < 0$ and the band is half-filled. This is where the renormalisation group method[11] comes in.

The idea is to find a mathematical relation between the original Hamiltonian and one for a different system which has a smaller bandwidth and different interactions between the particles. Frequently it is possible to "scale" in this way to weak coupling and so to solve the problem. In practice, this procedure can only be carried out approximately[11] and so is useful if the initial coupling is not too strong. But it is believed that at least the qualitative picture which emerges is more generally valid and, indeed, there is substantial agreement with strong coupling. The consequences of the field-theory models for phase transitions will be described in the lectures of Dr. Gutfreund.

3. INTERCHAIN COUPLING AND PHASE TRANSITIONS

Although there are no phase transitions in a purely one-dimensional system, a real material has coupling between the chains. As the temperature is lowered, the electrons act cooperatively over longer and longer distances within each chain and the interaction between chains is enhanced. Ultimately, this brings out the true three-dimensional nature of the system and allows a phase transition to take place.

This picture is particularly clear when interchain effects are weak, secondary, aspects of the material. It is inappropriate in the other extreme of an almost isotropic system. Real materials lie somewhere in between - hopping may be small but Coulomb forces often are not. This intermediate situation is difficult to deal with theoretically and is not well explored at present. It is not sufficient to think in terms of an anisotropic three-dimensional system because the characteristic one-dimensional effects do play an important role and are difficult to build in to such an approach. Some attempt has been made to use the renormalisation group method[18] but, in the theory of critical phenomena, this has not been a notoriously successful way of obtaining phase diagrams, although recently developed space renormalisation group methods[19] are more promising for this purpose. For these reasons, only weak inter-chain coupling will be considered here.

3.1 Weak Interchain Coupling

In this case, it is possible to use a self-consistent pertur-bation theory for interchain coupling provided the one-dimensional motion is treated accurately. The lowest order is mean-field theory in which each chain feels the average "Hartree" field produced by its neighbours[19] and, for the strong coupling limit[5] described in Sec. 2, this leads to a transition temperature

$$T_c \sim \frac{t_\parallel^2}{|U|} \left| \frac{t_\perp^2}{t_\parallel^2} + \frac{V_\perp |U|}{2t_\parallel^2} \right|^{\frac{1}{2-\theta-1}} \tag{3.1}$$

for a CDW instability (U < 0) and

$$T_c \sim \frac{t_\parallel^2}{U} \left(\frac{t_\perp^2}{t_\parallel^2} \right)^{\frac{1}{2-\theta}} \tag{3.2}$$

for superconductivity (U < 0) or an SDW state (U > 0 and a half-filled band). Here

$$\theta = \frac{1}{2} + \pi^{-1} \sin^{-1} \left(1 + \frac{V_\parallel |U|}{2t_\parallel^2} \right) \tag{3.3}$$

and it has been assumed that H_2 has only near-neighbour inter-actions $V_\parallel$ along a chain and $V_\perp$ between chains. From these equa-tions it can be seen that there is only a CDW transition when $t_\perp = 0$ but all three can take place if $V_\perp = 0$ but $t_\perp \neq 0$. Further-more, for superconductivity to occur at a higher temperature than a CDW transition (for U < 0), it is necessary that either $V_\perp < 0$ or $V_\parallel < 0$ or both, i.e., the electron-phonon interaction outweighs the Coulomb force between electrons on neighbouring sites.

The conclusions of the field-theory models, which will be given by Dr. Gutfreund are different in certain respects because the con-straint of having exactly one or exactly two electrons on every site is relaxed. This probably is of importance in a material like $(SN)_x$ which is superconducting (see Dr. Greene's lectures).

3.2 The CDW State and Phase Locking

It is worth describing the CDW state in more detail because it gives rise to observable lattice effects (see Sec. 4). If an

external electric field with components of wave vector $2k_F$ is
applied to the system, it drives a CDW on the individual chains.
At low temperatures, the amplitude is large because the range of
order is quite long. But, through the interchain coupling, each
chain sees an additional field due to its neighbours and the amplitude is built up even more. At T_c, the response to the external
field is divergent and a CDW state, given by Eq. (1.1), is established. The overall phase ϕ is free but the relative phases on
different chains are specified by $q_\perp$ - the component of q perpendicular to the chains. The value of $q_\perp$ is mainly determined by
the Coulomb interaction which causes regions of excess like charge
in the CDW to avoid each other. In a material like TTF-TCNQ, $q_\perp$
can vary with temperature to accommodate different rates of ordering of the two types of chain,[20,21] as found in the x-ray and
neutron scattering experiments described by Dr. Comès.

If there is a direction along which the component of q is
commensurate (i.e., a rational fraction of the component of the
reciprocal lattice vector G) the overall phase ϕ may become "locked"
to the lattice. In effect, the free energy depends upon ϕ because
there is a non-linear Bragg scattering of the CDW's and ϕ has to be
determined by minimisation. This mechanism can cause a component
of $q_\perp$ to lock at a commensurate value, so that it can take advantage of the extra free energy and it appears to occur in TTF-TCNQ.[20]

This is the simplest situation, but if q is almost commensurate
in any direction, the lowest free energy[20,22] is given not by a
constant ϕ but by a sequence of static solitons[20] which separate
commensurate regions. Previously[20] this has been imagined for $q_\perp$,
but it could also occur along a chain direction. For example,
TTF-TCNQ with $2k_F = 0.295G$ is perhaps close enough to a quarter-
filled band to allow this to happen. Then current may be carried
by solitons[23] moving through the static soliton lattice.

3.3 Impurities

Random impurities have pronounced effects in one-dimensional
systems. Localisation of the electrons has been extensively
studied[24] although it is clear that electron interactions, which
are usually omitted, are important and can change the conclusions
altogether. Impurities also have an influence on phase transitions,
acting in two ways which are best illustrated by examples. Consider
first the $U > 0$ strong coupling limit for a half-filled band, where
there is exactly one electron per site, and suppose the impurities
do not have available states for an electron. They are then non-
magnetic and move the antiferromagnetic transition to a lower temperature. On the other hand, if the impurities go in between chains,
they provide a random electric field which drives large fluctuations
of the phase of a CDW and destroys the long-range order.[25] It has
been suggested[26] that this mechanism is responsible for the absence

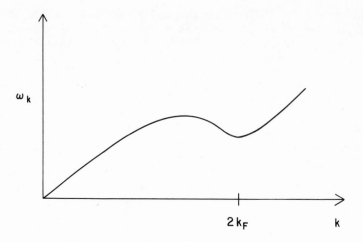

Fig. 9   Phonon spectrum with a Kohn anomaly.

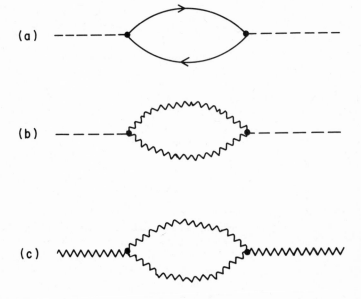

Fig. 10   Diagrams for second order contributions to the
          energy of phonons or charge-density waves. (a)
          phonon (dashed line) mixed with an electron-hole
          pair, (b) phonon mixed with two CDW's (wavy lines),
          (c) one CDW mixed with two CDW's.

of a CDW transition in KCP. (see the talks of Dr. Renker and Dr. Carneiro)

4. LATTICE DISTORTIONS AND PHONON ANOMALIES

The coupled electron-lattice problem has not so far been solved with the same accuracy as the electron gas models considered in Secs. 2 and 3. The principal difficulty is that the effective electron-electron interaction, produced by polarisation of the lattice, is retarded, because the lattice is unable to respond instantaneously to the motion of the electrons. This spoils many of the mathematical tricks which may be used when the electron Hamiltonian is static, and it was the reason for including the lattice-induced interaction as an effective time-independent contribution to V_{ij}. More important from an experimental point of view is the effect of the electrons on the lattice. The onset of a CDW state is associated with a (Kohn) anomaly in the phonon spectrum and a (Peierls) distortion of the lattice, which may be seen by x-ray and neutron scattering and provide some of the most direct evidence of the nature of the phase transitions in almost one-dimensional systems. These experiments will be described by Dr. Comès.

The electrons interact with lattice vibrations because $t_{\parallel}$, $t_{\perp}$ and V_{ij} depend upon the distance between molecules so that their values are modulated when a phonon passes by. Through this interaction, a static CDW in the electron gas creates a periodic potential in the lattice which causes a distortion of wave vector $2k_F$. But the lattice plays a more active role by reacting back on the electron gas and helping to drive the CDW, thereby increasing the transition temperature. This coupled phase change in the electrons and the lattice is known as a Peierls transition. Above T_c, the phonons reflect the impending transition by changes in their energy and damping for wave vectors near $2k_F$. The origin of this effect may be understood from Fig. 8, where the shaded area shows allowed energies of electron-hole pairs[27] with relative wave vector k. When the energy of a phonon (shown schematically by a dashed line) lies within the electron-hole spectrum, there is hybridisation - the phonon is damped because it can decay into an electron-hole pair and there is a dip in its energy as shown in Fig. 9. This is known as a Kohn anomaly. The initial stages may be described by second order perturbation theory pictured in Fig. 10a, which shows a phonon (dashed line) turning into an electron-hole pair through its modulation of $t_{\parallel}$ and $t_{\perp}$. As the temperature is lowered, the dip at $2k_F$ deepens continuing (in the simplest theories based on the process in Fig. 10a), until $\omega_k = 0$ at $k = 2k_F$ when it becomes energetically favourable to have static phonons at this wave vector - in other words there is a Peierls transition. In reality, the transition is not adequately described by second order perturbation theory and it need not take place by this "soft mode" mechanism.

A Kohn anomaly and lattice distortion have been observed[28] in
TTF-TCNQ but, at present, it is not known if the phonon becomes
completely soft.

The observation of a lattice distortion does not, in itself,
prove the existence of a CDW state in the electron gas for, if
there is magnetoelastic coupling, it could conceivably be a consequence of a SDW state. This is known to occur in insulators[29]
and is called a spin-Peierls transition. The possibility that a
similar thing may be happening in TTF-TCNQ will be discussed by
Dr. Torrance.

$4k_F$ Anomalies and Rotons in the Electron Gas

Perhaps the most interesting experiment in this field in the
past year has been the discovery[30] in TTF-TCNQ of a phonon anomaly
and lattice distortion with wave vector $4k_F$, in addition to those
previously seen at $2k_F$. The experimental situation will be described
in detail by Dr. Comès but, for the present purpose, the significant features are that the $4k_F$ phonon anomaly is present at room
temperature whereas there is nothing to be seen at $2k_F$ until 150K;
both lattice distortions are present at low temperatures.

In one sense, these experiments are not difficult to understand because, a period of $4k_F$ arises naturally when U is large and
positive, as shown in Fig. 3a for a quarter-filled band, which is
not too far from the situation in TTF-TCNQ. Of course this picture
is too static but it is not difficult to show[31] that when $U \to \infty$,
the period of the Peierls distortion is $4k_F$ even when hopping is
allowed.[32] In practice, the situation is not quite so simple, for
there is a lattice distortion together with a phonon anomaly at
$2k_F$ as well as at $4k_F$, and it appears to require the existence of
strong correlations between charge-density waves aided by a phonon
modulation of V_{ij} as will now be described.

In Sec. 2, it was shown that, for a half-filled band, $G = 4k_F$,
umklapp scattering of electron pairs (Fig. 4d) coupled the charge-
density waves and transformed the excitations into solitons. When
the band is not half-filled, this effect does not have the same
importance because the relevant electrons are not near to the Fermi
points, but it is reinstated if there is a superlattice distortion
with reciprocal vector $4k_F$ in the chain direction. This would
raise the elastic energy but, nevertheless, it would be profitable
if there were an even greater decrease in electronic energy as a
result of the umklapp-induced correlations. It is possible to work
out all of this in the field-theory model and to show[33] that there
will be a $4k_F$ superlattice distortion at low temperatures provided
β, which occurs in Eq. (2.9), satisfies $\beta^2 > 4\pi$. In terms of the
lattice model of Eqs. (2.2) to (2.5), this condition requires U
positive and larger than about $2t_\parallel$. Since $2k_F$ correlations are

present for a half-filled band, they are not removed by the formation of a $4k_F$ superlattice, although they may well be modified.
Above the transition, there is an effect on the phonons which is
similar to the Kohn anomaly at $2k_F$. The process which replaces
Fig. 10a is shown in Fig. 10b, where the wavy lines are charge-
density waves, and, indeed, for the relevant values of U, the
charge-density waves may be transformed into interacting Fermions
which take over the role played by the electrons in the $2k_F$
anomaly.[33] The main physical difference is that the electron-
phonon coupling comes from modulation of V_{ij}, enabling the phonons
to decay into two particles and two holes which become the intermediate charge-density waves in Fig. 10b, in virtue of their interaction.

In an interacting electron gas, it is possible for a CDW to
scatter into two CDW's to give the diagram shown in Fig. 10c. By
comparison with Fig. 10b, it can be seen that there will be a dip
in the CDW spectrum at wave vector $4k_F$. This is completely analogous[34] to a <u>roton</u> in liquid He^4 for which Feynman's wave function
mixes one and two phonon states in just the same way. In general,
the roton minimum occurs at wave vector $2\pi/a$, where a is the mean
spacing between particles, and for a one-dimensional system a =
L/N_e so, using Eq. (2.6), $2\pi/a = 4k_F$. As $U \rightarrow \omega$, this description
goes over to the strong coupling picture considered earlier.[34]
When the temperature is lowered, the dip in the charge-density
wave spectrum deepens and eventually there will be a static CDW,
which will be assisted by coupling to the lattice, as before. The
discussion of phase relations between the chains and phase locking
to the lattice both parallel and perpendicular to the chains,
given in Sec. 3.2, applies equally to the $4k_F$ distortion.

If this mechanism for the $4k_F$ anomaly is correct, it shows
directly that U is large and positive on at least one set of chains
(TTF or TCNQ) in TTF-TCNQ. To avoid this conclusion, it is necessary to find some other way of accounting for the experiments -
but the obvious possibility of coupling to two softened $2k_F$ phonons
is excluded by the different temperature dependences, coupling to
single-chain electronic collective modes is what has been discussed
here, and coupling to electronic modes from different chains should
be weaker and also requires large positive U. This argument,
together with the behaviour of the magnetic susceptibility,[35] which
does not decrease exponentially (as it should for attractive interactions (see Sec. 2)), strongly supports the conclusion that there
are repulsive interactions on at least one chain in TTF-TCNQ.

5. CONCLUSION

This account of the basic physics of nearly one-dimensional
conductors is far from complete, although many of the neglected
topics will be covered in the other lectures. Perhaps the most

significant omission is a discussion of the electrical conductivity, particularly through collective mechanisms related to the CDW state, but this is a subject in itself and it should also be taken together with the dynamical effects of phonons, a description of the ordered phases, and the inclusion of stronger interchain coupling to form a list of problems awaiting a more satisfactory solution.

REFERENCES AND FOOTNOTES

*   Work supported by Energy Research and Development Administration
1.   W. A. Little, Phys. Rev. A134, 1416 (1964).
2.   J. Bardeen, L. N. Cooper and J. R. Schrieffer, Phys. Rev. 108, 1175 (1957).
3.   A. J. Leggett, Rev. Mod. Phys. 47, 331 (1975).
4.   K. B. Efetov and A. I. Larkin, Zh. Eksp. Teor, Fiz. 69, 764 (1975) [Sov. Phys. JETP 42, 390 (1976)].
5.   V. J. Emery, Phys. Rev. B14,   (1976).
6.   J. des Cloizeaux and J. J. Pearson, Phys. Rev. 128, 2131 (1962); J. D. Johnson, S. Krinsky and B. M. McCoy, Phys. Rev. A8, 2526 (1973).
7.   Boson Representations were introduced by D. C. Mattis and E. H. Lieb, J. Math. Phys. 6, 304 (1965) and by K. D. Schotte and U. Schotte, Phys. Rev. 182, 479 (1969).
8.   A. Luther and I. Peschel, Phys. Rev. B9, 2911 (1974).
9.   A. Luther and V. J. Emery, Phys. Rev. Lett. 33, 589 (1974).
10. V. J. Emery, A. Luther and I. Peschel, Phys. Rev. B13, 1272 (1976); H. Gutfreund and R. A. Klemm, Phys. Rev. B14, 1073 (1976).
11. Yu. A. Bychkov, L. P. Gor'kov and I. E. Dzyaloshinsky, Zh. Eksp. Teor, Fiz. 50, 738 (1966) [Soviet Phys. JETP 23, 489 (1966)]; N. Menyhárd and J. Sólyom, J. Low Tempt. Phys. 12, 529 (1973); J. Sólyom, J. Low Temp. Phys. 12, 547 (1973).
12. The idea is that so many of the properties of "elementary" particles can be understood if they are made up of still mote fundamental fermions (quarks) but, so far, they have not been found in any experiments. This could be understood if the quarks were "confined" by their interactions so that it would be impossible to pull them out of the vacuum (ground state).
13. S. Coleman, Phys. Rev. D11, 2088 (1975).
14. A. Scott, F. Chu and D. McLaughlin, Proc. IEEE 61, 1443 (1973).
15. R. F. Dashen, B. Hasslacher and A. Neveu, Phys. Rev. D11, 3424 (1975); A. Luther, Phys. Rev. B14,   (1976).
16. T. D. Schultz, J. Math. Phys. 4, 666 (1963).
17. L. Mihály and J. Sólyom, J. Low Temp. Phys. 24, 579 (1976).
18. Th. Niemeijer and J. M. J. van Leeuwen, Phys. Rev. Lett. 31, 1411 (1973); Physica 71, 17 (1974).
19. D. J. Scalapino, Y. Imry and P. Pincus, Phys. Rev. B5, 2042 (1975); Y. Imry, P. Pincus and D. J. Scalapino, Phys. Rev. B12, 1978 (1975).
20. P. Bak and V. J. Emery, Phys. Rev. Lett. 36, 978 (1976).

21. T. D. Schultz and S. Etemad, Phys. Rev. B13, 4928 (1976); K. Šaub, S. Barisic and J. Friedel, Phys. Lett. 56A, 302 (1976).
22. W. L. McMillan, Phys. Rev. B14,   (1976).
23. M. J. Rice, A. R. Bishop, J. A. Krumhansl and S. E. Trullinger, Phys. Rev. Lett. 36, 432 (1976).
24. D. J. Thouless, Physics Reports 13C, 93 (1974).
25. P. Lacour-Gayet and G. Toulouse, J. Phys. (Paris) 35, 425 (1974); Y. Imry and S.-K. Ma, Phys. Rev. Lett. 35, 1399 (1975); H. G. Schuster, to be published.
26. L. J. Sham and B. R. Patton, Phys. Rev. B13, 3151 (1976).
27. This is easy to work out for a non-interacting system. Since a hole must be inside the Fermi sea and an electron outside, there is a restriction on the range of possible energies for each k.
28. For a list of references see the lectures of Dr. Comès.
29. J. W. Bray, H. R. Hart, Jr., L. V. Interrante, I. S. Jacobs, J. S. Kasper, G. D. Watkins, S. H. Wee and J. C. Bonner, Phys. Rev. Lett. 35, 744 (1975).
30. J. P. Pouget, S. K. Khanna, F. Denoyer, R. Comès, A. F. Garito and A. J. Heeger, Phys. Rev. Lett. 37, 437 (1976).
31. The argument is that the evaluation of expectation values can, by symmetry, always be arranged as integrals over the positions of the particles in a fixed order along the lines. Then the only restriction is that no more than one particle shall occupy a site, otherwise they are free. But then the answer is the same as for non-interacting, spinless, fermions which can have only one electron per state, and Eq. (2.6) becomes $N_e = \bar{k}_F L/\pi$. All of the singular effects take place at $2\bar{k}_F = 2\pi N_e/L$ which is just $4k_F$ by Eq. (2.6).
32. A. A. Ovchinnikov, Zh. Eksp. Teor. Fiz. 64, 342 (1973) [Sov. Phys. JETP 37, 176 (1973)]; J. Bernasconi, M. J. Rice, W. R. Schneider and S. Strässler, Phys. Rev. B12, 1090 (1975).
33. V. J. Emery, Phys. Rev. Lett. 37, 107 (1976).
34. V. J. Emery, to be published.
35. See the lectures by A. J. Heeger in this volume.

THE ORGANIC METALLIC STATE: SOME CHEMICAL ASPECTS

D. Cowan, P. Shu, C. Hu, W. Krug, T. Carruthers,
T. Poehler, and A. Bloch

Department of Chemistry, The Johns Hopkins University
Baltimore, Maryland 21218 U.S.A.

I. REVIEW OF DESIGN CONSTRAINTS[1]

Most crystalline organic compounds are insulators or semi-conductors with room temperature conductivities in the range of 10^{-9} to $10^{-14}\Omega^{-1}cm^{-1}$ and exhibit exponentially decreasing conductivity profiles as the temperature is decreased.[10] However in the 1960's Melby and the duPont group made the startling discovery that a few of the many TCNQ (fig. 2) radical ion salts they prepared and studied[11-17] had conductivities as high as $10^{2}\Omega^{-1}cm^{-1}$. While this is low compared with metals like copper ($\sigma_{RT}=5\times10^{6}\Omega^{-1}cm^{-1}$), it is remarkably high for an organic compound. Why do some of the TCNQ salts exhibit this high conductivity and can the conductivity of organic compounds be increased above $10^{2}\Omega^{-1}cm^{-1}$?

Certainly TCNQ is a good oxidizing agent; electron acceptor (A), and forms stable open-shell or free radical salts. However, there are numerous examples of insulating stable free radicals and free radical salts. Even an electron acceptor, such as chloranil, which has a planar structure and readily accepts an electron into and extended π-molecular orbital, structural characteristics shared with TCNQ, forms insulating or semiconducting radical ion salts.[10,18]

We can also note that all of the highly conducting TCNQ salts have a segregated stack....$A^{-n}A^{-n}A^{-n}A^{-n}$....and not a mixed cation --- anion stack.....$D^{+n}A^{-n}D^{+n}A^{-n}$....type of crystal structure.[19] One of the structural features that seems to facilitate this segregated type of stacking is an inhomogeneous charge and spin distribution. Both theory[20] and experiment[21] suggest that in TCNQ- the odd electron resides to a large extent on the terminal portions of the molecule containing the four electron attracting nitrile groups.

25

This gives rise to the "ring-double bond" overlap shown in fig. 1 for the TCNQ component in TTF-TCNQ.[22] Probably because of a similar inhomogeneous charge and spin distribution in $TTF^{\bullet +}$, an open-shell cation to be considered in some detail (see fig. 2), the "ring-double bond" overlap is also characteristic of the cation stacking in TTF-TCNQ.[22,23] The picture that emerges at this point is a stack of planar open-shell molecules arranged so as to allow relatively strong π-orbital overlap forming parallel conducting linear chains which are only weakly coupled.

The interplanar stacking distance (3.17 to 3.30Å) and the resulting π-orbital overlap integral between TCNQ ions indicates that the band widths in these solids are rather narrow, probably in the range of 0.1 to 0.6 of an electron volt.[24] One problem with a structure that gives rise to a narrow quasi-one-dimensional con-duction band is that it is subject to periodic lattic distortions of wave-vector $2k_F$ which open up a gap at the Fermi level producing an insulating or semiconducting material.[1] Most of the duPont TCNQ

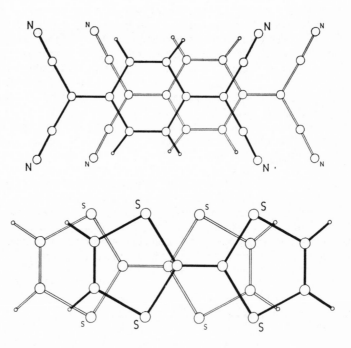

Fig. 1.   Molecular overlap in the columnar stacks of TTF and TCNQ in TTF-TCNQ (from ref. 22).

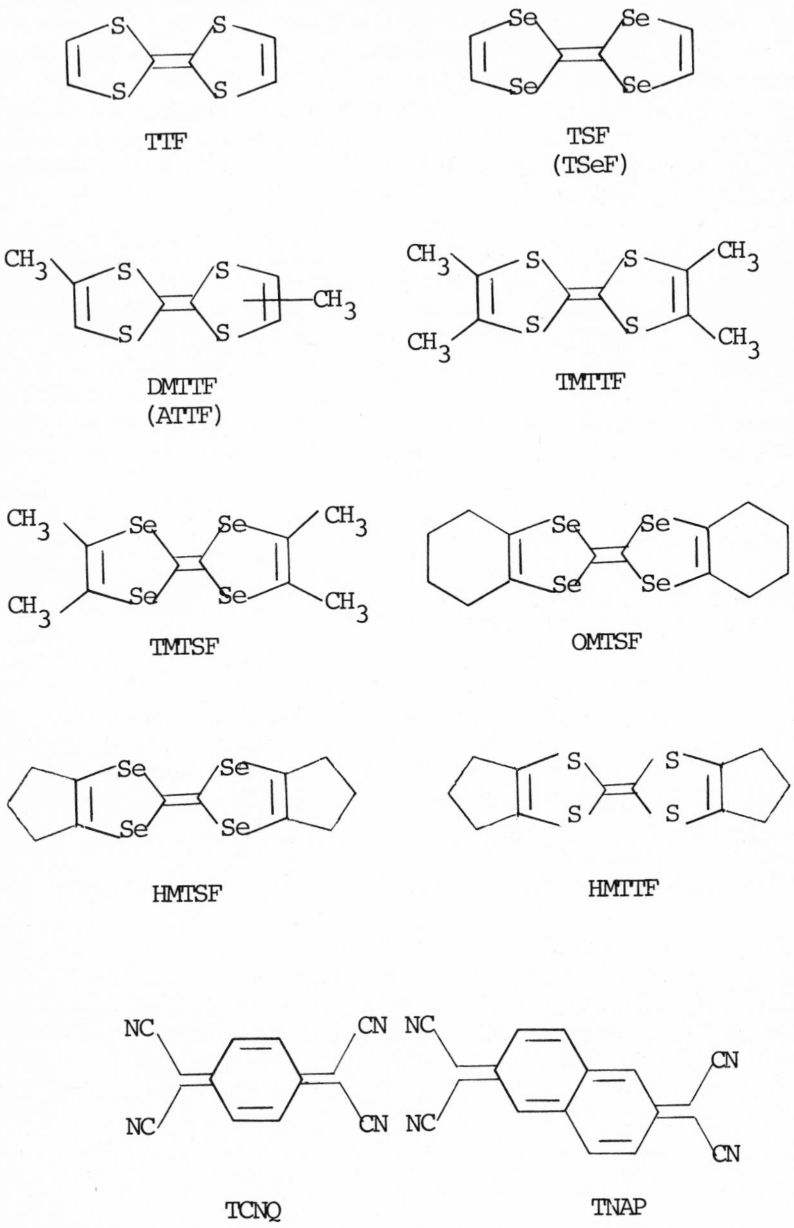

Fig. 2. Molecular structures of donor and acceptor molecules.

salts which have segregated stack structures also have periodically distorted linear chain structures and are insulators.[19] Only the highest conductivity TCNQ salts have both a segregated stack and a nondistorted linear chain structure.[19] The early notable members of this very exclusive class were N-methylphenazinium (NMP) TCNQ and quinolinium (Q) (TCNQ)$_2$, both with room temperature conductivities ~$10^2 \Omega^{-1} cm^{-1}$.[24,25] In both of these cases there is disorder in cation stack which tends not only to suppress the structural instability but also to localize the electronic wave functions.[26] The electrical conductivity, while relatively high, is limited by the conduction mechanism -- phonon assisted hopping among localized states.[2,26] Both NMP-TCNQ and Q(TCNQ)$_2$ have a very low conductivity maximum ($\sigma_{max}/\sigma_{RT} \approx$ 2-4) near room temperature and then go through a broad maximum and become activated as the temperature is lowered.

Cowan and co-workers first prepared the tetrathiafulvalene (TTF) complex of TCNQ on the premise that the Fermi surface for this solid should be more complicated than that for simple closed-shell cation salts of TCNQ because of the two bands resulting from the highest occupied molecular orbital (HOMO) of the donor (TTF) and the lowest unoccupied molecular orbital (LUMO) of the acceptor (TCNQ) with the attendant possibility that the metal-to-insulator transition temperature would be lowered and the conductivity increased.[23] Another structural feature that makes TTF-TCNQ an attractive candidate for study is the polarizability of the cation portion of this molecule. Le Blanc proposed that when the cation is relatively small and polarizable the conduction electron-cation chain exciton coupling can facilitate conduction by reducing the effective Coulomb repulsion (U_{eff}) between two electrons on the same site.[27,28]

The now-familiar four-probe d.c. conductivity profile for TTF-TCNQ as a function of temperature is shown in fig. 3. The average room temperature conductivity is about 500 $\Omega^{-1} cm^{-1}$ and the maximum conductivity is of order $10^4 \Omega^{-1} cm^{-1}$ near 59°K ($\sigma_{max}/\sigma_{RT} \approx$ 20). Certainly it appears to be more difficult to open up a gap across the entire Fermi surface since the metal-to-insulator transition temperature is so low. In addition if we assume that a mean free path λ of a carrier can be calculated from the one-dimensional metallic formula:[26,29]

$$\sigma = 2Ne^2\lambda/\pi\hbar$$

where N is the number of conducting chains per unit cross sectional area (2TTF, 2TCNQ per unit cell), then the mean free path λ at room temperature would be about 2Å, less than the b-axis lattice spacing of 3.82Å, but is at least 40Å at 59°K. While this is short when compared with the mean free path in copper at room temperature

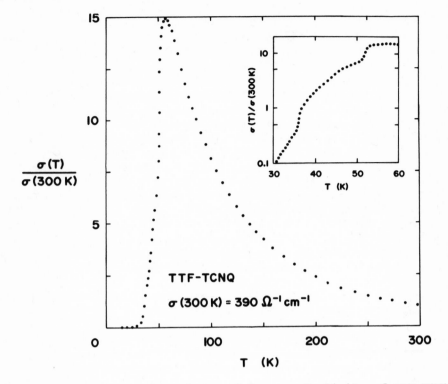

Fig. 3.   Temperature dependence of d.c. conductivity of TTF-TCNQ
          (45b).

(about 400Å), it is long enough to suggest that TTF-TCNQ deserves
the title of the first organic metal or semimetal.

In addition to the relatively high conductivity, the temperature range over which the conductivity increases as the temperature
decreases (300–59°K) is considerable. In the region 300–70°K the
resistivity can be fit to the following equation.[30,31]

$$\rho(T)/\rho(300K) = a + B\ T^{\gamma}$$

where $\gamma = 2.3 \pm 0.1$. This is a considerably strong temperature dependence than the usual T^1 dependence noted for metals such as copper
where this behavior is due to phonon scattering.

Several other interesting observations can be made regarding
this system. First, the phase transitions noted in the conductivity[32,33] (fig. 3), heat capacity[34] and low temperature X-ray and
neutron data[35,36] occur below the conductivity maximum. From

the conductivity data two transitions are easily discernible (53°
and 38°K) and a third (47°K) is apparent from a derivative plot
to be presented in the next paper. Inasmuch as the conductivity
maximum is at a higher temperature (T_{max}) than the first phase
transition (59 vs 53 K), it is reasonable to suppose that whatever
is driving the phase transition may be building up at higher temperatures and influencing the conductivity in the T_{max} region.[38,39]
Second, since the HOMO donor band and LUMO acceptor band need not
be half-filled, the intramolecular and intermolecular correlation
effects that are a result of electron-electron repulsion can be
reduced. This is roughly analogous to the mixed-valence type of
concept that would suggest that a complex salt $M^+(TCNQ)_2^-$, one added
electron per two TCNQ neutral molecules, should be more conducting
than the corresponding simple salt M^+TCNQ^- when all other factors
are held constant.[37] There is conductivity data on simple and complex TCNQ salts that can be interpreted in this fashion. However,
the problem is that more than one structural feature has always
been altered. The low temperature, three-dimensional superlattice
as determined from X-ray diffuse scattering for TTF-TCNQ provides a
measure of the fractional band filling.[35] For TTF-TCNQ about 0.58
electrons are transferred per TTF-TCNQ unit.

Recent work in our group has examined how a change of heteroatom (Se in place of S) and a change of substituents associated
with the donor portion (TTF) of the salt alters bandwidths, band
overlap and filling, and interchain coupling. Engler and colleagues[40,41,42] first prepared the unsubstituted selenium analogue
of TTF; TSF; and the mixed sulfur selenium compounds, DTDSF, while
our group[38,43,44] prepared the substituted selenium analogues TMTSF,
HMTSF, and OMTSF and a few others of less interest (see fig 2). The
first thing that we can note is that the room temperature value of
the electrical conductivity has been increased for each of the conducting salts. One is tempted to ascribe this to an increased bandwidth because of the selenium orbitals. However, the maximum conductivity for these new charge transfer salts is not greater that
the maximum conductivity of TTF-TCNQ. In fact, the normalized high
temperature conductivity profiles are all remarkably similar.[43] The
major exception is HMTSF-TCNQ which remains metallic from 300K to
6mK.[38] Electronically the donor portion of this molecule (HMTSF)
is indistinguishable from TMTSF and OMTSF. That is, the three
donors have essentially identical electronic absorption spectra and
the same solution oxidation potentials as determined by cyclic voltammetry.[38,44] However, the conducting form of TMTSF-TCNQ becomes
semiconducting below 70°K and OMTSF-TCNQ forms only a red insulating
phase.[45] Again the physics of these systems will be considered in
detail in the accompanying paper, but we can note here that the keys
to unlock this puzzle are contained in the crystal structures. They
suggest that HMTSF-TCNQ is more two-dimensional than any of the
other organic charge transfer salts produced to date.[46-48]

While TMTSF-TCNQ crystallizes in two phases, a red insulating form which is probably a DADA stack and a black conducting form, OMTSF-TCNQ has only been obtained in a red insulating form. However, when TNAP, an electron acceptor which is similar to TCNQ but which has a size comparable to OMTSF replaces TCNQ as the oxidant in the reaction with OMTSF, a black conducting phase of OMTSF-TNAP can be isolated ($\sigma_{RT} \approx 200\Omega^{-1}cm^{-1}$).[45,49]

Before considering the chemical synthesis of some of the donor molecules, we can summarize some of the organic metal design constraints as follows:

1. Stable open-shell (free-radical) salts

2. Planar molecules with delocalized π-molecular orbitals

3. Inhomogeneous charge and spin distribution

4. Segregated stacks of radical ions

5. Uniform stacks -- no periodic distortions

6. Both cation and anion open shell

7. Both cation and anion nominally divalent

8. Fractional charge transfer

9. Polarizable cation

10. No disorder (symmetric anions and cations)

11. Cation and anion of similar size

While the best organic metals at the present time have all of these features in common, the design of new donor and acceptor molecules based on these constraints does not guarantee one a good new organic metal. There is still unknown territory to be explored.

II. SYNTHESIS

Inasmuch as 1,4,5,8-tetrathiafulvalene (TTF) complexes have played such an important role in the recent development of organic conductors, we examine in the remaining portion of this paper synthetic routes for the preparation of substituted derivatives of TTF and TSF. To date we know of no successful attempts to make the tellurium analog TTeF.

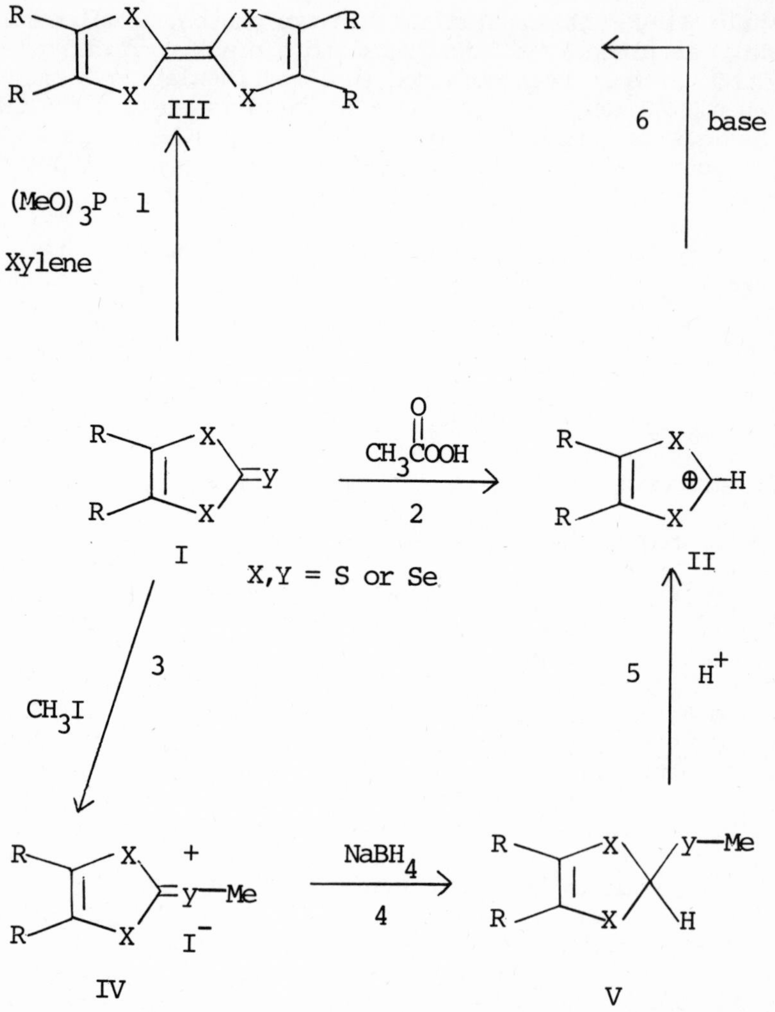

Fig. 4

A general scheme for the preparation of TTF and TSF compounds is given in fig. 4. While there are preparations which cannot be summarized by this scheme, more than 95% of the synthetic routes use 1,3-dithiole-2-thiones or their selenium counterparts as intermediates in the preparation of the desired electron donors. The earliest route used was described by Prinzbach, Berger and Luttringhaus[50] in 1965 and involves the peracid oxidation of 1,3-dithiole-2-thiones(I) to form 1,3-dithiolium salts(II) followed by deprotonation of this salt with a base like N-ethyldiisopropylamine to provide the desired tetrathiafulvalene(III). The intermediate

produced in the deprotonation reaction is presumed to be a
resonance stabilized nucleophilic carbene

This general route has been used to prepare TTF, DMTTF, TMTTF,
HMTTF and a number of di- and tetraaryl substituted TTF compounds.[50-56] This general scheme does not work with selenium
compounds.

An alternate route which is also useful with the sulfur containing compounds but has not as of yet proved very useful for
selenium compounds, involves methylation of the 1,3-dithiole-2-
thiones using a variety of methylating reagents followed by reduction of the 2-thiomethyl-1,3-dithiolium salt(IV) with sodium boro-
hydride to form the 2-S-methyl-1,3-dithiole and subsequent acid
cleavage to form the same 1,3-dithiolium salt(II) used in the prior
sequence. This sequence has been used by Wudl for the small scale
preparation of tetrathiafulvalene.[57] Fig. 5 describes an unpublished example of this type of synthesis that was recently completed
at Hopkins by Mr. Krug. 4,4´,5,5´-Tetra(thiomethoxy)tetrathia-
fulvalene $(CH_3S)_4TTF$ was first prepared by Chambers[58] using an
electrochemical reductive coupling of the ethylated 4,5-di(thiomethoxy)-1,3-dithiole-2-thione followed by a pyrolysis of the ortho-
thiooxalate to obtain the desired substituted $(CH_3S)_4TTF$. We find
the sequence shown in fig. 5 a very convenient route to this compound when more than a few milligrams of $(CH_3S)_4TTF$ are desired.
$(CH_3S)_4TTF$ cannot be prepared by the peracid oxidation route, tri-
methylphosphite desulfurization or photodesulfurization routes.
Other applications of the key electrochemically generated intermediate (I) are shown in fig. 6. A synthesis of (IV), a monomer
which when desulfurized would yield a rigid TTF-like polymer,
seemed plausible via (I).[59] From (I) and cyanogen bromide 4,5-di-
(thiocyanate)-1,3-dithiol-2-thione(II) was obtained which was
readily converted to the cyclic imidocarbonate(III). Attempts to
cleave (III) to obtain the desired compound (IV) are in progress
at the present time. Still unresolved complications arose in our
attempt to prepare (IV) via reaction of (I) with thiophosgene or
1,1´-thiocarbonyldiimidazole. A solution to this problem could
entail isolation of the intermediate (I) before subsequent reactions.

A second new example of the methylation-reduction-acid cleavage-
base deprotonation-scheme based on the work of C. Hu is presented in
fig. 7. Again in this case the strained 1,3-dithiole ring is very
sensitive to the reaction conditions and only the route shown gave

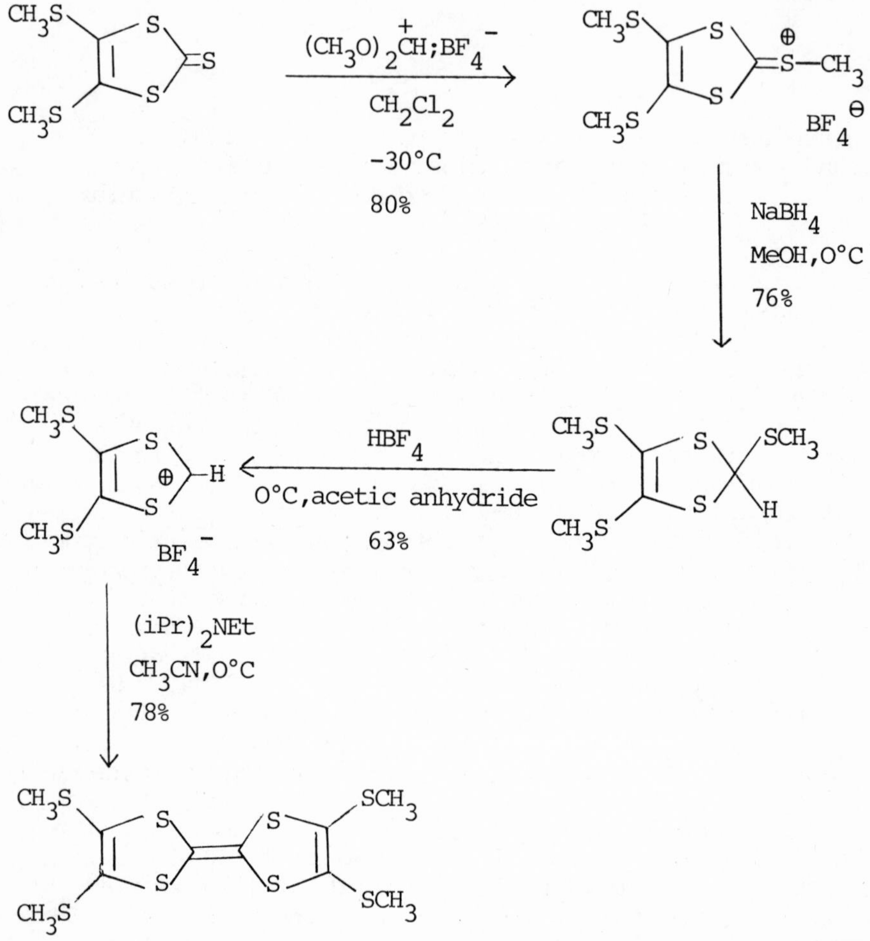

Fig. 5

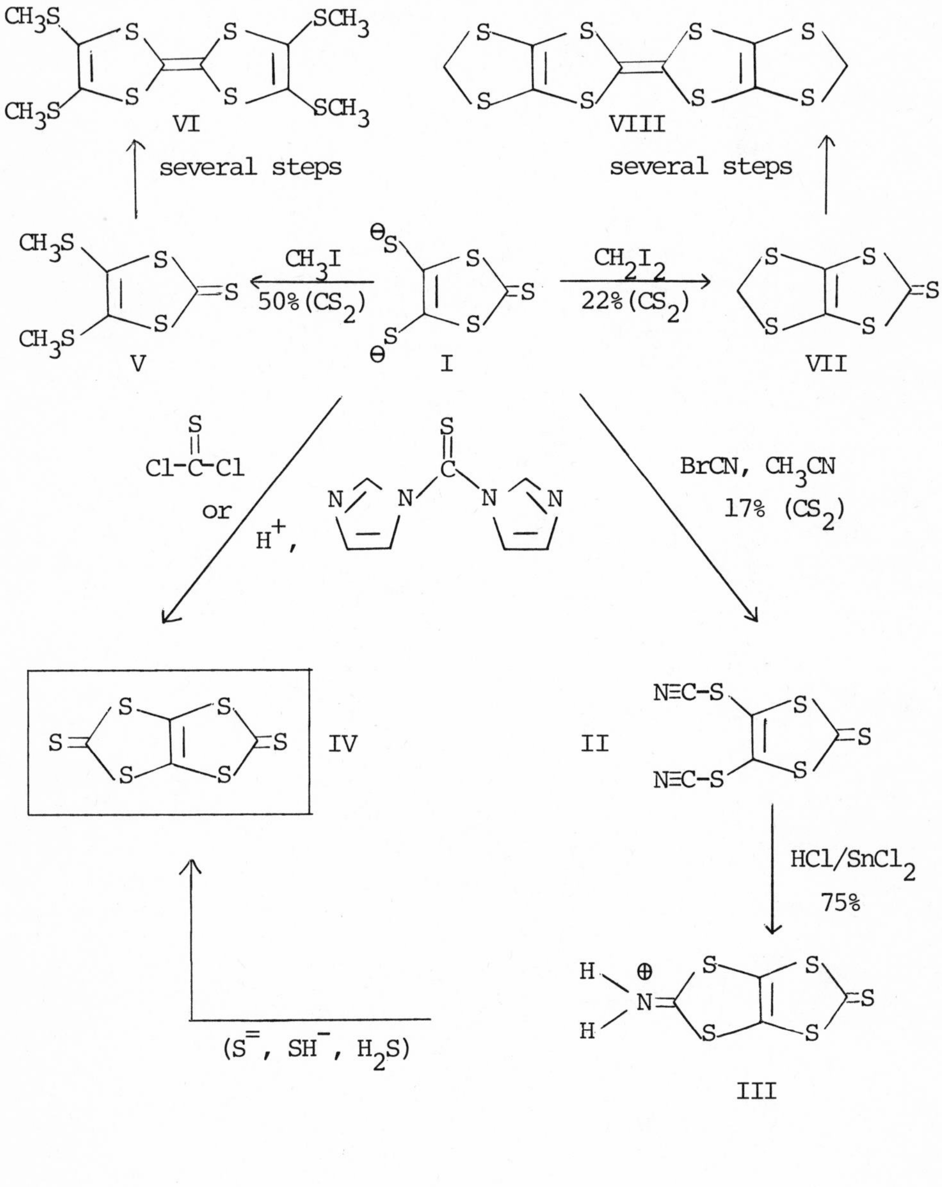
VI
several steps
VIII
several steps
CH3I
50% (CS2)
I
CH2I2
22% (CS2)
V
VII
Cl-C-Cl
or
H+,
BrCN, CH3CN
17% (CS2)
IV
II
HCl/SnCl2
75%
(S=, SH-, H2S)
III

Fig. 6

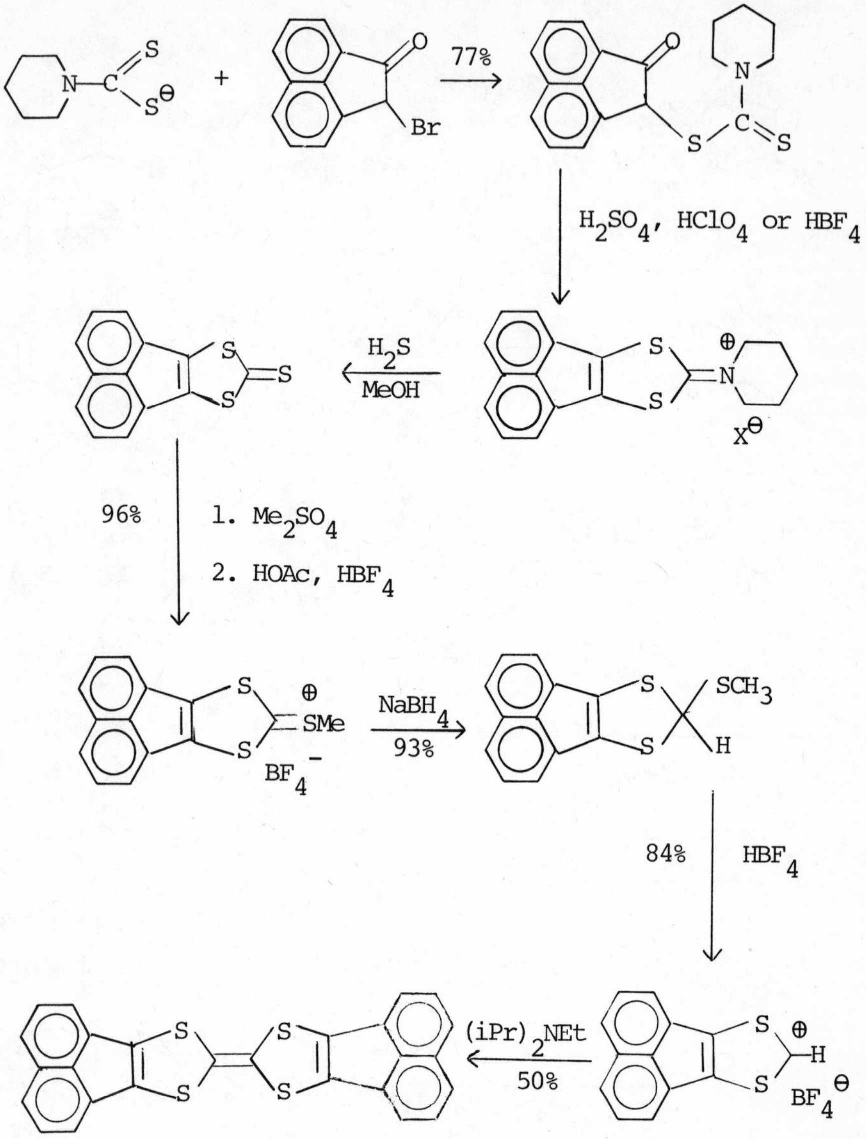

Fig. 7

satisfactory results. By preparing the aryl ring-fused TTF salts
of TCNQ, such as the acenaphtho-compound, it may be possible to
retain the stacking properties of TTF while varying the electron
density of the TTF portion of the molecule. Further, such compounds are obvious vehicles for the study of exciton-conduction
electron coupling.[60]

The only route for the coupling of selones that has proven
generally useful is the reaction with trivalent phosphorus compounds (see fig. 4). For sulfur containing compounds, the desul-
furization-coupling with trivalent phosphorus compounds has been
successful only when a strongly electron withdrawing substituent,
such as -CN, CF or -COOR is attached to the four- and/or five-
positions of the 1,3-dithiole-2-thione.[61] The 1,3-diselenole-2-
selones are more reactive than the thiones and coupling by this
route is useful for the unsubstituted 1,3-diselenole-2-selone and
for the selones with either electron withdrawing or donating substituents.[42,46,62] We have even been able to prepare the unsym-
metrically substituted TSF shown in fig. 8 by this route.

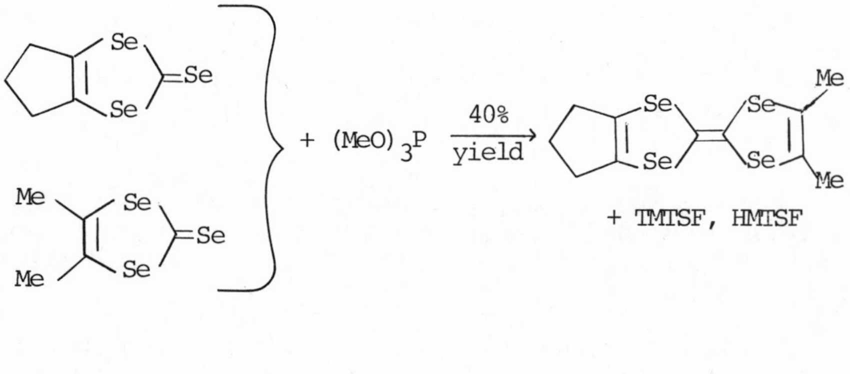

Fig. 8

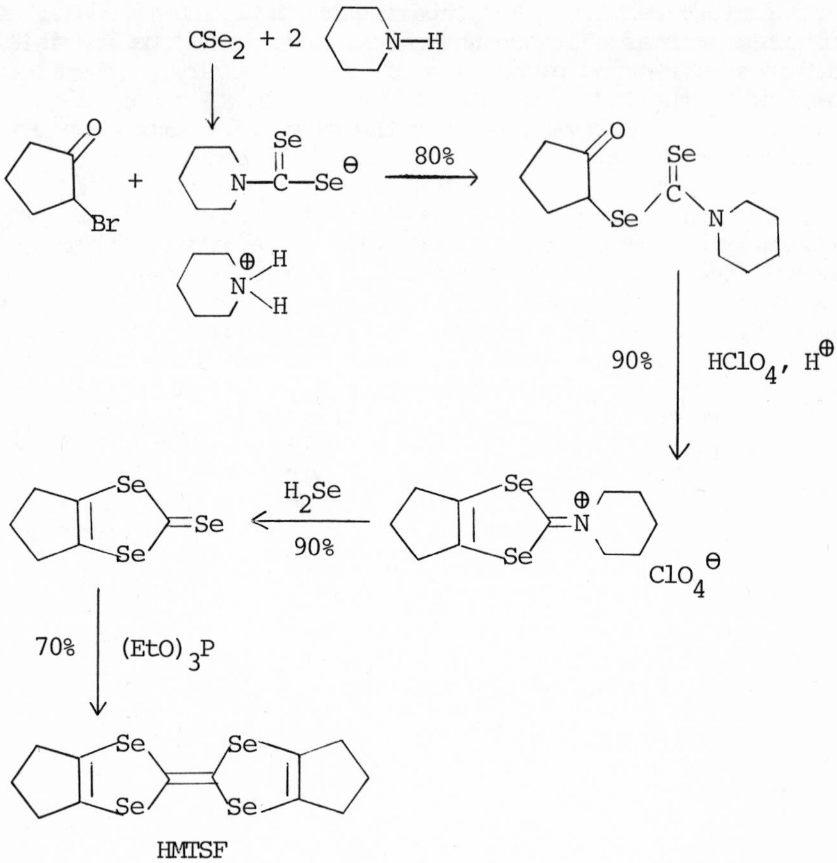

Fig. 9

Fig. 9 gives an example of the general scheme developed and utilized by Bechgaard and Shu for the synthesis of TMTSF, HMTSF, OMTSF, and a number of other di- and tetra-substituted derivatives of TSF.[43,44]

In the preparation of HMTSF, the first step involves nucleophilic displacement of a halogen from 2-bromocyclopentanone by the N,N-pentamethylenediselenocarbamate anion. Ring closure of the resulting oxo ester proceeded smoothly in concentrated sulfuric acid to yield the 2-(N,N-pentamethylenimino)-1,3-diselenolium salt. The aforementioned salt was subsequently treated with perchloric

or fluoroboric acid to yield the corresponding perchlorate or
fluoroborate salt, respectively. Cleavage of this salt with excess
H_2Se gave the anticipated 1,3-diselenole-2-selone which was coupled
by triethylphosphite to yield HMTSF. The only failure using this
general scheme came when we attempted to prepare the cyclobutyl
analogue of HMTSF. The ring closure of the oxo ester did not yield
I fig. 10 but apparently III instead.[45]

A new route for the preparation of tetraselenafulvalenes that
we have had some successwith recently is shown in fig. 11. This
route avoids the use of CSe_2 and uses the easier to prepare
selenaureas.[45]

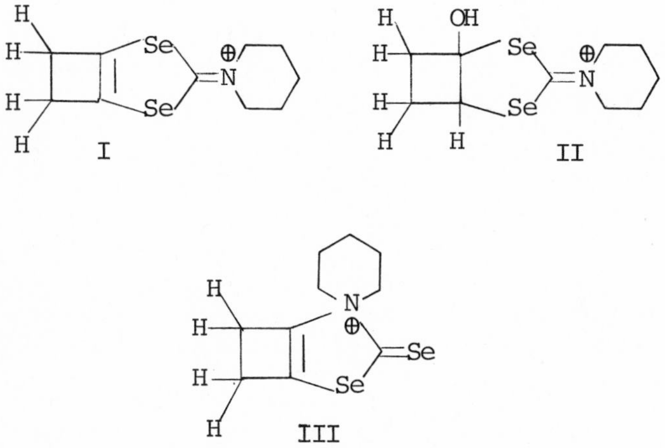

Fig. 10

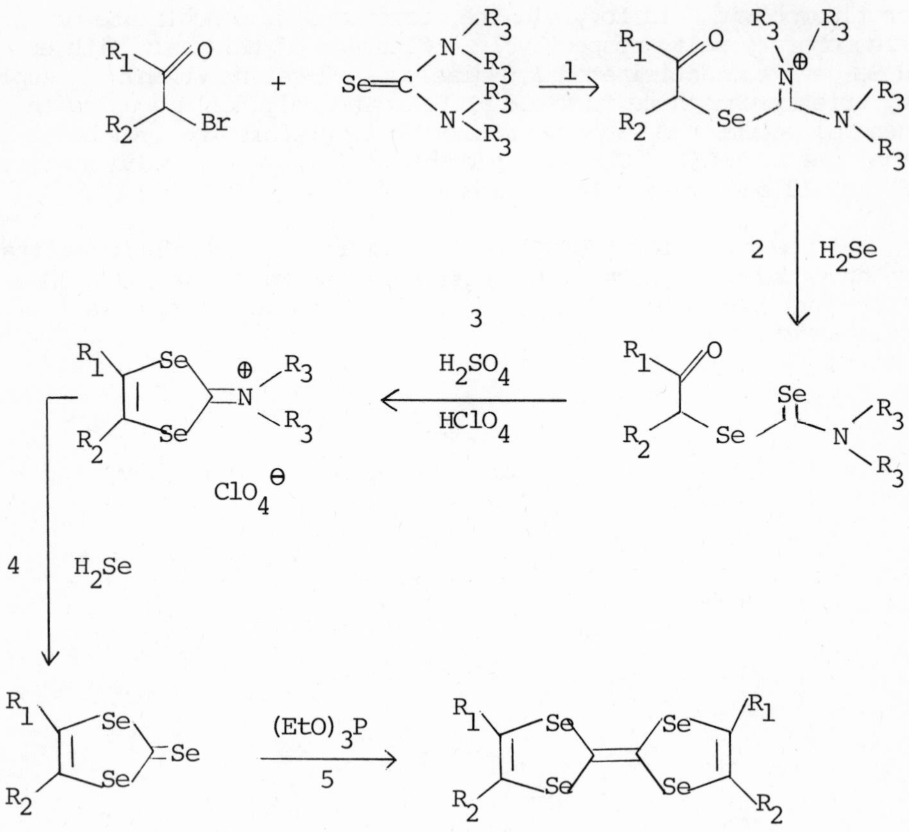

Fig. 11

III.   ACKNOWLEDGMENTS

This work was supported by the Materials Science Office, Advanced Research Projects Agency, Department of Defense. We thank J. Wilson, F. Wudl, R. Schumaker, K. Bechgaard, and E. Engler for useful comments regarding the chemistry of organic metals.

IV.   REFERENCES

1. A number of review papers have been published regarding organic conductors, see ref. 2-9, 18.

2. D. O. Cowan, C. LeVanda, J. Park, F. Kaufman, Accounts Chem. Res., 6, 1 (1973).

3. A. N. Bloch in "Energy and Charge Transfer in Organic Semiconductors," ed. K. Masuda and M. Silver, Plenum Press, New York, 1974, p. 159.

4. A. N. Bloch, D. O. Cowan, and T. O. Poehler in "Energy and Charge Transfer in Organic Semiconductors," ed. K. Masuda and M. Silver, Plenum Press, New York, 1974, p. 167.

5. D. O. Cowan, A. N. Bloch, T. Poehler, T. Kistenmacher, J. Ferraris, K. Bechgaard, R. Gemmer, C. Hu, P. Shu, W. Krug, R. Pyle, V. Walatka, T. Carruthers, T. Phillips, and R. Banks, Mol. Cryst. and Liq. Cryst., 32, 227 (1976).

6. E. M. Engler, Chemical Technology, 6, 274 (1976).

7. F. Wudl, G. A. Thomas, D. E. Schafer, and W. M. Walsh, Jr., Mol. Cryst. and Liq. Cryst., 32, 147 (1976).

8. A. F. Garito and A. J. Heeger, Accounts Chem. Res., 7, 232 (1974).

9. A. F. Garito and A. J. Heeger in "Low-Dimensional Cooperative Phenomena," ed. H. J. Keller, Plenum Press, New York, 1975, p. 89.

10. F. Gutmann and L. E. Lyons, "Organic Semiconductors," J. Wiley, New York, N.Y., 1967.

11. D. S. Acker and W. R. Hertler, J. Am. Chem. Soc., 84, 3370 (1962).

12. L. R. Melby, R. J. Harder, W. R. Hertler, W. Mahler, R. E. Benson, and W. E. Mochel, J. Am. Chem. Soc., 84, 3374 (1962).

13. W. J. Siemons, P. E. Bierstedt, and R. G. Kepler, J. Chem. Phys., 39, 3523 (1963).

14. L. R. Melby, Can. J. Chem., 43, 1448 (1965).

15. R. G. Kepler, J. Chem. Phys., 39, 3528 (1963).

16. D. B. Chesnut, J. Chem. Phys., 40, 405 (1964).

17. M. T. Jones and D. B. Chestnut, J. Chem. Phys., 38, 1311 (1963).

18. Z. G. Soos, Annu. Rev. Phys. Chem., 25, 121 (1974).

19. F. H. Herbstein in "Perspectives in Structural Chemistry,"
    Vol. IV, J. D. Dunity and J. A. Ibers, ed. Wiley, New York,
    N. Y., 1971, p. 166.

20. H. T. Jonkman, G. A. Van der Velde, and W. C. Nieuwpoort,
    Chem. Phys. Lett., 25, 62 (1974); H. T. Jonkman and J. Kom-
    mandeaur, Chem. Phys. Lett., 15, 496 (1972); F. Herman and
    I. P. Batra, Phys. Rev. Lett., 33, 94 (1974); D. Salahub,
    R. Messmer, and F. Herman, Phys. Rev. B, 13, 4252 (1976).

21. P. Reiger and G. Fraenkel, J. Chem. Phys., 37, 2795 (1962);
    J. Murgich and S. Pissanetzky, Chem. Phys. Lett., 18, 420
    (1973).

22. T. E. Phillips, T. J. Kistenmacher, J. P. Ferraris, and D. O.
    Cowan, Chem. Commun., 472 (1973); T. J. Kistenmacher, T. E.
    Phillips, and D. O. Cowan, Acta Cryst., B30, 763 (1974).

23. J. Ferraris, D. O. Cowan, V. Walatka, and J. Perlstein,
    J. Am. Chem. Soc., 95, 948 (1973); A. N. Bloch, J. P. Ferraris,
    D. O. Cowan, and T. O. Poehler, Solid State Commun., 13, 753
    (1973); D. E. Schafer, F. Wudl, G. A. Thomas, J. P. Ferraris,
    and D. O. Cowan, Solid State Commun., 14, 347 (1974).

24. A. J. Berlinsky, J. F. Carolan, and L. Weiler, Solid State
    Commun., 15, 795 (1974); U. Bernstein, P. M. Chaikin, and
    P. Pincus, Phys. Rev. Lett., 34, 271 (1975); V. K. S. Shante,
    et al., Bull. Am. Phys. Soc., 21, 287 (1976).

25. L. B. Coleman, J. A. Cohen, A. F. Garito, and A. J. Heeger,
    Phys. Rev. B, 7, 2122 (1973); J. H. Perlstein, M. J. Minot,
    V. Walatka, Mat. Res. Bull., 7, 309 (1972); L. I. Buravov,
    D. N. Fedutin, and I. F. Shchegolev, Zh. Eksp. Teor. Fiz.,
    59, 1125 (1971).

26. A. N. Bloch, R. B. Weisman, and C. M. Varma, Phys. Rev. Lett.,
    28, 753 (1972).

27. O. H. Le Blanc, J. Chem. Phys., 42, 4307 (1965).

28. P. M. Chaikin, A. F. Garito, and A. J. Heeger, J. Chem. Phys.,
    58, 2336 (1973).

29. G. A. Thomas, et al., Phys. Rev. B, 13, 5105 (1976).

30. R. P. Groff, A. Suna, and R. E. Merrifield, Phys. Rev. Lett., 33, 418 (1974). For a contrasting view, see M. J. Cohen, L. B. Coleman, A. F. Garito, and A. J. Heeger, Phys. Rev. B, 10, 1298 (1974).

31. While this equation gives the best conductivity fit for the entire temperature range (70 to 300°K), recent work by T. Carruthers indicates that this may be an oversimplication of the temperature dependence.

32. D. Jérome, W. Müller, and M. Weger, J. Physique Lett., 35, L-77 (1974).

33. S. Etemad, Phys. Rev. B, 13, 2254 (1976).

34. R. A. Craven, M. B. Salamon, G. DePasquali, R. M. Herman, G. Stucky, and A. Schultz, Phys. Rev. Lett., 32, 769 (1974).

35. F. Denoyer, F. Comes, A. F. Garito, and A. J. Heeger, Phys. Rev. Lett., 35, 445 (1975); S. Kagoshima, H. Anzai, K. Kajimura, and T. Ishiguro, J. Phys. Soc. Japan, 39, 1143 (1975); C. Weyl, E. M. Engler, S. Etemad, K. Bechgaard, G. Jehanno, Solid State Commun., 19, 925 (1976).

36. R. Comes, S. Shapiro. G. Shirane, A. Garito, and A. J. Heeger, Phys. Rev. Lett., 35, 1518 (1975); P. Bak and V. E. Emery, Phys. Rev. Lett., 36, 978 (1976).

37. See references 2, 13, 27, and E. B. Yagubsky, M. L. Khidekel, I. F. Shchegolev, L. I. Buravov, B. G. Gribov, M. K. Makova, Izv. Acad. Nauk, ser. chim., 9, 2124 (1968).

38. A. N. Bloch, D. O. Cowan, K. Bechgaard, R. E. Pyle, R. H. Banks, and T. O. Poehler, Phys. Rev. Lett., 34, 1561 (1975).

39. H. Fukuyama, T. M. Rice, and C. M. Varma, Phys. Rev. Lett., 33, 305 (1974).

40. E. M. Engler and V. V. Patel, J. Am. Chem. Soc., 96, 7376 (1974); S. Etemad, T. Penney, E. M. Engler, B. A. Scott, and P. E. Seiden, Phys. Rev. Lett., 34, 741 (1975).

41. Y. Tomkiewicz, E. M. Engler, and T. D. Schultz, Phys. Rev. Lett., 35, 456 (1975).

42. E. M. Engler and V. V. Patel, J. Org. Chem., 40, 387 (1975).

43. K. Bechgaard, D. O. Cowan, and A. N. Bloch, Chem. Commun., 937 (1974).

44. K. Bechgaard, D. O. Cowan, A. N. Bloch, and L. Henriksen,
     J. Org. Chem., $\underline{40}$, 746 (1975); K. Bechgaard, D. O. Cowan,
     A. N. Bloch, Mol. Cryst. Liq Cryst., $\underline{32}$, 227 (1976); J. Cooper,
     M. Weger, D. Jerome, D. Lefur, K. Bechgaard, A. Bloch, and
     D. Cowan, Solid State Commun., $\underline{19}$, 749 (1976).

45. (a) P. Shu, unpublished results; (b) T. Carruthers, unpublished
     results.

46. T. E. Phillips, T. J. Kistenmacher, A. N. Bloch, and D. O.
     Cowan, Chem. Commun., 334 (1976).

47. T. E. Phillips, T. J. Kistenmacher, A. N. Bloch, J. P. Ferraris,
     and D. O. Cowan, Acta Cryst., $\underline{B33}$ (1977).

48. K. Bechgaard, T. J. Kistenmacher, A. N. Bloch, and D. O. Cowan,
     Acta Cryst., $\underline{B33}$ (1977).

49. T. O. Poehler, unpublished results.

50. H. Prinzbach, H. Berger, and A. Lüttringhaus, Angew Chem.,
     Int. Ed. Engl., $\underline{4}$, 435 (1965).

51. D. L. Coffen, J. Q. Chambers, D. R. Williams, P. E. Garrett,
     and N. D. Canfield, J. Am. Chem. Soc., $\underline{93}$, 2258 (1971).

52. F. Wudl, G. M. Smith, and E. J. Hufnagel, Chem. Commun., 1453
     (1970).

53. S. Hünig, G. Kiesslich, H. Quast, and D. Scheutzow, Liebigs
     Ann. Chem., 310 (1973).

54. R. Schumaker, et al., Bull. Am. Phys. Soc., $\underline{20}$, 495 (1975).

55. J. P. Ferraris, T. O. Poehler, A. N. Bloch, and D. O. Cowan,
     Tet. Lett., $\underline{27}$, 2553 (1973).

56. A. Takamizawa, K. Hirai, Chem. and Pharm. Bull. (Japan), $\underline{17}$,
     1931 (1969).

57. F. Wudl and M. L. Kaplan, J. Org. Chem., $\underline{39}$, 3608 (1974); for
     another high yield route, see L. R. Melby, H. D. Hartzler, and
     W. A. Sheppard, J. Org. Chem., $\underline{39}$, 2456 (1974).

58. R. P. Moses and J. Q. Chambers, J. Am. Chem. Soc., $\underline{96}$, 945
     (1974); E. M. Engler, D. C. Green, and J. Q. Chambers, Chem.
     Commun., 148 (1976).

59. R. Schumaker (IBM) has recently prepared this compound,
     private communication.

60. W. A. Little in "Low-Dimensional Cooperative Phenomena," ed.
    H. J. Keller, Plenum Press, New York, 1975, p. 35.

61. W. Hartzler, J. Am. Chem. Soc., 92, 1412 (1970); 95, 4379
    (1973); M. G. Miles, J. D. Wilson, D. J. Dahm, and J. H. Wagen-
    knecht, Chem. Commun., 751 (1974); C. U. Pittman, Jr.,
    M. Narita, and Y. F. Liang, J. Org. Chem., 41, 2855 (1976).

62. M. V. Lakshmikantham and M. P. Cava, J. Org. Chem., 41, 879
    (1976).

THE ORGANIC METALLIC STATE: SOME PHYSICAL ASPECTS AND CHEMICAL TRENDS

A. N. Bloch, T. F. Carruthers, T. O. Poehler, and
D. O. Cowan

Department of Chemistry, The Johns Hopkins University
Baltimore, Maryland   21218   U.S.A.

I.   INTRODUCTION

To a degree unmatched among other intrinsic conductors, the electronic properties of organic charge-transfer salts[1] are subject to chemical control. Among the derivatives and analogs of TTF-TCNQ,[2-7] for example, even minor differences in molecular structure often manifest themselves as sharp contrasts in electrical behavior. Indeed, despite their general chemical and crystallographic similarities, such compounds span the full range from insulators to the best organic conductors known.

From an experimental point of view, this sensitivity to molecular detail presents an unusual opportunity. Certainly it complicates the interpretation of materials in this class, but it also multiplies the amount of information available and the possibilities for design. In effect, the chemical modifications at our disposal represent a crude set of controlled variables which can be adjusted systematically to probe the physics of the class as a whole. Some particularly fruitful examples are molecular substitutions chosen so as to alter the dimensionality of a conductor. To illustrate, we survey here the series of compounds progressing in this manner from TTF-TCNQ[2] to HMTSF-TCNQ,[6] the first organic whose conductivity remains metallic in magnitude as T→0. We shall find that the trends and distinctions within this sequence afford physical perspectives not readily apparent from the study of any single material, and suggest approaches toward optimization of the organic metallic state.

II.   GENERAL CONSIDERATIONS

High electrical conductivity can occur in a purely organic system only under special chemical conditions.[1] In the present context, the most significant is the constraint that a sizeable population of carriers can be energetically accessible at ordinary temperatures only if the basis states for conduction consist of extended π-molecular orbitals. At least two important consequences follow.

First, because π-π interactions are relatively weak, the carriers are confined to comparatively narrow bands. In fact, for the charge-transfer salts considered here, the dominant π-π overlaps are out-of-plane and the corresponding tight-binding band-width parameter (charge-transfer integral) $t_{||}$ is but of order 0.1 eV.[8,9] This is no larger than typical room-temperature scattering rates and intramolecular optical phonon frequencies, and is probably smaller than typical unscreened intramolecular coulomb correlation energies. Under these conditions, the electrical,[10] magnetic,[11] and optical[12] response need not resemble those of a simple metal, and their interpretation may require special care.

Second, because of the geometry of the π-orbitals, the band-widths and hence the electronic properties are inherently anisotropic. Nowhere is this effect more dramatic than in the conducting charge-transfer salts. These contain planar, open-shell molecular ions, stacked so as to form a periodic array of parallel conducting chains.[13] Insofar as the interchain charge-transfer integral $t_{\perp}$ is very much smaller than $t_{||}$, the propagation of carriers is confined effectively to one dimension.

Now, it is well known that physics in one dimension is pathological. As the one-dimensional limit is approached, we should expect the metallic state to become unstable[14] and fluctuation effects, static[15] and dynamic,[16] to become severe. Hence the electronic properties in this regime ought to be especially sensitive to any chemical modifications which relax or tighten the one-dimensional constraints.

Let us examine, briefly and schematically, some of the forms that this sensitivity can take. We know, for example, that nearly one-dimensional conductors must be particularly susceptible to the influence of impurities or structural disorder.[17,15] Indeed, it has been rigorously shown[18] that in the extreme one-dimensional limit, all eigenstates of an aribitrarily weak random potential are localized, so that conduction can occur only by thermally activated hopping.[19] What bearing does this result have upon a material which is three-dimensional, but strongly anisotropic in the sense we have described? Shante and Cohen[20] have found that extended states can certainly exist provided that the rms random potential δ is small enough, and $\bar{t}_\perp$ (the average value of $t_\perp$) large enough, that

$$zt_{||}\,\bar{t}_\perp >> \delta^2 \tag{1}$$

where z is the number of nearest neighbors in the transverse directions. Though approximate, Equation (1) has the correct behavior in the one-dimensional limit $t_\perp = 0$ and in the isotropic limit $t_{||} = t_\perp$, where it takes the form of the usual Anderson[21] criterion.

The same parameters also bear upon the persistence of the instabilities of the one-dimensional metal toward formation of charge-or spin-density waves. These are conveniently discussed in terms of the corresponding response functions, which in one dimension diverge at wavenumber $q = 2k_F$. When the flat sheets of Fermi surface at $\pm k_F$ are sufficiently blurred due to thermal or static disorder, or sufficiently curved due to non-zero interchain coupling, these divergences are alleviated and the instabilities suppressed. For concreteness we consider the charge-density wave instability, governed by the static density-density response function. For a non-interacting electron gas, its magnitude at $q = 2k_F$ is:

$$\chi(2k_F,0) \sim \ell n \left[\frac{\max(T,\bar{t}_\perp,\delta)}{E_F} \right] \tag{2}$$

provided that the argument of the logarithm is much smaller than unity. When the electron-electron interaction is taken into account, the logarithm becomes an inverse power law,[22] but this need not concern us here. The important point is that the electronic polarizability at $q = 2k_F$ will be anomalously large at low temperatures when

$$E_F >> z\bar{t}_\perp,\delta \tag{3}$$

If $E_F \sim t_{||}$, as is the case in most of these materials, Equations 1 and 3 can be combined to express the condition that $t_\perp$ be large enough for the states at the Fermi level to be extended, yet small enough for the response to display one-dimensional anomalies. Such a "quasi-one-dimensional metal" is possible when

$$t_{||} >> (zt_{||}\,\bar{t}_\perp)^{\frac{1}{2}} >> \delta \tag{4}$$

The conditions for the metallic state actually to give way to a static Peierls distortion[14] and insulating gap at low temperatures are more stringent. The gain in electronic energy favoring the distortion is lost if δ is so large as to place a sufficient population of localized states in the gap,[1b] or if $t_\perp$ is so large as to intro-duce sufficient dispersion of the gap across the Brillouin zone. Explicit calculations of the effect of disorder[23,1b] and of inter-chain charge-transfer interactions[24] confirm the intuition that a gap can be sustained at $T = 0$ if

$$\Delta_o(0) > \delta \qquad\qquad (5a)$$

and

$$\Delta_o(0) > z\bar{t}_\perp \qquad\qquad (5b)$$

where $2\Delta_o(0)$ is the gap at $T = 0$ calculated in the limit $\delta = t_\perp = 0$.

The same conclusion can be reached in a slightly different way. A non-zero Peierls gap at $T = 0$ implies that mean-field theory would predict a non-zero transition temperature T_P at which overscreening by the electrons drives the frequency of a $q = 2k_F$ phonon to zero. Employing (2) in the usual mean-field linear response theory,[25] we find for the frequency of that phonon:

$$\omega^2 = \omega_o^2 \, \lambda \, \ell n \left[\frac{\max(T, z\bar{t}_\perp, \delta)}{T_P} \right] \qquad\qquad (6)$$

where ω_o is the unscreened phonon frequency and λ the dimension-less electron-phonon coupling constant. Obviously ω does not vanish, and the transition does not occur, unless $z\bar{t}_\perp$ and δ are smaller than T_P. Since $T_P \leqslant T_{P_O}$, its value in the limit $t_\perp = \delta = 0$, this is essentially the statement of Equations (5).

We note that according to (6), appreciable softening of the phonon can occur even if the transition does not. This will come about when (3) is satisfied, but not (5a) (as may be the case in KCP[26,1b]) or (5b) (as may be the case in HMTSF-TCNQ).

Finally, we recall that since long-range order at finite temperatures cannot develop in a strictly one-dimensional system,[16] a phase transition can occur only as a result of the finite inter-chain coupling. Hence the actual transition will be observed at a temperature T_C not necessarily equal to T_P but dependent upon both T_P and the strength of the relevant interchain coupling.[27] In the case of a Peierls transition this probably consists largely of the coulomb coupling between charge-density waves on neighboring chains.[28] This observation suggests that it may be difficult to predict the effect of a given structural modification upon T_C. An increase in interchain separation, for example, may tend to reduce T_C by reducing the interchain coulomb interaction, but it may also tend to enhance T_C by reducing $t_\perp$, thereby enhancing T_P. Some flexibility is available, however, insofar as $t_\perp$ depends upon both interatomic distances and relative molecular orientations, whereas the coulomb coupling depends only upon the former.

III. MATERIALS

How are real organic conductors to be classified according to the scheme outlined in the previous section? Presumably TTF-TCNQ, in which the Peierls distortion has been directly observed,[29] satisfies both Equations (4) and (5). It is therefore of particular interest to seek systems in which these conditions are not met.

A. The Variation of δ: Disordered One-Dimensional Conductors

Organic conductors which simultaneously violate (1), (4), and (5a) have been identified for some time.[17] In the TCNQ salts of N-methyl phenazinium (NMP), acridinium, and quinolinium, dipolar cations are randomly oriented[30] in such a way that the conducting TCNQ chains experience $\delta \sim t_{||}$ and retain an undistorted structure. The electronic properties of these materials, which bespeak a dense set of localized states at the Fermi level, have been reviewed elsewhere from this perspective.[1b] Since then, at least two important confirming results have developed. First, Theodorou and Cohen[31] have shown that a substantial δ is necessary in order to account for the very low-temperature magnetic behavior of NMP-TCNQ, and have presented an elegant explanation in terms of a disordered Hubbard model. Second, Kobayashi[32] has found a more ordered variation of NMP-TCNQ in which the dipoles alternate regularly along the stacking axis. We have collected samples from five different laboratories,[33] including our own, and have found by X-ray analysis[34] that the degree of disorder varies widely from sample to sample depending upon the method of preparation. Theodorou and Cohen[31] have reached much the same conclusion by comparing magnetic susceptibility data obtained by different experimental groups. Our preliminary conductivity measurements[35] indicate that the ordered Kobayashi form is a semiconductor $[\sigma(300K)\sim 1\Omega^{-1}cm^{-1}; E_{act}\sim 0.1\text{ eV}]$, in agreement with simple band theory. The high conductivities usually found[36] occur only in the presence of sufficient disorder to destroy the insulating gap.

From the point of view of design, the most important lesson of these materials[1b,c] is that while violation of (5a) does stabilize the structure against a Peierls distortion, the concomitant violation of (1) localizes the electron states and limits conductivity to the diffusive range of a few hundred reciprocal ohm-centimeters.[17,37,1b] These constraints suggest that a more versatile class of conductors may result if Equation (4) is probed from the opposite direction -- the variation of $t_\perp$ while δ remains small.

B.   The Variation of $t_\perp$:   Modifications of TTF-TCNQ

Fortunately, TTF-TCNQ is a particularly suitable prototype
for such an enterprise, because of the amenability of the TTF donor
molecule to chemical modification. Some of the possibilities[1e] are
illustrated in Figure 1. Any or all of the sulfur heteroatoms can
be replaced with selenium, and the terminal protons can be replaced
with any of a wide variety of substituent groups. To study the
role of interchain coupling, we seek to hold the molecular electronic structure and intrachain stacking patterns as nearly constant as possible, while introducing steric factors chosen so as
to force changes in the crystal structure. In this respect our
work complements some by the IBM-Yorktown Heights group, which has
studied systematically the weaker but significant effects of adjusting molecular parameters within an isostructural series.[38]

IV.   STRUCTURES AND BAND STRUCTURES

Let us examine the interchain coupling problem for TTF-TCNQ
in some more detail. To appreciate the role of $t_\perp$, we consider
the band structure predicted by simple one-electron theory.[8,9]

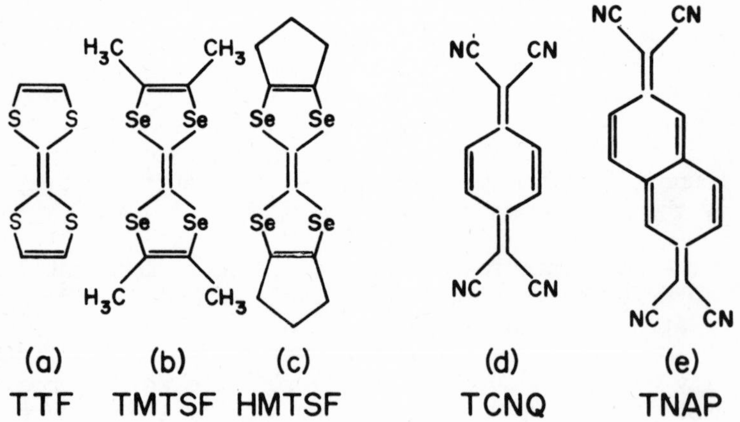

Figure 1 - Molecular structures of TTF, TCNQ, and some of their
analogs.

The essential features of our results were anticipated by the speculations of Cohen et al,[39] and independently by those of Bernstein et al,[40] though both invoked incorrect molecular physics.

The crystal structure of TTF-TCNQ,[41] shown in Figure 2, consists of parallel conducting chains of TTF (Figure 1a) and TCNQ (Figure 1d) molecular ions, separately stacked along the crystallographic b-axis. Contrary to our original expectations,[1c] the interchain stacking patterns and the nodal structures of the highest occupied molecular orbitals are such that the corresponding transfer integrals $t_{||}$ (denoted t_F for TTF and t_Q for TCNQ) are of opposite sign: $t_F > 0$ and $t_Q < 0$. Then (unless low-lying excited states play an improbably important role) the energy of states in the TTF-like band monotonically decreases, and that of states in the TCNQ-like band monotonically increases, as k_b is increased from zero. Since each formula unit contains exactly two conduction electrons, the system can be metallic only if these bands overlap in energy;[1c] in the absence of interchain coupling, they would cross precisely at the Fermi level.[39,40]

This degeneracy is lifted by the a-axis interaction between chains of unlike molecules; we denote the corresponding value of $t_\perp$ as t_{FQ}. If c-axis interchain coupling is neglected, the gap at the Fermi level proportional to t_{FQ} vanishes by symmetry at the zone boundary $k_a = \pi/a$, forming a zero-gap semiconductor. The c-axis coupling connects chains of like molecules ($t_\perp \equiv t_{QQ}'$, t_{FF}') and doubles the corresponding bands, forming pockets of Fermi surface in what is now a semimetal.

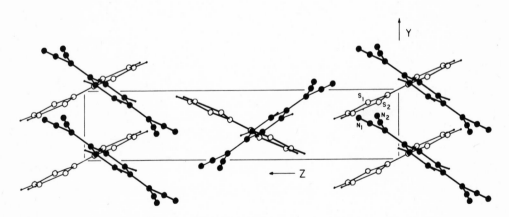

Figure 2 - Crystal structure of TTF-TCNQ projected along the a-axis.

Figure 3 presents the density of states N(E) for undistorted
TTF-TCNQ, calculated[8] in the tight-binding approximation using
Slater atomic orbitals and a Hartree potential adjusted to the
experimental[29] value of the charge transfer. Experience argues
that because of the artifically rapid decay of the Slater orbitals,
the absolute energies in the figure may be too small by a factor of
2, but that the relative values should be fairly trustworthy. Our
best estimate is that $t_Q:t_F:t_{QQ'}:t_{FQ}:t_{FF} \sim 1000:500:40:20:1$, with
$t_Q \sim 0.1$ eV. This is in fair agreement with the rough calculations
of Berlinsky et al,[9] but more consistent with the observed[42]
anisotropy of the conductivity.

From these figures arise several significant conclusions.
First, by comparison of $t_\perp$ with the experimental low-temperature
Peierls gap parameter $\Delta(0) \leqslant \Delta_0(0)$, we verify that Equation (5b)
is satisfied, though not strongly. As we show later in the paper,
our magnetic and electrical measurements both indicate $\Delta(0) \sim 100$K.

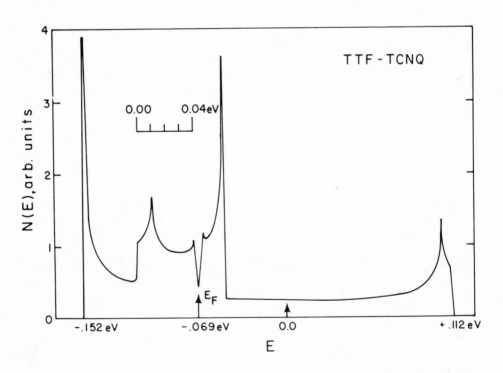

Figure 3 - Density of states for undistorted TTF-TCNQ, calculated
in one-electron tight-binding theory at T = 0.

Second, since $t_\perp < T$ above $T_C \sim 53K$, it is doubtful that covalent bonding between chains maintains any long-range interchain coherence, or that the calculated structure in $N(E)$ near E_F is meaningful, in the high-temperature metallic state.[39] Nevertheless, our results do suggest that states near the Fermi level should be especially sensitive to somewhat larger values of $t_\perp$; that, other factors being equal, an increase in $t_\perp$ by less than an order of magnitude should be sufficient to suppress the Peierls transition; and that at least at low temperatures the resulting material should behave as a semimetal. As we shall see, in large measure this prospect appears to be realized in HMTSF-TCNQ.[43]

To see how an adjustment of $t_\perp$ can be effected, we consider the interchain coupling on a molecular level. The perspective of Figure 2 emphasizes that in TTF-TCNQ, the coupling t_{FQ} (as well as the a-axis interchain coulomb interaction) occurs largely through the four short S$\cdots$N contracts between TTF and TCNQ molecules separated by $\tfrac{1}{2}a+\tfrac{1}{2}b$. As Table I indicates, the S$\cdots$N separations of 3.20 and 3.25Å are substantially shorter than the nominal van der Waals separations, and comparable with **intrachain** interplanar stacking distances. Even so, t_{FQ} remains small because of the geometry of the π-molecular orbitals and the relative tilts of the TTF and TCNQ molecular planes.[41]

The c-axis coupling can better be appreciated from the perspective of Figure 4a, which views the structure along the stacking b-axis. The strongest interaction, $t_{QQ'} > t_{FQ}$, occurs with Z=2 between TCNQ molecules separated by $\tfrac{1}{2}b+\tfrac{1}{2}c$ and is mediated by the short contacts between cyano groups (Table I). In contrast, the shortest TTF-TTF c-axis contacts involve only the terminal protons, so that $t_{FF'}$ is negligible.

The situation is drastically altered when these protons are replaced by electronically inert but physically bulky substituent groups. As Figure 4b demonstrates, even the tetramethyl derivative TMTTF[3] is simply too large to be accommodated by the crystal structure of TTF-TCNQ. The methylation does produce a slight increase in stacking distance[44] (and probably in charge transfer[45]), but the principal effect is to force a rotation of each stack about its axis.

The change in the pattern of interchain contacts (Table I) from Figure 4a to Figure 4b is striking. The strongest interchain coupling in TTF-TCNQ, the c-axis N$\cdots$N interaction, has altogether disappeared: each TCNQ cyano group now faces the inert methyl substitutents of the cation. In place of the four short a-axis S$\cdots$N contacts in TTF-TCNQ, we now have only two, and at 3.45Å these are appreciably longer than the van der Waals separation.

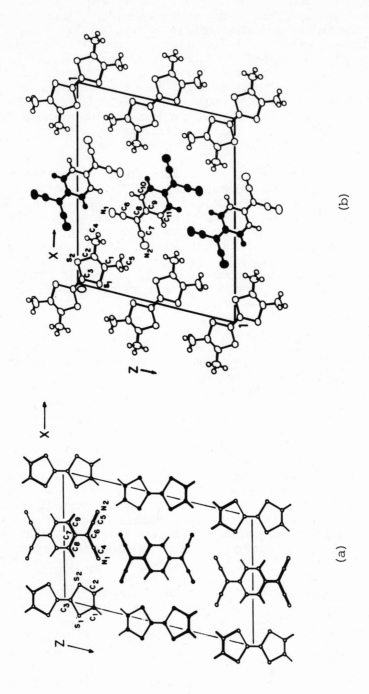

Figure 4 – Comparison of the crystal structures of (a) TTF–TCNQ and (b) TMTTF–TCNQ, projected along their conducting axes.

TABLE I

Electronically Significant Interchain Contacts

	Contact	Number	Distance, Å	van der Waals Distance, Å	Ref.
TTF-TCNQ	S··N (a)	2	3.20	3.35	41
	S··N (a)	2	3.25	3.35	
	C··N (c)	2	3.29	3.10	
TMTTF-TCNQ	S··N	2	3.45	3.35	44
	S··S	2	3.66	3.70	
TMTSF-TCNQ	Se··N	2	3.36	3.50	46
	Se··Se	2	4.00	4.00	
HMTSF-TCNQ	Se··N	4	3.10	3.50	47

A new pair of S•••S contacts does appear, but they, too, are comparable with the van der Waals spacing. Thus, its crystallography
suggests that TMTTF-TCNQ is considerably more "one-dimensional"
than its TTF parent compound. We shall find, in fact, that its
interchain coupling is the weakest of any material in the series
we have studied. (Perhaps because these interactions are so weak,
several distinct crystalline forms of TMTTF-TCNQ have been
observed.[44,46] The one described here, though different from
that originally reported in Ref. 3, is by far the most prevalent.[44])

When the sulfur heteroatoms in TMTTF are replaced with
selenium to form TMTSF[5] (Figure 1b), the arrangement of interchain
contacts in the TCNQ salt (Figure 5a) is qualitatively unchanged.[47]
However, because of the greater spatial extent of Se and the shorter
Se•••N separations (Table I), $t_\perp$ is likely to be considerably
larger. The TMTSF-TCNQ structure[47] is also the simplest in the
series in that it is the only one containing but one cation-anion
molecular pair per unit cell.

This material is of special interest for another reason. By
any spectroscopic or electrochemical measure,[6,48] the TMTSF molecule is electronically identical with its hexamethylene analog
HMTSF,[6] Figure 1c. Further, the intrachain molecular overlaps
and stacking distances in their TCNQ salts[47,49] are practically
indistinguishable. Hence, TMTSF-TCNQ and HMTSF-TCNQ can differ
only in their respective interchain couplings, and comparison of
the two materials promises an exceptionally sharp differentiation
of interchain from intrachain and molecular effects.

The contrast in interchain coupling is emphasized in Figure 5.
Just as the TMTTF and TMTSF salts are sterically precluded from
duplicating the structures of their unsubstituted parent compounds
(Figure 4), so the still larger HMTSF molecule cannot enter the
structure of Figure 5a, and HMTSF-TCNQ adopts that of Figure 5b
instead.[49] In the process the number z of short Se•••N contacts
per molecule is increased from 2 to 4, and their length is shortened to a remarkable 3.10A, considerably shorter than even the
S•••N distances in TTF-TCNQ (Table I). Further, the familiar
herringbone alternation between the stacks in TTF-TCNQ[41] (Figure 2),
preserved throughout the rest of the series,[44,47] is replaced in
HMTSF-TCNQ by the geometry of Figure 6. This arrangement is far
more conducive to direct Se•••N π-bonding. The clear inference
is that in HMTSF-TCNQ, the (b-axis) interchain coupling $(zt)_{FQ}$ is
substantially larger than the corresponding parameter for any other
material in the series.

Along the a-axis, on the other hand, the coupling must be
exceptionally weak. Here unlike molecules again alternate, but
the electron-rich TCNQ cyano groups face only the saturated outer

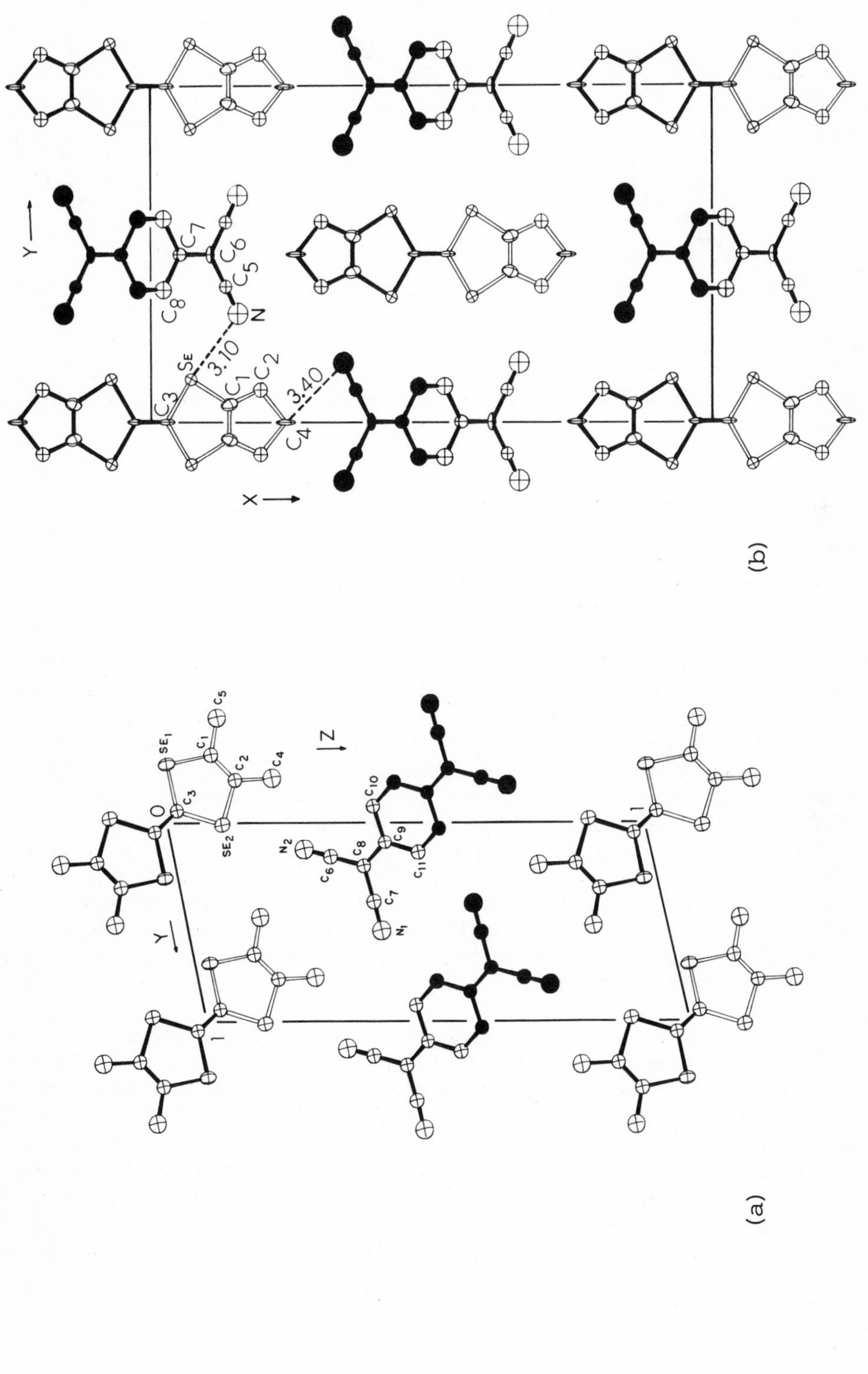

Figure 5 – Structures of (a) TMTSF–TCNQ and (b) HMTSF–TCNQ, projected along their conducting axes.

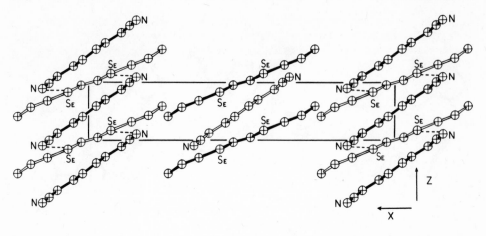

Figure 6 - Crystal structure of HMTSF-TCNQ, projected along the b-axis.

rings of the HMTSF cations. As Figures 5 and 6 indicate, these
rings form crude electrical insulators between the π-electron
systems of adjacent b-c layers, and also increase their spacing
so that even the interlayer coulomb interaction is diminished.
Indeed, so weak is the a-axis coupling that the structure is substantially disordered in this direction, as strong diffuse streaking
in X-ray photographs attests.[49] Electron micrographs[50] show evidence of macroscopic layering in the same direction. The ordered
structure of Figures 5b and 6 corresponds, in fact, to only 80%
of the total heavy-atom electron density in the crystal, the
remainder corresponding to some b-c layers displaced along the
stacking c-axis, and some composed of molecules tilted in the
direction opposite the one shown. (Unlike that discussed in
Section 2, this form of disorder affects neither the periodicity
nor the extent of the wavefunctions along the conducting chains.
It may, of course, influence the form of the density of states and
the onset of phase transitions.)

Their crystallography strongly suggests, then, that HMTSF-TCNQ
is the most "two-dimensional," and TMTTF-TCNQ the most "one-dimensional," of the conductors in this series. To map the intermediate
régime, it may be useful to make reference to some of the materials
developed and studied in other laboratories. Of these, TSF-TCNQ,[4]
the selenium analog of TTF-TCNQ, is isostructural with the parent
compound,[38] as are the solid solutions of the two. The disulfur-
diselenium analog DTDSF-TCNQ[38] also shares this structure. On the
other hand, the structure of HMTSF-TCNQ is not quite reproduced in

its sulfur analog HMTTF-TCNQ;[51] compared with the former, the interchain coupling in the latter should be geometrically similar but considerably weaker along the b-axis, and somewhat stronger along the a-axis. The structure of the dimethyl isomeric mixture (cis,trans)-DMTTF-TCNQ (ATTF-TCNQ)[52] is unknown.

Based upon this information, we should expect on chemical and crystallographic grounds that in effective "dimensionality" these conductors should rank as follows:

$$HMTSF- > TSF- > TMTSF-TCNQ \qquad (7a)$$

and

$$HMTTF- > TTF- > TMTTF-TCNQ \qquad (7b)$$

with each selenium compound substantially higher than its sulfur counterpart. (Further information is required to compare, say, TMTSF-TCNQ and HMTTF-TCNQ.) In the next section we inquire to what extent these trends are reflected in the electronic properties of the materials.

V.   EXPERIMENTAL RESULTS

Of the substantial body of experimental data now available for these materials, we concentrate upon those aspects most directly pertinent to the trends discussed above. Some of this information is summarized in Table II.

A.   Spin-Resonance Linewidths

In contrast with other materials of high conductivity, organic conductors in this class typically display sharp, well-defined spin-resonance signals.[52] The associated g-values have been exploited as a measure of local susceptibilities,[53] and the anisotropic line-shape[54] and intensity[55] as measures of microwave conductivity. Our immediate concern, however, is with the spin-resonance linewidth. In collaboration with Tomkiewicz and Schultz, we have shown that this quantity probes the effective dimensionality of a conductor.[56,57]

We recall that in an isotropic conductor, the electronic spin-lattice relaxation time T_1 is usually governed by two closely related processes. One is the phonon modulation of the spin-orbit coupling, studied by Overhauser.[58] The other was treated by Elliott,[59] who observed that an electronic wavefunction predominantly of one spin also contains a term of opposite spin, first-order in the spin-orbit coupling; and that the ordinary electron-phonon interaction scatters between the large component of one

TABLE II

Properties of 1:1 TCNQ Salts of:

	R	X	σ_{RT} (cm$^{-1}\Omega^{-1}$)	σ_{max}/σ_{RT}	T_{max}	T_c	$\Delta g \times 10^4$ (300K)	$\Delta H,G$ (300K)	
TTF	-H	S	500	20	59	53,47,38	20,40,-2	6	JHU
DMTTF	2-H,2-CH$_3$	S	50	25	~50	~35 (broad)	37,2	5	JHU/Penn
TMTTF	-CH$_3$	S	350	15	60	34	37,37,1	3-4	JHU
HMTTF	-(CH$_2$)$_3$-	S	500	4	80	50,40	~40	11	IBM(SJ)
DTDSF	-H	2S,2Se	500	7	64	~45 (broad)	~100	250	IBM(Y)
TSF	-H	Se	800	12	40	28	~100	500-650	IBM(Y)
TMTSF	-CH$_3$	Se	1,200	6	61	57	88,-30	100	JHU
HMTSF	-(CH$_2$)$_3$-	Se	2,000	3.5	No Transition		Not Observable		JHU

spinor and the small component of another. Yafet[60] has shown that at all temperatures, the two mechanisms in combination produce a linewidth $\Delta H \sim T_1^{-1} \sim (\delta g)^2 \tau_{ph}^{-1}$, where δg is the average g-shift and τ_{ph}^{-1} the scattering rate characterizing the phonon part of the resistivity.

Our point[56],[57] is that in a purely one-dimensional metal, these mechanisms are ineffective. Here the only possible scattering wavevectors lie near $q=0$ or $q=2k_F$. In the former case, energy conservation precludes a single phonon of energy qv_s from creating or destroying an electron-hole pair of energy qv_F. (If the phonon dispersion is three-dimensional, the process may be energetically allowed, but the matrix element[60] is never larger than order q^2.) Spin relaxation via backscattering through $q=2k_F$, on the other hand, would simultaneously reverse spin and linear momentum. This is forbidden by time-reversal symmetry,[59],[60] and the matrix element is proportional to $|q-2k_F|$. Hence the spin resonance line for a one-dimensional metal can be sharp even where that for an isotropic metal of the same τ_{ph} would be too broad to observe.

In nearly one-dimensional conductors such as those described here, the situation is somewhat modified. As $t_\perp$ is increased, so are the phase space available for spin-flip scattering and hence the linewidth. The most important processes involve interchain scattering with $q_{||} \sim 2k_F$. A rough calculation, taking account of the phase-space cutoffs imposed by conservation of energy, yields in the high-temperature limit:

$$\Delta H \sim (\delta g)^2 \, \tau_{ph}^{-1} \left(\frac{t_\perp}{t_{||}}\right)^2 \left(\frac{\max[t_\perp, \omega_{ph}]}{t_\perp}\right), \qquad T \gg t_\perp, \omega_{ph}, T_c \quad (8)$$

where ω_{ph} is the frequency of a phonon at $q=2k_F$. Admittedly this simple expression does not describe the temperature dependence of the linewidth, which increases with cooling in some of the compounds[52],[56] and decreases in others.[56] At lower temperatures the problem becomes considerably more complicated, particularly when phonon softening is taken into account. Nevertheless, at room temperature the dependence on $(\delta g)^2$ is experimentally established,[57] and the linewidths listed in Table II suggest precisely the trends anticipated in Equations (7). These are reinforced if the room temperature conductivities σ_{RT}, also listed in the table, are a measure of τ_{ph}.

Particularly encouraging is the result that HMTSF-TCNQ, which we have taken to possess the highest dimensionality in the series, is the only member of the class in which no spin-resonance signal is observed. In our experiment this implies a linewidth substantially larger than 4000 G, at least throughout the temperature range 4-300 K. Our interpretation of both the technique and the material is buttressed by a comparison with TMTSF-TCNQ, the compound which, we have argued, differs from HMTSF-TCNQ only in interchain coupling. The appearance of a well-defined (100 G) line in TMTSF-TCNQ is strong evidence that the extra broadening in the HMTSF salt is not an intramolecular or intrachain effect. Further, we note that observable spin resonance is also absent in the semimetallic normal state of the polymer $(SN)_x$.

Our conclusions in this section are largely consistent with those drawn by Jerome and Weger[61] from the frequency dependence of nuclear spin-lattice relaxation rates. However, their results imply a much larger interchain coupling for TMTTF-TCNQ than do ours, which are of course more consistent with the crystal structure.[44] At present this discrepancy is unexplained, unless either by the multiple crystalline forms of the material,[44,46] or else by the intervention of an additional spin relaxation mechanism in either the NMR or ESR experiments.

B.   Static Magnetic Susceptibilities

Qualitatively, the temperature-dependent magnetic suscepti-bility $\chi(T)$ for each member of the series has same the general shape as the plots in Figure 7, becoming increasingly diamagnetic almost linearly with decreasing temperature above the metal-to-insulator transition temperature T_c, and quickly dropping and flat-tening below. Some features of the data are summarized in Table III. Except in the case of HMTSF-TCNQ,[64] the low-temperature dia-magnetic limit $\chi(0)$ is very close to the core diamagnetism calcu-lated from Pascal's constants or determined experimentally from separate measurements of the neutral molecules. If this contribu-tion is treated as temperature-independent and subtracted from the data, the resulting room-temperature spin susceptibilities $\Delta\chi$ are all consistent in magnitude with a simple Pauli paramagnetism, except possibly for the case of TTF-TCNQ itself.[65] Unless the bandwidth parameters are really almost as small as indicated in Figure 3, $\Delta\chi$ is 20-30% larger than calculated.[65,8] This is con-sistent with a modest enhancement due to electron-electron inter-actions,[65] or simply with a departure from cosinusoidal dispersion of the one-electron conduction bands due to configuration inter-action with excited molecular states.[8,66]

In any case, we have verified by explicit calculation[8] that the undistorted band structure alone cannot account for the temper-ature dependence of $\Delta\chi$, which is not fully understood. We emphasize that a correct explanation must be adequate to explain the nearly identical behavior of $\Delta\chi(T)/\Delta\chi(300K)$ above T_c for all members of the class.[62-64] In the context of discussions of the magnitude of the effective electron-electron interaction,[65,67,22] it is inter-esting that this behavior extends to TTF-TCNQ, which exhibits[29] a $4k_F$ phonon softening from room temperature down, and an onset of

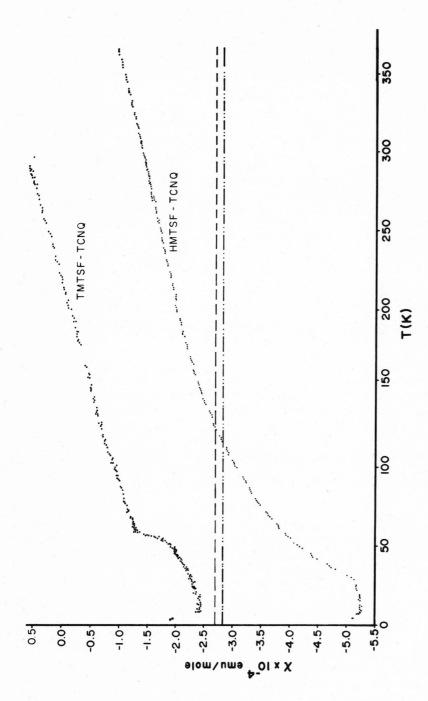

Figure 7 – Temperature dependence of the static magnetic susceptibilities of TMTSF-TCNQ and HMTSF-TCNQ. The experimental core diamagnetism of TMTSF$^\circ$ (---) and HMTSF$^\circ$ (-·-·-·-) + TCNQ$^\circ$ are shown for comparison.

TABLE III

Static Magnetic Susceptibilities, 10^6 emu/mole

	χ_{core}(calc)	χ_{core}(exp)	$\chi(0)$	$\Delta\chi=\chi(300K)-\chi(0)$	Ref.
TTF-TCNQ	-193	-221	-205	+605	62
DMTTF-TCNQ	-214	—	-197	+340	62
TMTTF-TCNQ	-238	—	-257	+480	62
TSF-TCNQ	-249	—	-263	+300	63
TMTSF-TCNQ	-273	-270	-287	+333	64
HMTSF-TCNQ	-285	-298	-524	+380	64

$2k_F$ softening near 150K; to TSF-TCNQ, which exhibits[68] $2k_F$ softening below 230K and no $4k_F$ softening; and to HMTSF-TCNQ, which
exhibits[68] $2k_F$ softening even at room temperature, but no anomaly
at $4k_F$.

Despite the ubiquity of the temperature dependence, it is
apparent from the figure and the table that the magnitude of $\chi(T)$
in HMTSF-TCNQ is anomalous. Here $\chi(0)$ is nearly twice the molecular core value, and the material remains diamagnetic even above
room temperature.[64,69] Comparison with TMTSF-TCNQ (Figure 7) is
again instructive: since the molecular and intrachain electronic
structures of the two materials are identical, the excess diamagnetism in the HMTSF salt must arise from the difference in interchain coupling. Consistent with the discussion of Section 3, we
assign it to formation of the large intermolecular coherent orbits
characteristic of a semimetal.[64,69]

It is tempting to suggest[69,61] that this orbital diamagnetism
is present only at low temperatures, where long-range interchain
coherence is fully established. Certainly that would be consistent
with the magnetoresistance,[70,61] which becomes appreciable only
below 100K. On the other hand, such behavior would also require
that the room-temperature spin susceptibility of HMTSF-TCNQ be
anomalously small for the series -- about 40% of the value for
TMTSF-TCNQ (Table 3). But if the three-dimensional electronic
structure near E_F were really obliterated by thermal fluctuations,
the room-temperature susceptibilities of the two materials ought
to be identical. Such reasoning, combined with the very nearly
parallel high-temperature $\chi(T)$ curves for the TMTSF and HMTSF salts
(Figure 7), persuades us that at least 70% of the low-temperature
excess diamagnetism of HMTSF-TCNQ is very probably present at room
temperature, at that the remaining diamagnetic contribution begins
to accumulate with cooling below ~140K, where the two susceptibility
curves cease to be parallel. This is close to the temperature at
which the normalized conductivity[6] of HMTSF-TCNQ begins to flatten
and fall below those of the other compounds, as discussed in the
next section.

Finally, we note that Figure 7 also reveals a sharp break in
the slope of $\chi(T)$ for HMTSF-TCNQ near 32K, and that heat capacity
measurements[64] confirm a phase transition at this temperature. The
transition obviously does not lead to an insulating state, but it
is reflected as a sharp change in the slope of the temperature-
dependent conductivity.[71] Hence studies of the conductivity under
pressure[70,61] can be interpreted as indicating that the transition
is continuously suppressed to lower temperatures with increasing P,
and disappears entirely below 2Kbar.

C. Conductivities and Excess Noise

Along the stacking axis, the room-temperature d.c. conductivities of compounds in this class (Table II) are near the lower threshold of the metallic regime: they correspond to nominal mean free paths[17,72] ranging from about one lattice constant in the sulfur compounds to about five in HMTSF-TCNQ. With cooling, the conductivity rises rapidly, then typically peaks at a temperature T_{max} and decreases perceptibly before under- going single or multiple phase transitions to a low-temperature semiconducting state. The sole exception in this series[73] is HMTSF-TCNQ, in which the maximum is extremely broad, the metal- to-insulator transition does not occur (notwithstanding the unexplained phase transition mentioned in the previous section), and the metallic magnitude of the conductivity persists at least to 6mK.[74]

The temperature-dependent resistivities of the four materials we have singled out for extended discussion are compared in Figure 8. For convenience we consider the various temperature regions separately.

1. High Temperatures $T > T_{max}$. It has been widely observed[75,76] that above T_{max}, the resistivity of TTF-TCNQ can be fit to the empirical formula

$$\rho(T) = \rho_o + b\, T^{\gamma} \tag{9}$$

where ρ_o is a sample-dependent constant that presumably arises from crystal imperfections. We find this fit to be universal for the series,[6,38,63] with the value of b varying among different com- pounds but the constant γ common to them all. The dependence on ρ_o and b can be eliminated from (9) by normalizing $\rho(T) - \rho_o$ to its value at 300K, as in Figure 8b. Plotted in this way, the high- temperature resistivities of the four compounds are practically indistinguishable from one another, with $\gamma = 2.3 \pm 0.1$.

We emphasize that apart from the importance of ρ_o,[75,76] no direct physical significance need be attached to the precise form of the empirical expression (9): it is but the simplest of several functional forms that fit the data equally well.[71] Nevertheless, the universality of the fit does imply that at high temperatures all of these materials conduct by the same mechanism. It follows that the temperature dependence of the conductivity, like that of the magnetic susceptibility,[62-64] is sensitive neither to the strength of the interchain coupling nor to the development of $2k_F$ or $4k_F$ phonon anomalies.[29,68]

This last observation, together with the short coherence lengths observed in diffuse X-ray scattering,[29,68] the modest ap- parent mean free paths,[72] and the experimental validity of the

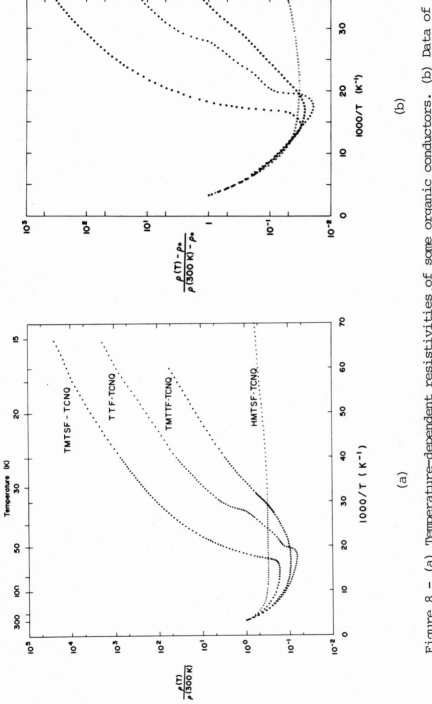

Figure 8 – (a) Temperature-dependent resistivities of some organic conductors. (b) Data of Figure 8a, corrected for sample-dependent constant ρ_0 and normalized to values at 300K.

Wiedemann-Franz Law for TTF-TCNQ,[77] militates against the hypothesis[78,42] that a current-carrying collective mode contributes appreciably to conduction at these temperatures. It is sometimes argued that such a mechanism would be required to explain the rapid rise in conductivity if, as the form of $\chi(T)$ might suggest,[62] the density of single-particle states $N(E_F)$ is falling with decreasing temperature. Such reasoning is not necessarily sound. For example, we observe that even in the weak-scattering limit, the theorem[79] that σ is independent of $N(E_F)$ in the metallic regime does not apply to a nearly one-dimensional metal. Here the single-particle conductivity calculated[17,72] from the usual kinetic equations would depend only upon the cross-sectional density of chains and the effective mean free path Λ. In Born approximation,[79] $\Lambda = v_F\tau \propto [N(E_F)]^{-2}$; and since $\sigma \propto \Lambda$, the conductivity rises as $N(E_F)$ falls, at least until $N(E_F)$ becomes so small that the metallic description is invalid. Such a rise can occur in one dimension because, in contrast with the isotropic case,[79] the increases in v_F and τ with decreasing $N(E_F)$ are not compensated by any decrease in the area of the Fermi "surface." One is of course reluctant to take these simple expressions seriously for the present materials (although the relation $\rho(T) \propto T[\chi(T)]^2$, deduced in this way for $\tau = \tau_{ph}$ above the Debye temperature, is in surprisingly good agreement with the data) but they do illustrate that an inverse relation between σ and $N(E_F)$ is not paradoxical in this geometry.

The precise mechanism for the high-temperature resistivity is still in question. A strong temperature dependence can certainly arise from the strong energy dependence of $N(E)$ near E_F implied by the band-structure calculations.[39,40,6,8] A T^2 contribution is also predicted for one dimension by calculations[80] of carrier propagation in the dynamically disordered environment created by incoherent multiphonon excitations at high temperatures. Optical experiments[81] indicate substantial coupling of the electrons to intramolecular optical modes; their contribution to the d.c. resistivity is strongly temperature-dependent in any dimension. Nor can the possibility[82] of a T^2 term due to strong electron-hole scattering be lightly dismissed. At present our work does not distinguish unambiguously among such effects, but studies of the variation of b [Equation (9)] with chemical trends in $t_{||}$, ionic masses, and charge transfer are in progress.

2. The Region of Phase Transitions. The data suggest that collective effects precursive to the phase transition may assume importance as the temperature is reduced to the vicinity of T_{max}. In the case of TTF-TCNQ, for example, the resistivity begins to rise at $T_{max} \sim 59$ K and has increased by about 50% before a second-order phase transition occurs at $T_c \sim 53$ K.[83,29] With diminishing sample quality, T_c is unchanged but the increase in resistivity commences at higher temperatures and lower conductivities;[42]

conversely, in highly purified samples[84] we have observed T_{max} as low as ~56 K. Such behavior is consistent with the theoretical result[85,86] that the resistivity, and in particular the impurity scattering, can be dynamically enhanced through the divergence of the electronic polarizability at $q = 2k_F$ as $T \to T_c$. The greater the impurity concentration, the larger would be this fluctuation contribution to the resistivity, and the higher the temperature at which it rises above the contribution (9) to form the conductivity maximum.

The development of this part of the resistivity with temperature is apparent in the derivative plot[87] of Figure 9, where the rate of rise of $\partial(\ln \rho)/\partial(T^{-1})$ with cooling begins to accelerate as early as 70 K. Horn and co-workers[88] were the first to demonstrate that the enhanced scattering rate leads to a critical divergence in $\partial\rho/\partial T$ as $T \to T_c$ from above; this behavior is also evident in the figure.

It is now well known that TTF-TCNQ undergoes at least three closely spaced phase transitions. On the basis of neutron scattering studies, Bak and Emery[89] have plausibly identified these as the three-dimensional ordering of one set of chains at ~53 K; the ordering of the other set at ~47 K, along with the onset of a continuous shearing of the superlattice so as to lower the interchain coulomb energy; and a first-order locking of the shear with transverse period 4a near 38 K. Etemad[38c] is responsible for the first convincing evidence that the 38 K transition is reflected in the conductivity. Figure 9 reproduces this result, and reveals[90] a weak anomaly at 47 K as well.

The phase transitions and their influence on the conduction can be studied in another way. One of us (T. F. C.) has observed that when an order parameter is coupled to the resistivity, its fluctuations should be manifested as fluctuations in the resistance, or excess noise.

Formally, we have described[91] the bulk electrical noise in a conductor by expanding the voltage autocorrelation function in powers of a steady applied potential V_o. The zero-order term simply states the Nyquist theorem, describing the fluctuations in current at constant resistance, or equilibrium Johnson noise[92]. The next non-vanishing term is quadratic in V_o, involves four long-wavelength density fluctuation operators, and describes the fluctuations in resistance at constant current, or excess noise.

Experimentally, the excess noise is the difference between the mean-square voltage fluctuations $<|\delta V|^2>$ measured in the presence and absence of a quiet d. c. voltage $V_o = I_o R$. We determine it using a four-probe technique designed to eliminate the relatively small excess contact noise.

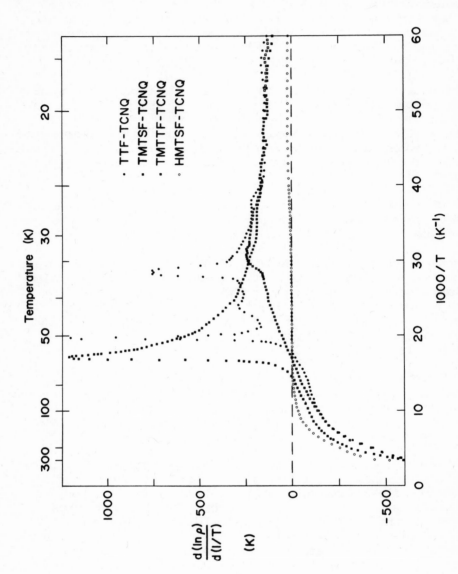

Figure 9 – Temperature derivatives of the data of Figure 8.

For TTF-TCNQ, we find that the spectral dependence of the excess noise is 1/f at least from 1 to 10,000 Hz at all temperatures. Of more immediate interest here, however, is the temperature dependence of the broadband excess noise power, shown in Figures 10 and 11 for two samples of TTF-TCNQ.

Above T_{max}, the excess noise power varies little despite the strong temperature dependence of $\rho(T)$ and $\partial\rho/\partial T$. In contrast, it is evident from Figure 10 that the rise in resistivity below $T_{max} \sim 59K$ is accompanied by a strong buildup in the mean-square resistance fluctuations, $<|\delta R|^2> = <|\delta V|^2>/I_o^2$. Indeed, the noise power begins to diverge critically as the second-order phase transition near 53K is approached. This divergence is even stronger than that in $(\partial\rho/\partial T)^2$.

Analysis[71] reveals that for a parallel set of independently conducting chains, the excess noise power is proportional to A_c, the cross-sectional coherence area. Hence, it may be that critical exponents in the noise power directly reflect the growth of the transverse coherence length at the three-dimensional ordering temperature. Our data, however, are not yet of sufficient quality to evaluate this proposition quantitatively.

Below 53K, $<|\delta R|^2>$ eventually becomes nearly proportional to R^2, as in a semiconductor.[92] In particular, the discontinuity in $\rho(T)$ at the first-order 38K transition also is reflected in $<|\delta R|^2>$, Figure 10. But in the normalized noise power $\delta P/P = <|\delta R|^2>/R^2$ (Figure 11), only the 53K structure appears. In other words, critical resistance fluctuations appear at the second-order,[85,88] but not the first-order transition. There is also some indication of weak structure in Figure 11 near 47K, but because of the scatter in the data this observation remains tentative.

We have emphasized the complexity of the metal-to-insulator transformation in TTF-TCNQ so as to contrast it with the other materials in our series. We have already remarked in Section II that one should expect to find no detailed systematic variation of T_c (Table II) with our qualitative estimates of dimensionality, because of the competition between the tendencies of the interchain coupling to suppress T_c and to mediate the three-dimensional ordering at T_c. More interesting is the variety of behavior within the series once the region of T_c is reached.

For example, in TMTTF-TCNQ, whose structure and spin resonance suggest the weakest interchain coupling in the series, a critical divergence in $\partial\rho/\partial T$ (Figure 9) does not appear at all.

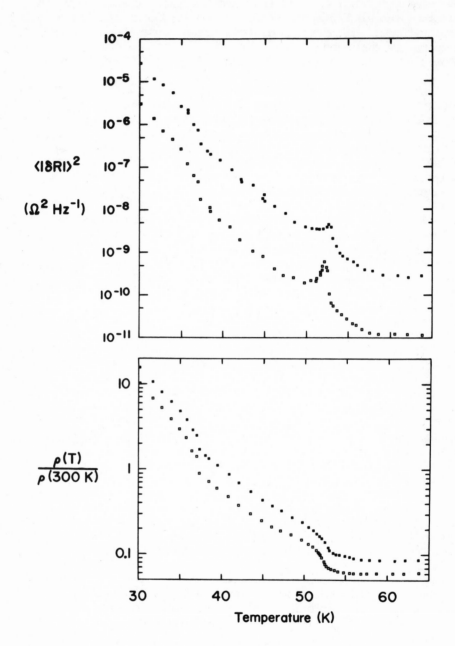

Figure 10 - Mean-square broadband (1-1000 Hz) resistance
fluctuations and d.c. resistivity of two samples of TTF-TCNQ.

The position of $T_{max} \sim 65K$ is typical for the series, but the maximum (Figure 8) is broad and the region of what we have called fluctuation resistivity is greatly extended, as in a system more one-dimensional[85] than TTF-TCNQ. Only near 34K does the resistivity show any sign of a phase transition, and this takes the form of a weak discontinuity in $\partial\rho/\partial T$, surprisingly reminiscent of the 47K resistance anomaly in TTF-TCNQ (Figure 9). Indeed, the entire TMTTF-TCNQ derivative curve strongly suggests the TTF-TCNQ and TMTSF-TCNQ curves with their critical divergences subtracted. Apparently the growth of one-dimensional fluctuations with cooling below T_{max} in this material does not lead to three-dimensional ordering as directly as in its more strongly coupled analogs. More precise conclusions await studies of the diffuse x-ray scattering and excess noise.

A critical divergence does occur in $(\partial\rho/\partial T)$ for TMTSF-TCNQ, whose interchain coupling we have taken to be of intermediate strength. This, the simplest structure[47] in the series with but one molecule pair per unit cell and one pair of short Se$\cdots$N

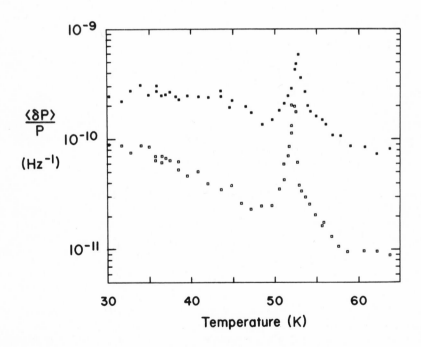

Figure 11 – Normalized broadband excess noise power $\delta P/P = \langle|\delta R|^2\rangle/R^2$ for the samples of Figure 9.

contacts, exhibits only one phase transition, with T_c = 57K.
It is apparent from Figure 9 and especially from Figure 8 that
the divergence of $\partial\rho/\partial T$ as $T \to T_{c+}$ is stronger than that in
TTF-TCNQ, but that the development of the gap below T_c is very
substantially slower. In this respect TMTSF-TCNQ differs from
its unsubstituted analog TSF-TCNQ, in which the divergence is
sharper than in the TTF salt on both sides of the transition.[88]
The differences in critical behavior among the three materials
appear difficult to rationalize fully in terms of the simple
theory of Horn et al.[88]

In the higher-dimensional analog HMTSF-TCNQ,[6] the situation
is of course entirely different. Here the Peierls transition
is suppressed and the resistivity merely flattens below 120K,
remaining essentially featureless on the scale of Figures 8 and
9 throughout this intermediate temperature region. Even at the
32K phase transition[64] there is only a weak discontinuity in
$\partial\rho/\partial T$ (which becomes apparent when Figure 9 is expanded by a
factor of ten). As we remarked earlier, the coincidence of the
flattening of the resistivity and the augmentation of the excess
orbital diamagnetism[64] suggests that both may simply arise from
establishment of the full semimetallic band structure with
cooling. Preliminary excess noise studies,[71] however, reveal a
weak structure near 120K, indicating that matters may be more
complicated.

3. <u>Low Temperatures</u>. At the lowest temperatures, the plots
of Figure 8 and especially Figure 9 reveal at least two remarkable
features. First, as described earlier, the conductivity of
HMTSF-TCNQ[6] never becomes activated, but falls slowly toward
a T = 0 intercept of order $10^3\Omega^{-1}cm^{-1}$; this behavior persists at
least to 6mK.[74] Unlike the results at higher temperatures, this
drop in conductivity is sample-dependent,[6] and becomes very mild
indeed in the best specimens.[93] Whether this temperature
dependence is really due to fluctuation effects as we originally
speculated[6] (based in part upon our misunderstanding of the
crystal structure[49]), or whether to the energy dependence of the
semimetallic density of states,[6,8,70] is not at present clear.
The situation is complicated by the identification of the 32K
phase transition,[64,71] and by indications in the heat capacity,[64]
thermoelectric power,[6] conductivity,[71] and excess noise[71] of a
possible second, even weaker phase transition near 12K.

The second important low-temperature feature of Figure 9 is
the puzzling similarity among the apparent activation energies
in the semiconducting states of TTF-, TMTTF-, and TMTSF-TCNQ.
The low-temperature values of $\partial\ln\rho/\partial(T^{-1}) \sim 100 - 120$K for these
materials are also in the range of Etemad's[38c] results for TSF-
and DTDSF-TCNQ. It appears, then, that the activation energy is

little influenced by significant variations in the parameters --
$t_{||}$, $zt_{\perp}$, E_F, λ, etc. -- that normally govern a Peierls transition.
This persistence is particularly surprising in light of differences
in the low-temperature magnetic g-values, which indicate that the
states near the band edge are cation-like in some of the materials
(such as TTF-TCNQ[52] and TMTTF-TCNQ[50,56]) but anion-like in others
(such as TMTSF-TCNQ[50,56]).

At present the insensitivity of the apparent activation
energy to otherwise well-established chemical trends is unexplained.
It is true that our findings for TTF-TCNQ contradict those of
Etemad,[38C] but they do agree with measurements by Horn[94] on
samples collected from several different laboratories.

The possibility that our results are extrinsic cannot of
course be excluded absolutely (although the similarity among
different materials would then be no easier to explain), but we
consider the prospect unlikely. In the case of TTF-TCNQ, for
example, our purification procedures[84] have been successful to
the extent that $\chi(T)$ remains activated and the g-tensor
temperature-independent in the semiconducting regime at least
down to 4K.[50]

The complete absence of low-temperature Curie behavior in our
TTF-TCNQ samples invites further study of the energy gaps. In a
simple, non-magnetic, small-gap semiconductor, an accurate
determination of the activation energy requires knowledge of the
pre-exponential factor. This factor is much more difficult to
establish a priori for the conductivity, where it involves both
the mobility and a number density, than for the magnetic
susceptibility, where it involves the number density alone. Here
the correct temperature dependence of the prefactor in the one-
dimensional case is $T^{-1/2}$ where the spins are intrinsic, and T^{-1}
where they originate from the thermal ionization of dilute non-
magnetic impurities. Analyzed in this way, our data for $\chi(T)$
below the 38K phase transition can be expressed as the sum of
two activated terms, one of activation energy $\Delta_1 = 104K$
and prefactor consistent with an intrinsic gap, and the other of
activation energy $\Delta_2 = 10.2K$ and prefactor consistent with about
50 ppm in non-magnetic defects or impurities. When published[95,53]
static[62] and spin-resonance[53,95] susceptibility data are treated
in the same way, Δ_2 is uncertain because of the Curie contribution,
but Δ_1 is in excellent agreement with our result.

These same activation energies describe our conductivity data
for these samples. Indeed, we can go farther. We have also
measured[96] the current-voltage characteristics of semiconducting
TTF-TCNQ at higher fields, using a pulsed technique designed to
eliminate sample heating effects. The results do not include
a negative differential resistance region as found in d.c.

measurements,[97,98,96] but are strongly non-ohmic. To describe
the data, we assume that a steady state is established and adapt
standard hot-electron theory[99] to the case of a small gap, nearly
one-dimensional semiconductor. We obtain excellent fits to the
entire set of I-V curves, from 4K to 38K and from V = 0 to field
saturation, by using the values for Δ_1, Δ_2, and the impurity
concentration that were obtained from the magnetic susceptibility.
The fits require that momentum relaxation occur via elastic
collisions with impurities or defects, but that energy relaxation
occur principally via low-lying optical phonon modes. Since we
take the phonon frequencies from neutron data[100] and the band
width parameters are fairly well established,[8] the only adjustable
parameter in the theory is the power of T in the low-field
mobility μ_0. The best fit to the conductivity in the low-field limit
gives $\mu_0 \propto T^{3/2}$ almost exactly; the family of high-field curves are
then fit with no adjustable parameters whatever. The average low-
field mobility ranges from $\sim$ 40 to $\sim 10^3$ cm^2/V-sec between 4 and
38K; these values are not overly large considering that $\Delta_1/t_{||}$ is
small and that the effective mass near the band edge is
m* $\sim \hbar^2 b^{-2}(\Delta_1/4t_{||}{}^2) \sim 0.1$ m.

Taken together, these results bear important implications for
current discussions of the low-temperature character of TTF-TCNQ,
and presumably of the other materials as well. First, the smallness
of the implied gap $2\Delta_1 \sim$ 208K suggests that the mean-field
transition temperature T_p may not be so far above T_c as some
theoretical treatments[28] might predict. Second, the success of
the magnetic activation energies in fitting the whole temperature-
field conductivity surface is exceptionally strong evidence that
the magnetic and semiconducting gaps in TTF-TCNQ are identical.
Hence our results flatly contradict the assertion,[65] based upon
the data of Ref. 38c, that a set of spin excitations lies below
the semiconducting gap. Finally, we find no persuasive evidence
that phase soliton excitations[101,98,22] play any appreciable
role in low-temperature transport. The combined magnetic and
electric data would require that the soliton carry a spin, and
in any case the non-ohmic conductivities are well described
by standard semiconductor theory. Indeed, at low temperatures
we find nothing to distinguish TTF-TCNQ from a very ordinary, if
very anisotropic, small-gap semiconductor.

D. Some Comments on the Optical Conductivity of TTF-TCNQ.

The relatively small intrinsic low-temperature gaps deduced
in the last section from d.c. conductivity and magnetic data
invite comparison with the structure in published optical data
for TTF-TCNQ.[102] The far-infrared optical conductivity shows a
threshold near 300 cm^{-1} and rises to a peak at $\sim$1000cm,$^{-1}$ and
the latter figure is sometimes interpreted as the Peierls gap.

We disagree. On physical grounds, it is inconceivable that a gap of order 0.1 eV could arise simply from the weak Peierls modulation of $t_{||}$ which is itself of order 0.1 eV. On spectroscopic grounds, it is more appropriate to assign the "gap" to the absorption threshold at low temperatures than to the absorption peak.

Even then, the apparent optical gap of ~ 300 cm^{-1} is nearly twice as large as our estimates. This may, of course, simply reflect the difference between a direct and an indirect gap. There is evidence, however, that the 300 cm^{-1} absorption "edge" in the optical conductivity is not a simple semiconducting gap. First, the edge remains practically temperature-independent even up to the room-temperature metallic state. Second, Torrance and Bloch[81] have demonstrated, both experimentally and theoretically, the presence at higher frequencies of strong Fano[103] anti-resonances in the optical conductivity due to the coupling to the electronic continuum of discrete, totally symmetric intra-molecular optical phonon modes. Such modes also occur in the far infrared, and the calculated couplings suggest a strong, nearly temperature-independent dip in the optical conductivity very close to the apparent 300 cm^{-1} absorption edge. The coupling to these low-frequency phonons must also renormalize the electronic ground state,[104] so that the conductivity resembles that of a polaronic system. In fact, simple polaron theory is remarkably successful in describing the optical conductivity of TTF-TCNQ.[105]

At least one other aspect of the high-frequency response of TTF-TCNQ bears upon some of the discussion in this volume. The plasmon spectrum has been experimentally determined,[106] and extends without Landau damping across the entire Brillouin zone. The dispersion is very well described within the random phase approximation.[107] However, when a modest amount of exchange is included in such calculations for nearly one-dimensional systems, the dispersion is severely modified and Landau damping is intro-duced at large wavevectors.[108] Hence rationalization of the plasmon spectrum may be a significant burden for descriptions of TTF-TCNQ[65,61,22,67] in which the short-range electron-electron interaction is taken to be large.

VI.   CONCLUDING REMARKS

We summarize our survey as follows. The extraordinary instabilities and fluctuation effects characteristic of a hypothetical one-dimensional metal are quenched rapidly as the effective dimensionality is increased. Hence we should expect that the physics of a nearly one-dimensional conductor should be especially sensitive to the effective interchain coupling.

This is particularly true in two-band organic systems based upon the prototype TTF-TCNQ, where the one-electron band structure in the undistorted state is nominally semimetallic, and the shape of the Fermi surface and density of states at the Fermi level are dominated by interchain charge-transfer integrals.

These observations suggest a means of systematically adjusting the electronic properties. By adding saturated substituent groups to the molecular components of such systems, we are able to control molecular size and interchain spacing without appreciably affecting the electronic structures of the molecule, the interchain stacking patterns and distances, or the effective radii of the conducting strands. We find that while the details of the interchain coupling apparently have little effect upon the temperature dependences of the high-temperature conductivity and magnetic susceptibility, or upon the low-temperature semiconducting gap, they do dominate the evolution of the metal-to-insulator transition. As the dimensionality is lowered, for example, the multiple phase transitions of TTF-TCNQ give way to a single very weak transition (TMTTF-TCNQ), with indications that the width of the critical region may approach the magnitude of the transition temperature itself. Conversely, with increased interchain coupling the metal-to-insulator transition is suppressed and the semimetallic band structure fully established. The result is HMTSF-TCNQ, whose electronic properties appear midway between those of TTF-TCNQ and a conventional semimetal. The crucial role of dimensionality in this system is emphasized by contrast with the more conventional TMTSF-TCNQ, which differs from HMTSF-TCNQ only in interchain coupling.

The development of HMTSF-TCNQ has not only underscored the physical importance of higher dimensionality, but illuminated the chemical means for attaining it. For example, preliminary studies indicate that the analogous series of compounds based upon the electron acceptor TNAP (Figure 1e) displays a wider range of electrical behavior than does the TCNQ sequence. It may not be overly optimistic to look forward to the emergence of a class of higher-dimensional organic conductors which is much broader and electrically more versatile than the materials we have so far developed.

VII. ACKNOWLEDGEMENTS

Much of this paper is based upon work supported by a grant from the Materials Science Office, Advanced Research Projects Agency, Department of Defense. We are very grateful to F. J. Di Salvo, to T. J. Kistenmacher, and to V. K. S. Shante for permission to present Figures 7, 6, and 3, respectively, prior to publication elsewhere.

VIII.   REFERENCES

1. Reviews include (a) I. F. Shchegolev, Phys. Stat. Sol. (A) 12, 9 (1972); (b) A. N. Bloch, in Energy and Charge Transfer in Organic Semiconductors, edited by K. Masuda and M. Silver (Plenum Press, New York, 1974), p. 159; (c) A. N. Bloch, D. O. Cowan, and T. O. Poehler, in Energy and Charge Transfer in Organic Semiconductors, edited by K. Masuda and M. Silver (Plenum Press, New York, 1974), p. 167; (d) A. F. Garito and A. J. Heeger, Accounts Chem. Res. 7, 232 (1974); (e) A. J. Berlinsky, Contemp. Phys. 17, 331 (1976); (f) D. O. Cowan, P. Shu, C. Hu, W. Krug, T. Carruthers, T. Poehler, and A. Bloch, this volume.

2. J. Ferraris, D. O. Cowan, V. Walatka, and J. H. Perlstein, J. Amer. Chem. Soc. 95, 498 (1973).

3. J. P. Ferraris, T. O. Poehler, A. N. Bloch, and D. O. Cowan, Tet. Lett. 27, 2553 (1976).

4. E. M. Engler and V. V. Patel, J. Amer. Chem. Soc. 96, 7376 (1974).

5. K. Bechgaard, D. O. Cowan, and A. N. Bloch, Chem. Comm. 1974, 937.

6. A. N. Bloch, D. O. Cowan, K. Bechgaard, R. E. Pyle, and R. H. Banks, Phys. Rev. Lett. 34, 1561 (1975).

7. R. L. Greene, J. J. Mayerle, R. Schumaker, G. Castro, P. M. Chaikin, S. Etemad, and S. J. La Placa, preprint.

8. V. K. S. Shante, A. N. Bloch, D. O. Cowan, W. M. Lee, S. Choi, and M. H. Cohen, Bull. Amer. Phys. Soc. 21, 287 (1976); and to be published.

9. A. J. Berlinsky, J. F. Carolan, and L. Weiler, Sol. St. Comm. 15, 795 (1974).

10. See, for example, M. H. Cohen, J. Non-Cryst.Sol. 2, 432 (1970).

11. Z. G. Soos, Ann. Rev. Phys. Chem. 25, 121 (1974).

12. P. F. Williams and A. N. Bloch, Phys. Rev. B10, 1097 (1974).

13. For a review, see F. H. Herbstein, in Perspectives in Structural Chemistry, edited by J. D. Dunitz and J. A. Ibers (Wiley, New York, 1972), Vol. IV, p. 166.

14. R. E. Peierls, Quantum Theory of Solids, (Oxford Press, London, 1955), p. 108.

15. See, for example, B. I. Halperin, Adv. Chem. Phys. 13, 123 (1966).

16. L. D. Landau and E. M. Lifschitz, Statistical Physics (Pergamon Press, New York, 1958), p. 482.

17. A. N. Bloch, R. B. Weisman, and C. M. Varma, Phys. Rev. Lett. 28, 753 (1972).

18. R. E. Borland, Proc. Roy. Soc. A274, 529 (1963).

19. N. F. Mott and W. D. Twose, Phil. Mag. 10, 107 (1961).

20. V. K. S. Shante and M. H. Cohen, unpublished.

21. P. W. Anderson, Phys. Rev. 109, 1492 (1958).

22. See, for example, V. J. Emery, this volume.

23. P. Sen and C. M. Varma, Sol. St. Comm. 15, 1905 (1974).

24. G. Beni, Sol. St. Comm. 15, 269 (1974).

25. See, for example, M. J. Rice and S. Strassler, Sol. St. Comm. 13, 125 (1973).

26. A. N. Bloch and R. B. Weisman, in Extended Interactions Between Metal Ions in Transition Metal Complexes, edited by L. V. Interrante (American Chemical Society, Washington, D. C., (1974), p. 356.

27. D. J. Scalapino, Y. Imry, and P. Pincus, Phys. Rev. B11, 2042 (1975).

28. See, for example, P. A. Lee, T. M. Rice, and P. W. Anderson, Phys. Rev. Lett. 31, 462 (1973).

29. F. Denoyer, R. Comes, A. F. Garito, and A. J. Heeger, Phys. Rev. Lett. 35, 445 (1975); R. Comes, S. M. Shapiro, G. Shirane, A. F. Garito, and A. J. Heeger, Phys. Rev. Lett. 35, 1518 (1975); S. Kagoshima, H. Anzai, K. Majimura, and T. Ishigoro, J. Phys. Soc. Japan 39, 1143 (1975); R. Comes, this volume.

30. See, for example, C. J. Fritchie, Acta Cryst. 20, 892 (1966).

31. G. Theodorou and M. H. Cohen, Phys. Rev. Lett. 37, 1014 (1976).

32. H. Kobayashi, Bull. Chem. Soc. Japan 48, 1373 (1975).

33. We thank M. A. Butler, R. Gemmer, B. Morosin, Z. Soos, and Y. Tomkiewicz, respectively, for supplying these samples.

34. T. J. Kistenmacher, unpublished.

35. P. Leung and L. Kraft, unpublished.

36. A. J. Epstein, S. Etemad, A. F. Garito, and A. J. Heeger, Phys. Rev. B5, 952 (1972).

37. A. N. Bloch and C. M. Varma, J. Physics C6, 1849 (1973).

38. (a) S. Etemad, T. Penney, E. M. Engler, B. A. Scott, and P. E. Seiden, Phys. Rev. Lett. 34, 741 (1975); (b) Y. Tomkiewicz, E. M. Engler, and T. D. Schultz, Phys. Rev. Lett. 35, 456 (1975); (c) S. Etemad, Phys. Rev. B13, 2254 (1976).

39. M. H. Cohen, J. A. Hertz, P. M. Horn, and K. S. Shante, Int. J. Quant. Chem. Symp., No. 8, 491 (1974).

40. U. Bernstein, P. M. Chaikin, and P. Pincus, Phys. Rev. Lett. 34, 271 (1975).

41. T. J. Kistenmacher, T. E. Phillips, and D. O. Cowan, Acta Cryst. B30, 763 (1974).

42. M. J. Cohen, L. B. Coleman, A. F. Garito, and A. J. Heeger, Phys. Rev. B10, 1298 (1974).

43. A similar interpretation of this material has been conceived independently by M. Weger, Sol. St. Comm. 19, 1149 (1976). See also Refs. 61 and 70.

44. T. E. Phillips, T. J. Kistenmacher, A. N. Bloch, J. P. Ferraris, and D. O. Cowan, Acta. Cryst. (in press).

45. M. A. Butler, J. P. Ferraris, A. N. Bloch, and D. O. Cowan, Chem. Phys. Lett. 24, 600 (1974).

46. T. J. Kistenmacher, T. E. Phillips, D. O. Cowan, J. P. Ferraris, and A. N. Bloch, Acta Cryst. B32, 539 (1976).

47. K. Bechgaard, T. J. Kistenmacher, A. N. Bloch, and D. O. Cowan, Acta. Cryst. (in press).

48. K. Bechgaard, D. O. Cowan, and A. N. Bloch, Mol. Cryst. and Liq. Cryst. 32, 237 (1976).

49. T. E. Phillips, T. J. Kistenmacher, A. N. Bloch, and D. O. Cowan, J. C. S. Chem. Comm. 1976, 334 (1976).

50. T. O. Poehler, unpublished.

51. L. B. Coleman, M. J. Cohen, D. J. Sandman, F. G. Yamagishi, A. F. Garito, and A. J. Heeger, Sol. St. Comm. 12, 1125 (1973).

52. Y. Tomkiewicz, B. A. Scott, L. J. Tao, and R. S. Title, Phys. Rev. Lett. 32, 1363 (1974).

53. Y. Tomkiewicz, A. R. Taranko, and J. B. Torrance, Phys. Rev. Lett. 36, 751 (1976).

54. S. K. Khanna, E. Ehrenfreund, A. F. Garito, and A. J. Heeger, Phys. Rev. B10, 2205 (1974); A. H. Kahn, J. Appl. Phys. 46, 4965 (1975); V. V. Walatka, A. N. Bloch, J. Bohandy, and T. O. Poehler, unpublished.

55. Y. Tomkiewicz, D. Garrod, A. R. Taranko, and A. N. Bloch, preprint.

56. T. O. Poehler, J. Bohandy, A. N. Bloch, and D. O. Cowan, Bull. Amer. Phys. Soc. 21, 287 (1976).

57. Y. Tomkiewicz, T. D. Shultz, E. M. Engler, A. R. Taranko, and A. N. Bloch, Bull. Amer. Phys. Soc. 21, 287 (1976). See also Ref. 38b.

58. A. W. Overhauser, Phys. Rev. 89, 689 (1953).

59. R. J. Elliot, Phys. Rev. 96, 266 (1954).

60. Y. Yafet, Sol. St. Phys. 14, 1 (1963).

61. (a) D. Jerome and M. Weger, this volume; (b) G. Soda, D. Jerome, M. Weger, J. M. Fabre, and L. Giral, Sol. St. Comm. 18, 1417 (1976).

62. J. C. Scott, A. F. Garito, and A. J. Heeger, Phys. Rev. B10, 3131 (1974).

63. S. Etemad, private communication.

64. F. J. DiSalvo, W. A. Reed, F. Hsu, A. N. Bloch, and D. O. Cowan, unpublished.

65. J. B. Torrance, this volume.

66. See, for example, A. Karpfen, J. Ladik, G. Stollhoff, and P. Fulde, Chem. Phys. Lett. 31, 291 (1975).

67. P. A. Lee, T. M. Rice, and R. A. Klemm, preprint.

68. C. Weyl, E. M. Engler, S. Etemad, K. Bechgaard, and G. Jehanno, Sol. St. Comm. 19, 925 (1976).

69. The data for HMTSF-TCNQ (Ref. 64), first presented at the 1975 March meeting of the American Physical Society (Denver, Col.), have been essentially reproduced by G. Soda, D. Jerome, M. Weger, K. Bechgaard, and E. Pederson, Sol. St. Comm. 19, (in press); see also Ref. 61. Our results and interpretation

were communicated directly to these authors in early 1976.

70. J. R. Cooper, M. Weger, D. Jerome, D. Lefur, K. Bechgaard, A. N. Bloch, and D. O. Cowan, Sol. St. Comm. 19, 749 (1976).

71. T. Carruthers, unpublished.

72. G. A. Thomas, et al., Phys. Rev. B13, 5105 (1976).

73. Substantial residual conductivities as T→0 have also been observed in HMTSF-TNAP (Ref. 6) and in the halides of tetrathiatetracene, TTT (E. Perez-Albuerne, work presented at the 1976 March meeting of the American Physical Society [Atlanta, Ga.]; I. F. Schegolev, work presented at the Conference on Organic Conductors and Semiconductors, Siofok, Hungary, Sept. 1976.)

74. R. L. Greene, unpublished.

75. R. P. Groff, A. Suna, and R. E. Merrifield, Phys. Rev. Lett. 33, 418 (1974).

76. D. E. Schafer, F. Wudl, G. A. Thomas, J. P. Ferraris, and D. O. Cowan, Sol. St. Comm. 14, 347 (1974).

77. M. B. Salamon, J. W. Bray, G. DePasquali, R. A. Craven, R. Herman, G. Stucky, and A. Schultz, Phys. Rev. B11, 619 (1975).

78. J. Bardeen, Sol. St. Comm. 13, 357 (1973).

79. For a lucid physical explanation, see N. F. Mott, Phil. Mag. 13, 989 (1966).

80. A. Madhukar and M. H. Cohen, preprint; A. A. Gogolin, V. I. Mel'nikov, and E. I. Rashba, Abstracts of the Conference on Organic Conductors and Semiconductors, Siofok, Hungary, (1976), p. 15.

81. J. B. Torrance, E. E. Simonyi, and A. N. Bloch, Bull. Amer. Phys. Soc. 20, 497 (1975), and to be published.

82. See, for example, P. E. Seiden and D. Cabib, Phys. Rev. B13, 1846 (1976).

83. R. A. Craven, M. B. Salamon, G. DePasquali, R. M. Herman, G. Stucky, and A. Schultz, Phys. Rev. Lett. 32, 769 (1974).

84. R. V. Gemmer, D. O. Cowan, T. O. Poehler, A. N. Bloch, and R. H. Banks, J. Org. Chem. 40, 3544 (1975).

85. H. Fukuyama, T. M. Rice, and C. M. Varma, Phys. Rev. Lett. 33, 305 (1974).

86. A. Luther and V. J. Emery, Phys. Rev. Lett. 33, 589 (1974).

87. T. Carruthers, A. N. Bloch, and D. O. Cowan, Bull. Amer. Phys. Soc. 21, 313 (1976), and to be published.

88. P. M. Horn and D. Rimai, Phys. Rev. Lett. 36, 809 (1976); and P. M. Horn and D. Guidotti, preprint.

89. P. Bak and V. J. Emery, Phys. Rev. Lett. 36, 978 (1976).

90. The 47K anomaly in the conductivity, though absent in the data of Ref. 38c, has now been observed independently in several laboratories. To our knowledge it was first reported by us (Ref. 87), and separately by P. Horn and D. Guidotti, at the 1976 March Meeting of the American Physical Society (Atlanta, Ga.).

91. V. K. S. Shante, A. N. Bloch, T. Carruthers, and D. O. Cowan, Bull. Amer. Phys. Soc. 21, 313 (1976), and to be published.

92. See, for example, Aldert van der Ziel, Noise: Sources, Characterization, Measurement (Prentice-Hall, Englewood Cliffs, N.J., 1970).

93. K. Bechgaard and B. S. Jensen, work presented at this conference.

94. P. Horn, private communication.

95. J. E. Gulley and J. F. Weiher, Phys. Rev. Lett. 34, 1061 (1975).

96. T. O. Poehler, R. M. Somers, A. N. Bloch, and D. O. Cowan, preprint.

97. H. Kahlert, Sol. St. Comm. 17, 1161 (1975); K. Seeger, Sol. St. Comm. 19, 245 (1976).

98. M. J. Cohen, P. R. Newman, and A. J. Heeger, preprint.

99. See, for example, K. Seeger, Semiconductor Physics (Springer-Verlag, New York, 1973).

100. H. A. Mook and C. R. Watson, preprint.

101. M. J. Rice, A. R. Bishop, J. A. Krumhansl, and S. E. Trullinger, Phys. Rev. Lett. 36, 432 (1976).

102. D. B. Tanner, C. S. Jacobsen, A. F. Garito, and A. J. Heeger, Phys. Rev. Lett. 32, 1301 (1974); 33, 1559 (1974); Phys. Rev. B13, 3381 (1976).

103. U. Fano, Phys. Rev. 124, 1886 (1961).

104. This effect is distinct from the stabilization of the Peierls gap by the softening of intramolecular optical modes suggested by M. J. Rice, C. B. Duke, and N. O. Lipari [Sol. St. Comm. 17, 1089 (1975)]. In fact their calculation is flawed in its assumption that the phases of the intramolecular and intermolecular soft modes are arbitrary. The correct theory [A. Madhukar, Chem. Phys. Lett. 27, 606 (1974)] recognizes that the former modes build up charge around the lattice sites and the latter in the bonding regions; hence the two instabilities compete.

105. H. Hinkelmann and H. G. Reik, Sol. St. Comm. 16, 567 (1975).

106. J. J. Ritsko, D. J. Sandman, A. J. Epstein, P. C. Gibbons, S. E. Schnatterly, and J. Fields, Phys. Rev. Lett. 34, 1330 (1975).

107. P. F. Williams and A. N. Bloch, Phys. Rev. Lett. 36, 64 (1976).

108. G. Giulliani, E. Tosatti, and M. P. Tosi, Lett. Nuovo Cimento 16, 385 (1976).

CHARGE DENSITY WAVE PHENOMENA IN TTF-TCNQ AND RELATED ORGANIC CONDUCTORS[*]

A. J. Heeger

Department of Physics and Laboratory for Research
on the Structure of Matter, University of Pennsylvania,
Philadelphia, Pennsylvania, 19174, U.S.A.

INTRODUCTION

Although it has long been known that a one-dimensional (1 D) metal is intrinsically unstable to the formation of charge- or spin-density waves,[1-3] the study of real physical metallic systems which are, in effect, truly one-dimensional in electronic structure has emerged only in the past few years.[4] The chainlike stacking of[5,6] planar molecules in the TCNQ salts with the implied π-electron overlap was suggestive of highly anisotropic properties and, in a rough sense, of quasi-one-dimensionality. However, it is important to recognize the difference between "anisotropy" and "one-dimensionality" since clearly all experimental chain-like systems form in three dimensional structures with, perhaps, weak coupling between individual chains. Understanding of the reality of the 1 D limit came with quantitative measurements of the anisotropy of the optical[7-13] and transport[14-16] properties of 1 D conductors.[17] The observation in polarized optical reflectance of a plasma edge for E parallel to the conducting axis and a flat frequency-independent reflectivity through the far IR for E

[*]Work at the University of Pennsylvania supported by the National Science Foundation through Grant DMR-74-22923 and through the Laboratory for Research on the Structure of Matter (DMR-72-03025), and by the Advanced Research Projects Agency through Grant DAHC-15-72-C-0174.

perpendicular to the conducting axis implies that even though the conductivity perpendicular to the chain is finite, this transverse transport does not arise from metallic behavior; i. e. , $\omega_p^{||} \tau_{||} > 1$ and $\omega_p^{\perp} \tau_{\perp} < 1$, where ω_p and τ are the plasma frequency and scattering time. Moreover, the measurements of the magnitude of the transverse components[14, 15, 16] of the conductivity tensor yield an estimated mean free path $l_{\perp}$ of order 10^{-3} of a lattice constant, whereas in the parallel direction the mean free path as estimated from the dc or the optical conductivity near the plasma edge implies metallic propagation. Thus $l_{||} > a_{||}$ and $l_{\perp} \ll a_{\perp}$, where $a_{||}$ and $a_{\perp}$ are the lattice constants parallel to and perpendicular to the principal conducting axis. The implication is that the interaction of the electrons with the lattice can be sufficiently strong to localize electrons on individual chains with diffusive hopping[15,16] between chains. This localization converts a metal with an anisotropic band structure into an array of 1 D metallic chains.[18]

In this sense TTF-TCNQ, KCP and a growing class of related compounds are one-dimensional metals and can be expected to exhibit the unique phenomena associated with the mathematical 1 D instabilities. On the other hand, poly(sulfur-nitride), $(SN)_x$, although chain-like in structure,[19, 20] has large enough interchain coupling to prevent the 1 D localization. As a result, the optical properties,[21-25] electron energy loss results,[26] and ultimately the absence of a Peierls transition[27-29] indicate that $(SN)_x$ is not a 1 D metal, but an anisotropic three dimensional metal.

We will focus in this paper on the charge density wave phenomena associated with the Peierls instability in the TTF-TCNQ system bringing in experimental results from related compounds where available and appropriate. TTF-TCNQ is a one-dimensional metal at high temperatures which undergoes a metal-insulator transition[30] at 54K to a high-dielectric-constant[16, 31, 32] non-magnetic[30, 33] ground state. The crystal and molecular structures are shown in Fig. 1.[6] Recent x-ray scattering[34, 35] and elastic-neutron-scattering[36] studies demonstrated the existence below 54K of a low-temperature incommensurate super-lattice having a periodicity of 3. 4b in the chain direction. In the conducting state above 54K the x-ray results showed 1 D diffuse scattering[34,35,37,38] consistent with a 1 D lattice distortion or a phonon anomaly. Inelastic neutron-scattering studies[39, 40] of the phonon spectrum revealed a Kohn anomaly at 0. 295b at room temperature which becomes stronger[40] with decreasing temperature. These structural studies, therefore, established the existence of the charge-

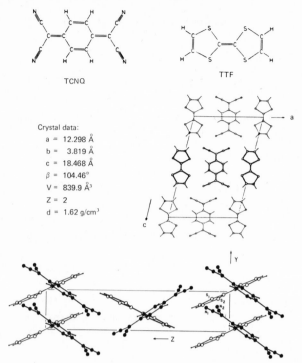

Crystal data:
a = 12.298 Å
b = 3.819 Å
c = 18.468 Å
β = 104.46°
V = 839.9 Å³
Z = 2
d = 1.62 g/cm³

Fig. 1. Molecular constituents and parallel chain structure of TTF-TCNQ$^{(ref. 6)}$. x corresponds to a, y to b, and z to c. The c^{*}-direction is normal to a and b.

density-wave ground state arising from the Peierls instability and the associated giant Kohn anomaly in the fluctuation regime above 54K, and provided detailed information on the temperature dependence.

The Peierls instability is a soft-mode structural metal-insulator transition driven by the 1 D electronic system. It is useful to view the instability in terms of electronic dielectric response function which relates the induced charge density to an external perturbing potential.

$$[\epsilon^{-1}(q,\omega) - 1] = (4\pi e^{2}/q^{2})[\rho(q,\omega)/V_{ext}(q,\omega)] \tag{1}$$

The dielectric response function can be written in the following form

$$\epsilon^{-1}(q,\omega) - 1 = \frac{4\pi e^{2}}{q^{2}} F(q,\omega, T) \tag{2}$$

where $F(k, \omega, T)$ is the temperature dependent Lindhard function.

$$F(q, \omega, T) = \sum_p \frac{n_{p+q} - n_p}{\epsilon_p - \epsilon_{p+q} + \hbar\omega} \tag{3}$$

where n_p is the Fermi distribution function for state p. The Lindhard function is the fundamental response function of the metallic electron sea and is shown in Fig. 2a for 1, 2, and 3 dimensional systems. In 3-D, the Lindhard function varies smoothly with an infinite slope at $q = 2k_F$ where the perturbing wave vector spans the Fermi surface such that $\epsilon_p - \epsilon_{p+2k_F} = 0$. In 2D the response function has a cusp and a discontinuous derivative at $q = 2k_F$. In 1 D, since the Fermi surface is a plane, the nesting condition $\epsilon_p - \epsilon_{p+2k_F} = 0$ is satisfied over the entire Fermi surface resulting in a logarithmic singularity. With increasing temperature, the smearing of the distribution function in the numerator of eq. 3 smooths out the singularity resulting in a sharp peak at $2k_F$ which progressively weakens with increasing temperature (Fig. 2b).

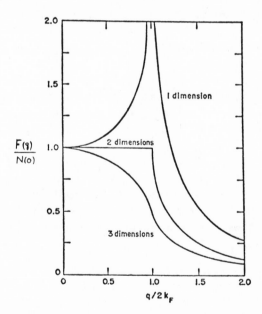

Fig. 2a. Lindhard function $F_{(q, o, o)}$ for 1, 2, and 3 dimensional electron gas (normalized to the density of states at the Fermi level).

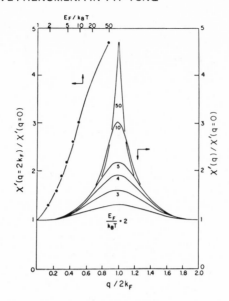

Fig. 2b.   Temperature dependent Lindhard function for the tight-binding band.   The calculations were carried out for the half-filled band.   The logarithmic dependence of $F(2k_F)$ as a function of (E_F/kT) is shown on the left-hand side.

   Equation (2) is a free electron result consistent with the simple Fröhlich hamiltonian neglecting long range Coulomb interactions. To obtain the proper long wavelength, high frequency behavior (plasmons), and metallic screening, one must include the long range electron-electron Coulomb repulsion.   However, for our purposes here equation (2) is adequate, since we will be interested in local response to lattice distortions for which charge neutrality is maintained within the unit cell.   Combining eq. (2) and (3)

$$\rho(q,\omega) \;=\; F(q,\omega) \, V_{ext}(q,\omega) \tag{4}$$

where $F(q,\omega)$ is the Lindhard function shown in Figure 1b for $\omega \simeq 0$ and assuming a tight-binding band structure.

   The Peierls instability arises from the divergent response of the one-dimensional metal to an infinitesimal driving potential at $q = 2k_F$ as a result of the logarithmic singularity in $F(q, T)$ as $T \rightarrow 0$.   Such an external potential arises from a distortion in the crystal.   Consider, for example, the half-filled band case, and a distortion δ where the lattice constant goes from a to $a + \delta$ or $a - \delta$

on alternating sites. Such a distortion doubles the unit cell making
the Brillouin zone of width $2\pi/2a = \pi/a = 2k_F$. The existence,
then, of the doubled zone implies

$$V_G = V_{\pi/a} = V_{2k_F} \neq 0 \tag{5}$$

As a result, the electronic energy is lowered by an amount

$$\Delta E_{electronic} = \tfrac{1}{2} \rho(2k_F) V(2k_F)$$

$$= \frac{|e|}{2} F(2k_F, \omega) V^2(2k_F) \tag{6}$$

More generally, we can write

$$\Delta E_{el}(q) = -\tfrac{1}{2} F(q, \omega) V_q^2 \tag{7}$$

where V_q is an external potential, for example, imposed by a
phonon in the crystal. In this context, equation (4) is written in
the spirit of the Born-Oppenheimer approximation, i. e., the
electronic system is assumed to adiabatically follow the much
slower nuclear motion.

We therefore consider a system described by the Hamiltonian

$$H = H_o + H_{e-e} + H_{e-p} \tag{8}$$

where H_o represents the unperturbed electron and phonon
subsystems

$$H_o = \sum_{k, \sigma} \epsilon_{k\sigma} c_{k\sigma}^+ c_{k\sigma} + \sum_q \hbar \omega_q^o (n_q + \tfrac{1}{2}) \tag{9}$$

and $\epsilon_{k\sigma}$ is the tight binding one-electron band structure. H_{e-e}
and H_{e-p} represent electron-electron and electron-phonon coupling terms, respectively. We assume the effective electron-
electron interaction term, H_{e-e}, has been reduced by intramolecular correlation, metallic screening and correlation, and general
polarizability effects to the point where Coulomb effects can be
neglected (at frequencies well below both the plasma frequency of
the resulting metal and the dominant singlet molecular π-π^*
transitions responsible for the polarizability). In this limit, one
has the simple Fröhlich Hamiltonian

$$\mathcal{H} = \sum_{k,\sigma} \epsilon_{k\sigma} c^+_{k\sigma} c_{k\sigma} + \sum_{k,q} g(q)(a_q + a^+_{-q}) c^+_{k+q} + \sum_q \hbar w^o_q (n_q + \tfrac{1}{2}) \quad (10)$$

where $g(q)$ is the coupling constant, $c^+_{k\sigma}$ $(c_{k\sigma})$ is the electron creation (annihilation) operator, and $a^+_q (a_q)$ is the phonon creation
(annihilation) operator.

To calculate the phonon spectrum including the effect of the
electron-phonon coupling we utilize equation (7) and write

$$H_{phonon} = \tfrac{1}{2} K \sum_\ell (U_{\ell+1} - U_\ell)^2 - \frac{|e|}{2} \sum_q F(q) V^2_q + \tfrac{1}{2} M \sum_\ell |\dot{U}_\ell|^2. \quad (11)$$

The first term is the usual "spring constant" potential energy contribution with U_ℓ being the displacement of the ℓth molecule from
equilibrium, and the second term is essentially a phonon self-
energy contribution arising from the electronic polarizability.
The final term is the kinetic energy contribution.

Imagine imposing a periodic distortion (wave number q) on
the lattice as in a phonon lattice wave. The resulting V_q can be
obtained from the form of the electron-phonon coupling. If the
nearby cores are displaced, the electrons at the ℓth site see a
potential

$$V_\ell = \tfrac{1}{2} \gamma [(U_{\ell+1} - U_\ell) + (U_{\ell-1} - U_\ell)] \quad (12)$$

The resulting V_q from a periodic wave is

$$V_q = \gamma \left\{ \frac{1}{N} \sum_\ell e^{-iq\ell} [(U_{\ell+1} - U_\ell) + (U_{\ell-1} - U_\ell)] \right\} \quad (13)$$

$$= \gamma \left(\frac{2\hbar}{NM w_q}\right)^{\tfrac{1}{2}} \sin^2 \tfrac{1}{2} q (a^+_{-q} + a_q) \quad (14)$$

The phonon spectrum is readily calculated from eq. 11

$$H_{phonon} = \sum_q \hbar w_q (n_q + \tfrac{1}{2}) \quad (15)$$

where

$$\omega_q^2 = (\omega_q^0)^2 \left[1 - \frac{\gamma^2 |e|}{KN} \sin^2(\tfrac{1}{2}qa) \, F(q, T, \omega_q) \right] \tag{16}$$

and

$$\omega_q^0 = 2\sqrt{K/m} \, \sin\tfrac{1}{2}qa.$$

Because of the peak in $F(q, T)$ at $q = 2k_F$ (see Fig. 1b) the calculated phonon spectrum shows a relatively sharp anomaly near
$q = 2k_F$. This giant Kohn anomaly in the phonon spectrum ultimately arises from the perfect nesting of the Fermi surface in the
1 D metal. [41]

The mean field transition occurs when a solution for the
screened phonon frequency appears at zero frequency, for in this
case one has a condensation of $2k_F$ phonons, i.e., a true distortion in the crystal. Thus, the equation for the mean field Peierls
transition temperature, T_p, is given for the simple acoustic mode
case ($\hbar\omega \ll E_F$) by[42, 43]

$$\left[1 - \frac{\gamma^2 |e|}{KN} \sin^2 k_F \, F(2k_F, T_p) \right] = 0 \tag{17}$$

giving

$$kT_p \simeq E_F \, e^{-1/\lambda} \tag{18}$$

where λ is the dimensionless electron-phonon coupling constant,
$\lambda = (\frac{2g^2}{\hbar\omega_q}) N_{(o)}$, where $g_{(q)}^2 = \gamma^2 \, 2\hbar/M\omega_q$ and $N_{(o)}$ is the density of
states at the Fermi energy.

The mean field temperature defined in eq. 18 should not be
viewed as a phase transition temperature, since the one-dimensional fluctuations shift the actual phase transition down to $T = 0\,K$
in absence of interchain coupling. Rather, $k_B T_p^{MF}$ defines the
characteristic energy scale below which the electrons and $2k_F$
phonons are strongly coupled; there exists short range order and
a dynamical distortion with finite coherence length, $\xi(T)$, but no
static distortion. A static distortion will appear in the structure
as a result of pinning[44, 45] the dynamical distortion by a phase
dependent potential which may arise, for example, from interchain Coulomb coupling or commensurability of the distortion with
the underlying lattice.

The Peierls' Instability in TTF-TCNQ: Structural
Aspects and Phonon Softening

The analysis of the previous section defines the structural
aspects of the Peierls' instability. At low temperatures a well-
defined 3d superlattice is expected with a superperiod along the
1 D chain direction of λ_s where

$$\frac{2\pi}{\lambda_s} = 2k_F \tag{19}$$

For a simple tight-binding band, $2k_F = \pi(f/b)$ where $0 \leq f \leq 1$ is
the fractional band filling; i.e., the number of electrons per molecule. Thus the period of the superlattice determines the charge
transfer, $f = 2b/\lambda_s$. The position of the superlattice satellite
peaks in the a^* and c^* directions will depend on the detailed interchain coupling mechanism.

At higher temperatures, one expects a loss of transverse
order leaving 1 D scattering with little or no correlation between
adjacent chains of the phase of the modulation in the chain direction.
This is the temperature regime of the giant Kohn anomaly; the
softening of the phonon spectrum at $q = 2k_F$. At still higher temperatures, the Kohn anomaly will gradually weaken as the peak in
the Lindhard function becomes progressively less pronounced.

Diffuse x-ray scattering studies[34, 35, 37, 38] above and below
54K and detailed neutron elastic scattering[36] measurements below
54K have provided the crucial structural evidence of the Peierls'
instability in TTF-TCNQ. These data are described in detail in
the accompanying article by Comès. We briefly summarize the
principal results here in the context of the related electronic
properties.

The x-ray and neutron scattering results have established
unambiguously the appearance of new satellite Bragg peaks below
54K implying a modulated structure. The satellites appear at
incommensurate positions along the chain or b^* direction with a
periodicity of $0.295b^*$. From eq. 19, we obtain $2k_F = 0.295b^*$
with the superlattice period $\lambda_s = 3.39b$. The charge transfer is
therefore $f = 0.59$ electrons.[46]

Further characterization of these $2k_F$ satellites was carried
out in the work of Comès et al.[36] These experiments established

that the two previously observed electronic transitions at 38 K[47, 48] and 54 K[49-52] have clear structural counterparts. Below 38 K, satellites appear at the points ($\pm 0.25a^*$, $\pm 0.295b^*$ c^*) in agreement with the x-ray data of Kagoshima et al.[35] At 38 K, the modulation in the a^* direction becomes incommensurate and displays an abrupt increase gradually reaching the value $0.5a^*$ at 49 K and remaining at this value until the 54 K transition.[36] Below 38 K all satellite widths are resolution limited indicative of long range order. However, the satellite peaks do show progressive broadening in the transition region as T approaches 54 K.

Above 54 K, 1 D precursor effects attributed to the giant Kohn anomaly were observed in the x-ray studies. Well-defined diffuse sheets occur at .295b* indicating that $2k_F$, and hence the charge transfer, remains constant through the ordering transition. From an estimate of the width of the $2k_F$ lines, the coherence length of the fluctuating distortion is about 50 lattice constants at 55 K.[34] With increasing temperature the $2k_F$ diffuse lines weaken in intensity with considerable broadening.[37, 38] The residual $2k_F$ scattering at room temperature[38] is broad and barely observable and suggests a coherence length of a few lattice constants. A plot of the peak intensity at $2k_F$ as a function of temperature is shown in Fig. 3a. The points were obtained from the data of Kagoshima et al.[38]. Although the error bars are large the strength of the

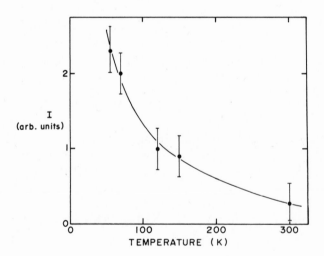

Fig. 3a. X-ray scattering intensity at $2k_F$ vs temperature. The points are taken from the data of Kagoshima et al. (ref. 38).

$2k_F$ scattering clearly increases with decreasing temperature. This is qualitatively consistent with the results of Comès et al. using photographic techniques. [37]

The intensity modulation of the diffuse scattering perpendicular to b^* above 54 K arises from the form factor of the relatively large molecules involved. Comès et al. [37] have shown that the intensity distribution of the diffuse scattering agrees with the calculated scattering from uncorrelated waves on the individual molecular stacks. The observation of diffuse scattering sheets by Kagoshima et al. [38] is in agreement with this conclusion. This provides a direct confirmation of the truly 1 D character of the scattering above 54 K. The dynamical distortions on individual chains are uncorrelated and have no phase coherence relative to one another.

Two independent inelastic neutron scattering investigations have reported observations of a giant Kohn anomaly at $2k_F =$ $0.295b^*$. Mook et al. [39] have described a phonon anomaly visible and sharp up to room temperature while Shirane et al. [40] have reported an anomaly which develops clearly only below 150 K. However, the temperature dependence of the $2k_F$ scattering as observed in the diffuse x-ray studies [37,38] described above is in complete agreement with the results of Shirane et al. [40] The strong temperature dependence (see Fig. 3a) rules out the possibility of a sharp phonon anomaly at room temperature, which would require enhanced phonon amplitude at $0.295b^*$.

Thus the giant Kohn anomaly predicted for the 1 D Peierls' instability has been observed in TTF-TCNQ. The results of Shirane et al. [40] show a broad and weak anomaly at room temperature developing into sharp structure below 150 K and becoming more pronounced as the transition at 54 K is approached. Below 150 K the anomaly develops near the transverse branch at $2k_F$.

In addition to the $2k_F$ scattering, x-ray studies [37,38] have revealed diffuse scattering at $0.59b^*$ which is equivalent to $4k_F$. This indication of two anomalies has been interpreted [37] as arising directly from the two chain structure of TTF-TCNQ with independent TTF and TCNQ chains. In this picture, one type of chain behaves as a normal 1 D metal and gives rise to the expected $2k_F$ Kohn anomaly at $0.295b^*$ and the other type of stack, because of electron interactions, gives rise to the second anomaly. Repulsive interactions between electrons in the strong coupling limit on one

molecular specie would require single occupancy of each momentum state and thus would double the Fermi wave vector giving rise
to the $4k_F$ anomaly. It is, however, unlikely that the strong coupling limit is applicable in view of the magnetic susceptibility[33] and
nuclear relaxation[53,54] results. Using weaker coupling, Emery[55]
has shown that a $4k_F$ anomaly can arise from phonon modulation
of the electron-electron interaction giving rise to simultaneous
scattering of two electrons across the Fermi surface with a momentum change of $4k_F$. Finally, it has been suggested[56] that the
scattering at $0.59b^*$ is the $2k_F$ anomaly with the $0.295b^*$ scattering arising from spin wave-phonon interaction at k_F. Although
an interesting possibility, the latter interpretation would require
a charge transfer in excess of unity and therefore appears unlikely
due to the relatively large Coulomb energy expected for static
double occupancy.

In summary, the major structural features are the observation of the incommensurate charge density wave ground state, the
transitional region between 38 K and 54 K, and the observation of
1 D precursor effects associated with the giant Kohn anomaly above
54 K. In this conducting region, the dynamical distortion has a
well-defined periodicity ($\lambda_s = 3.4b$) and a relatively long, temperature dependent, coherence length. An instantaneous snapshot of
the distortion in the fluctuation regime would look schematically
as shown in Fig. 3b. The lifetime of the dynamical distortion
would be relatively long on the scale of electronic times; certainly
greater than $1/\omega_{ph}$.

Just as the x-rays scatter from the finite coherence length
dynamical distortion, so, too, would the electronic Bloch waves
scatter as a result of the electron-phonon interaction. Thus the
increasing $2k_F$ x-ray scattering with decreasing temperature
(Fig. 3a) directly implies a corresponding increase in electron
$2k_F$ back-scattering across the Fermi surface with a reduction in
the electron transport mobility and the formation of a pseudogap
in the electronic density of states. This argument leads to the
conclusion that the single particle contribution to the electrical
conductivity decreases as the temperature is lowered.

The observation of the incommensurate superlattice is of
fundamental importance, for it implies that the phase of the distortion is arbitrary and only becomes fixed by higher-order
effects or the relatively weak interchain coupling. As a result,
so long as[57-59]

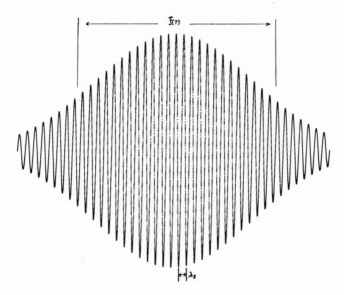

Fig. 3b. "Snapshot" of fluctuation of the dynamical distortion with wavelength λ_s and coherence length $\xi(T)$.

$$k_B T > [\xi(T)/b] V_o(\varphi) \tag{20}$$

where $\xi(T)$ is the longitudinal coherence length, b is the lattice constant, and $V_o(\varphi)$ is the amplitude of the periodic pinning potential, the fluctuating charge-density wave can move to form a current-carrying state; the collective mode transport first envisioned by Fröhlich.[2,60]

The strongly increasing electrical conductivity of TTF-TCNQ in the presence of increased $2k_F$ scattering therefore provides the fundamental evidence of collective mode transport in this 1 D conductor. In the following sections experimental results of studies of the electronic and magnetic properties will be summarized which provide evidence of the pseudogap expected from the observed $2k_F$ scattering.

The Pseudogap in the Density of States: Determination of the Local Susceptibility

Theoretical studies[61] have shown that the Peierls'-Fröhlich state in the 1 D fluctuation regime below the mean field transition temperature, T_p^{MF}, is characterized by a pseudo-gap in the electronic density of states resulting directly from the $2k_F$ scattering from the dynamical distortion. In the regime, $(T_{pin} < T < T_p^{MF})$ the magnetic susceptibility is given by

$$\chi = 2\mu_B^2 N(o) \tag{21}$$

and the density of states in the pseudo-gap is given by[61]

$$N(o) = N_o \frac{\hbar v_F \xi^{-1}(T)}{2\Delta} \tag{22}$$

where N_o is the unperturbed band density of states, v_F is the Fermi velocity, $\xi(T)$ is the longitudinal coherence length and $\Delta \equiv \langle \Delta^2 \rangle^{\frac{1}{2}}$ is the energy gap ($2\Delta = 3.5 k_B T_p^{MF}$). Thus the temperature dependence of the susceptibility can provide information on the pseudogap and the temperature dependence of the coherence length.

The local susceptibilities on the individual TTF and TCNQ chains were determined[54] through a study of the Knight shift associated with the ^{13}C labeled CN group located on the TCNQ molecule. The Knight shift in metals is given by $K = -\mu_B^{-1} H_{hf} \chi_i$ where H_{hf} is the average hyperfine field per electron and χ_i is the Pauli (local) susceptibility per molecule. The Knight shift thus provides a direct measure of the time-averaged local magnetic field at the TCNQ molecule and thereby determines the temperature dependence of the TCNQ chain local magnetic susceptibility. When combined with the total susceptibility, χ_T, the data can be analyzed to obtain the individual chain contributions (χ_Q for the TCNQ and χ_F for the TTF chains). The result of this decomposition is shown in Fig. 4.

The results indicate that at approximately 54K a real energy gap opens in the electronic excitation spectrum of the TCNQ chains, and the susceptibility drops rapidly to zero. At higher temperatures in the 1 D fluctuation regime, the density of states and the corresponding susceptibility are determined by the coherence length for the fluctuating charge density wave as described by equations 21 and 22. Within the model, the gap is relatively well formed with $\xi(T) \gg b$ so that the density of states is smaller than the simple band structure value, and the fluctuating charge

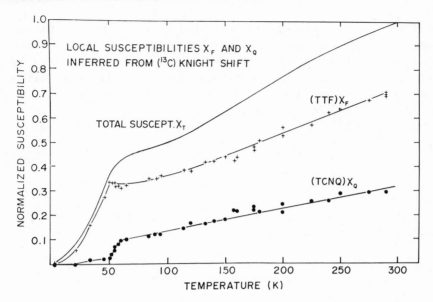

Fig. 4.   Normalized local susceptibilities on the TTF (χ_F) and TCNQ (χ_Q) chains.[54] The results were obtained from the ^{13} C Knight shift studies together with the static magnetic susceptibility.

density wave slides via time dependent phase to contribute to the dc conductivity;[62]

$$j_{\varphi} = N_s e(2k_F)^{-1} \dot{\varphi} \tag{23}$$

where φ is the phase of the order parameter and N_s is the density of condensed electrons. Infrared experiments[12-14] on polycrystalline and single crystal samples have confirmed the existence of the pseudogap at temperatures above 54K. However, the infrared conductivity studies do not yield direct information on the density of states for $\hbar\omega < 2\Delta$ since the mobility may be limited by strong scattering of the electrons by fluctuations associated with the dynamic Peierls' distortion.

The interpretation of the dc conductivity as resulting from collective mode motion implies that the order parameter should be complex with amplitude and phase as independent parameters.[62] The coherence length for the complex order parameter case can be calculated using the theory of Scalapino, Sears and Ferrell.[63] These authors treat the complex Ginzburg-Landau field starting with the free energy functional $F = \int dx [a|\Psi|^2 + b|\Psi|^4 + c|d\Psi/dx|^2]$. Where $\Psi = \Psi_0 e^{i\varphi}$ is the complex order parameter; and a, b and c are phenomenological constants to be obtained from microscopic theory for application to a particular problem. The statistical mechanics is solved by transformation to an equivalent one-particle (anharmonic oscillator) quantum mechanics problem with Hamiltonian; $H = -(4\beta^2 c)^{-1} d^2/d\Psi^2 + a|\Psi|^2 + b|\Psi|^4$ where $\beta = (k_B T)^{-1}$. With this transformation, Scalapino, et al. show[63]

$$\xi(T)^{-1} = \beta[E_1 - E_0] \tag{24}$$

where E_0 and E_1 are the ground state and first excited state energies of the Hamiltonian.

We consider the case $T \ll T_p^{MF}$ where T_p^{MF} is the mean field Peierls' temperature defined by

$$3.5 k_B T_p^{MF} = 2\Delta = 2|\Psi_0(o)| \tag{25}$$

where $\Psi_0(o) = \langle \Psi_0(T = 0) \rangle$ and $2\Delta = E_g$ is the energy gap. For this temperature regime, the free energy has formed a distinct minimum for $|\Psi| \neq 0$ but with arbitrary phase; the Fröhlich state. Since amplitude fluctuations are more energetically costly, the continuous phase-only case will be considered for Ψ complex, $\Psi = |\Psi_0(o)| e^{i\varphi}$ and only phase fluctuations are included. The Hamiltonian then takes the particularly simple form of a rigid rotor $H = -[4\beta^2 c|\Psi_0(o)|^2]^{-1} d^2/d\varphi^2$

$$E_i = \frac{1}{4\beta^2 c} \frac{1}{|\Psi_0^2(o)|} m^2 \tag{26}$$

where $m = 0, 1, \ldots$. We have initially neglected possible φ-dependent pinning potentials (e.g., impurities, commensurability, interchain Coulomb coupling, etc.) which would tend to fix φ.

The coherence length is therefore given by (eq. 24)

$$\xi(T)^{-1} = \frac{k_B T}{4c|\Psi_0(o)^2|} \tag{27}$$

The quantity $c|\Psi_o^2(o)|$ can be obtained from microscopic theory. Rice et al.[64] have shown that to obtain the linear phase phonons[61] the potential energy for compression of the charge density wave should be of the form

$$E^p_{compression} = (\tfrac{1}{2}mv_F^2 N_s)(2k_F)^{-2}\left|\frac{d\varphi}{dx}\right|^2.$$

Therefore,

$$c\left|\frac{d\Psi}{dx}\right|^2 = c\left|\Psi_o^2(o)\right|\left|\frac{d\varphi}{dx}\right|^2 = (\tfrac{1}{2}mv_F^2)N_s(2k_F)^{-2}\left|\frac{d\varphi}{dx}\right| \tag{28}$$

$N_s = N_e/L = f/b$ is the condensate density and f is the fractional charge transfer.

$$4c\left|\Psi_o(o)\right|^2 = (\tfrac{1}{2}mv_F^2)\frac{f}{b}k_F^{-2} \tag{29}$$

and

$$\xi^{-1}(T) = \frac{kT}{\tfrac{1}{2}mv_F^2}k_F^2\frac{b}{f} \tag{30}$$

Finally, using eq. (21) and (22)[65]

$$\chi(T) = \frac{\pi}{2}\frac{kT}{\Delta}\chi_p \simeq \chi_p\frac{T}{T_p^{MF}} \tag{31}$$

where $\chi_p = 2\mu_B^2 N_o$ is the bare Pauli susceptibility.

The TCNQ chain susceptibility, $\chi_Q(T)$, from Fig. 4 is re-plotted in Fig. 5 along with the theoretical curve from eq. (31) assuming $T_p^{MF} \simeq 300\,K$ ($3.5\,k_B T_p^{MF} \simeq 0.1$ eV) and $\chi_p \simeq 2\mu_B^2/W$ with a band width $W \simeq 0.35$ eV in rough agreement with earlier experimental and calculated results.[66] The measured susceptibility increases approximately linearly with increasing temperature with magnitude in good agreement with theory. As $T \to T_p^{MF}$, amplitude fluctuations are expected to become important so that $\xi(T)$ would decrease more rapidly than T^{-1}. The observed linearity of $\chi(T)$ all the way to room temperature may result from a partial cancellation of the small increase[63] due to amplitude

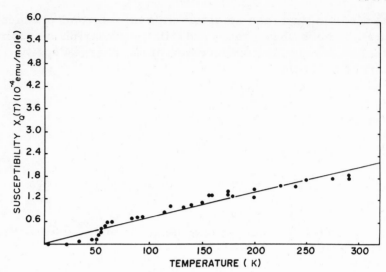

Fig. 5. TCNQ chain susceptibility, χ_Q, versus temperature.[54]
Data points were obtained directly from the [13]C Knight shift
analysis, solid line represents the temperature dependence of the
density of states resulting from the finite coherence length complex order parameter.

fluctuations balanced by the decrease arising from the temperature
dependence of the narrow band Pauli susceptibility. The density
of states as measured by the local susceptibility is consistent with
the complex order parameter description in the 1 D fluctuation
regime above 54 K. Below 54 K, the CDW becomes pinned giving
the large low frequency dielectric constant. The coherence length
grows rapidly in the pinned regime causing the susceptibility to
decrease dramatically, as observed.

At the pinning temperature, 54 K, the longitudinal coherence
length as calculated from eq. 30 is approximately 60 lattice constants. Taking into account the expected temperature dependence
due to the combined phase (eq. 30) and amplitude fluctuations, [63]
the coherence length at room temperature should be only a few
lattice constants (approximately five). These values, inferred
indirectly from the temperature dependent density of states, are
in agreement with estimates obtained from the diffuse x-ray and
neutron scattering results as described above.

The temperature dependence of $\chi_Q(T)$ may be represented
quantitatively by the Peierls'-Fröhlich model with a <u>complex</u>

order parameter describing the broken symmetry state for $T > 54K$ which becomes pinned when the phase is essentially locked below 54K. Systematic studies of the thermoelectric power in TTF-TCNQ,[51,52] TSeF-TCNQ,[67] and dilute alloys $(TTF)_{1-x}(TSeF)_x$-TCNQ[67] demonstrated that the conductivity in the "metallic" regime is dominated by the TCNQ chain. Thus, the dc transport occurs on the TCNQ chains in the temperature regime where the CDW fluctuations are building up with a relatively long coherence length which in turn arises from the growth of the complex order parameter. The CDW is therefore mobile and contributes to the total conductivity. Given that $\xi(T)/b \gg 1$, the $2k_F$ back-scattering should be strong so that the single particle contribution to the conductivity should be small, as argued above. Thus the Fröhlich CDW collective mode appears to be the primary low frequency transport mechanism in TTF-TCNQ above 54K.

The Pseudogap: Optical Properties

The optical properties and frequency dependent dielectric function expected for the Peierls'-Fröhlich charge density wave conductor have been described in detail in earlier publications.[12,17] The $2k_F$ scattering of the electron Bloch waves in the vicinity of k_F leads to a pseudogap in the density of states which becomes a real gap when the CDW phase is pinned at low temperature.

In the fluctuation regime, as shown in the previous section, the finite coherence length leads to states within the gap. However, because the coherence length is considerably greater than a lattice constant, these states are strongly scattered and hence effectively localized for $\hbar\omega > 2\Delta$. The conductivity in this region would therefore be diffusive with very low mobility and should go to zero as the coherence length grows long. For frequencies $\hbar\omega > 2\Delta$, the optical conductivity would be unaffected by the $2k_F$ scattering and hence should show the high frequency Drude-like behavior expected for a metal, in the ideal case. The expected increase in conductivity at $\hbar\omega \sim 2\Delta$, in semiconductor terms, is due to the interband transition across the Peierls' gap. Finally at very low frequencies oscillator strength associated with the Fröhlich CDW collective mode should dominate.

A schematic diagram of $\sigma_1(\omega)$ vs ω showing the characteristic features of the collective mode and single-particle excitations is shown in Fig. 6. In the conducting state the collective mode is centered at zero frequency with a width determined by the collective mode lifetime, $\tau_c(T)$. The pseudo-gap leads to a broad minimum in $\sigma_1(\omega)$ with the single-particle (semiconductor interband) transitions showing up at higher frequencies. Pinning of the collective mode will shift the oscillator strength into the infrared.

The fundamental signature of the CDW collective mode Fröhlich state is therefore relatively high electrical conductivity at dc and microwave frequencies in the presence of a pseudo-gap in the electronic excitation spectrum. The low frequency conductivity is expected to be extremely sensitive to pinning by defects and impurities. These are precisely the features observed in TTF-TCNQ for T > 54K.

Polarized reflectance data obtained from single crystals are shown in Fig. 7a[12] and 7b[68]. Fig. 7a shows the room temperature results for E∥b and E∥a in the range 50-10,000 cm^{-1}. The data below 300 cm^{-1} were obtained from unpolarized reflectance from a carefully aligned mosaic of single crystals mounted on thin gold wires. In this long wavelength region, $\Re$ (E∥b) was obtained by assuming $\Re$ (E∥a) is constant at the 300 cm^{-1} value so that $\Re$ (E∥b) = $2\Re_{meas}$ - $\Re$ (E∥a) (polarized data in the far IR are summarized below).

The single crystal measurements were repeated[68] in independent experiments and extended to lower temperatures as given

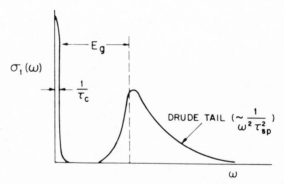

Fig. 6. Schematic representation of the frequency-dependent conductivity of the Peierls'-Fröhlich conductor showing the collective-mode peak and the single particle oscillator strength at higher frequencies.

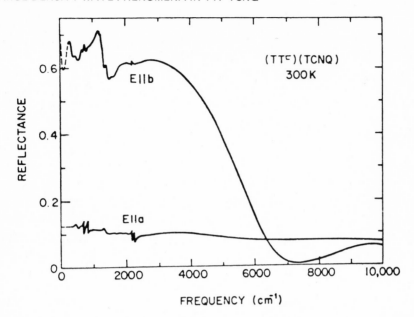

Fig. 7a.   Polarized reflectance of TTF-TCNQ single crystals over the whole region, 50-10,000 cm^{-1}. The dashed lines below 350 cm^{-1} indicate data derived from the unpolarized reflectance of an oriented mosaic (ref. 12).

in Fig. 7b where the data for E∥b from 1000 cm^{-1} to 8000 cm^{-1} are shown.   The plasma edge sharpens up with decreasing temperature consistent with the earlier results on single crystals[10] and thin films.[12,13] In addition, the absolute value of the long wavelength reflectance increases significantly from about 60-65% at room temperature to about 80% at 15K with intermediate values at 160K. This temperature dependent increase in reflectance indicates that the relatively low reflectance is intrinsic and not the result, for example, of imperfect surface scattering.

Single crystal polarized reflectance measurements have been extended into the far IR.[69] Results for two representative temperatures; 4.2K in the insulating, high dielectric constant (pinned) regime and 100K in the conducting (unpinned) regime are shown in Fig. 8.   Data obtained from different runs and on completely different mosaic samples were in good agreement.

At low frequencies ($\bar{\nu} < 20$ cm^{-1}) the polarized 100K reflectance turns rapidly upward heading toward 100% as required for a

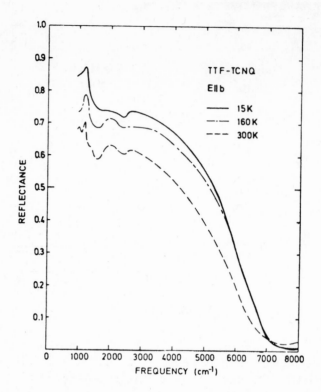

Fig. 7b. b-axis reflectance of TTF-TCNQ single crystals at selected temperatures (ref. 68).

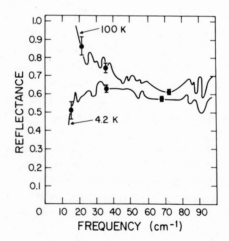

Fig. 8. Polarized ($\underline{E} \parallel \underline{b}$) reflectance of single crystal mosaic of TTF-TCNQ at 100 K and 4.2 K (resolution 2 cm^{-1}) (ref. 69).

system with large dc conductivity. This observation of nearly complete reflectance at the lowest frequency is of critical impor-tance for it demonstrates that the 65%-70% value in the region above 20 cm^{-1} is intrinsic and not the result of experimental dif-ficulties such as diffuse scattering, small particle effects, inter-rupted strands, misaligned mosaic, etc. The corresponding structure (decrease) in the 4.2 K reflectance below 20 cm^{-1} will be discussed below.

The relatively low b-axis reflectance at long wavelengths independently implies the existence of an energy gap at both tem-peratures. The reflectance expected for a simple Drude metal with a plasma edge near 7000 cm^{-1} and a dc conductivity of order 5000 Ω^{-1}-cm^{-1} would be in excess of 90% for all frequencies below 100 cm^{-1}. Therefore, the low reflectance ($\sim$70%) at 100 K above 20 cm^{-1} implies a very small FIR conductivity in the gap region. An estimate of the dielectric constant and conductivity in the gap can be obtained by comparison of the low and high temperature data. At 4.2 K TTF-TCNQ is an insulator; therefore, the low frequency reflectance is determined completely by the real part of the dielectric constant $\epsilon_1(w)$. Using the standard relation

$$\mathcal{R}(w) = \frac{(\epsilon_1^{\frac{1}{2}} - 1)^2}{(\epsilon_1^{\frac{1}{2}} + 1)^2} \tag{32}$$

gives the value ϵ_1 (20 cm^{-1} $< \bar{\nu} <$ 80 cm^{-1}) $\approx$ 90 in excellent agree-ment with the thin film results and the room temperature single crystal results obtained from a full Kramers-Kronig analysis of $\mathcal{R}(w)$. The small overall increase in reflectance above 30 cm^{-1} at 100 K compared to that at 4.2 K can be used to set an upper limit on the FIR conductivity in the gap region. Expanding the full ex-pression for reflectance in the limit $\epsilon_2/\epsilon_1 < 1$ one obtains

$$\mathcal{R}(w) \simeq \left(\frac{\epsilon_1^{\frac{1}{2}} - 1}{\epsilon_1^{\frac{1}{2}} + 1}\right)^2 [1 + 2\epsilon_2^2/\epsilon_1^{5/2}] \tag{33}$$

which leads to an estimate of the conductivity in the gap region at 100 K of σ^b (50 cm^{-1}) $\simeq$ 50 $(\Omega$-cm$)^{-1}$; two orders of magnitude below the dc and microwave values.

Using this result, the frequency dependent conductivity at long wavelengths is plotted in Fig. 9. The far IR data together with the typical dc[15] and microwave data[70] at 100 K are shown. The solid curve is from an approximate analysis of the E∥b reflectance using a single Lorentzian oscillator centered at zero frequency and a residual conductivity in the far IR of 50 $(\Omega\text{-cm})^{-1}$. These results provide clear evidence of the pseudo-gap present in the conducting regime above 54 K.

The value obtained for $\epsilon_1(\omega)$ in the FIR can be used to obtain an estimate of the Peierls' gap using the standard expression for the single particle contribution[16, 61, 71]

$$\epsilon_1^{sp} = \epsilon_0^b + \frac{2}{3}\frac{\omega_p^2}{\omega_g^2} \tag{34}$$

With $\hbar\omega_p = 1.2$ eV and $\epsilon_0^b \simeq 2.5$, we obtain $\hbar\omega_g \simeq 0.1$ eV in good agreement with the results of the full Kramers-Kronig analysis of the single crystal[12] and thin film[13] reflectance data as sketched in the inset to Fig. 9. The full frequency dependence of $\sigma_1^b(\omega)$ shows a maximum at 1000 cm^{-1} with a smooth Drude-like frequency dependence observed at higher frequencies. The corresponding behavior of $\epsilon_1^b(\omega)$ is like that of a semiconductor with transitions across the energy gap sufficiently strong to give negative values between 1000 and 6000 cm^{-1}. At lower frequencies ϵ_1^b is positive with magnitude of approximately 100 in the gap region. From the maximum in $\sigma_1^b(\omega)$ and the zero-crossing in $\epsilon_1^b(\omega)$, we infer a pseudogap with magnitude $\hbar\omega_g \simeq 0.1$ eV (10^3cm^{-1}).

The results in the conducting regime are therefore qualitatively consistent with the x-ray observations of strong 2k$_F$ scattering from the dynamical distortion with relative long coherence length.

The Transition Region: 38 K < T < 54 K

Two transformations, at 54 K and 38 K, in TTF-TCNQ have been widely studied with driect structural evidence coming from x-ray and neutron studies. Principally by analyzing transport measurements, Etemad[48] suggested the 54 K transformation results from ordering on TCNQ chains, and the 38 K transformation,

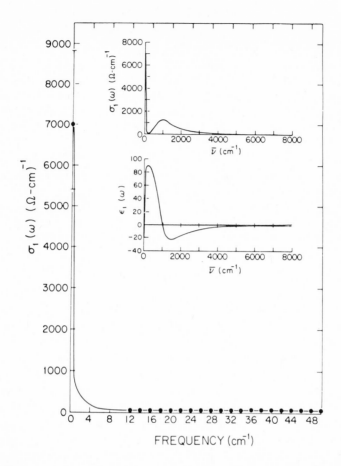

Fig. 9.   Frequency dependence of $\sigma_1 (\omega)$ below 50 cm^{-1} at 100 K.
Insert: Sketch of full frequency dependence of $\sigma_1 (\omega)$ and $\epsilon_1 (\omega)$ at
100 K.

on TTF chains.   From elastic neutron studies, Comès, et al.[36]
showed that the region 38 K $<$ T $<$ 54 K is characterized by an
incommensurate superlattice where the a-axis modulation changes
continuously from 2a near 54 K to 4a with a discontinuous step at
38 K.   On the basis of their Ginzburg-Landau treatment of the two
sets of coupled chains, Bak and Emery[72] have shown the neutron
results for TTF-TCNQ are consistent with three transitions.
They suggested long range order sets in on one of the chains at
54 K and on the second chain near 49 K with the a-axis modulation
remaining at 2a between 54 and 49 K.   From 49 K to 38 K both
chains continuously order with respect to one another, finally

locking discontinuously to 4a at 38K. The 38K transition is first-order with hysteresis observed in both the structural[36, 38] and transport studies. [73]

The Bak-Emery theory assumes that the TTF and TCNQ chains have scale temperatures which are large compared to the transition temperature so that their coherence lengths are long, and the nearly static lattice distortions may be regarded as complex order parameters in terms of which the free energy may be expanded. These assumptions appear to be valid as indicated by the structural, optical and magnetic data summarized above.

Although the Bak-Emery analysis of the structural data provided the first evidence of the 49K transition, the independent ordering of the TCNQ and TTF chains had been inferred from transport[48] and spin resonance data. [74, 75] The 49K transition is clearly observable in the local susceptibility data as obtained from the [13]C Knight shift studies. [54] These Knight shift data as shown in Fig. 10 directly identify the transitions; the TCNQ chains order at 54K and the TTF chains near 49K.

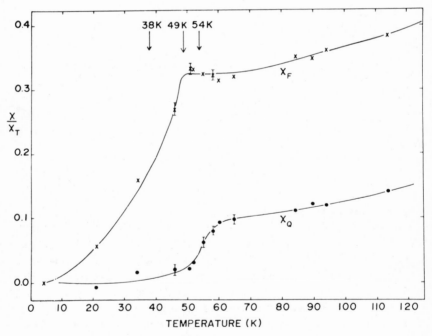

Fig. 10. Temperature dependence of the TTF and TCNQ chain local susceptibilities shown in detail (ref. 54).

In the region $49\,K < T < 54\,K$, where only the TCNQ chains
are ordered, the superlattice periodicity is $\frac{1}{2}a^*$ [36] implying that
the two TCNQ chains within the unit cell are antiphase ordered.
With the onset of TTF chain order at $49\,K$ the periodicity becomes
incommensurate, [36] $q + \frac{1}{2}a^*$, with q varying continuously in the
region $38\,K < T < 54\,K$ and jumping discontinuously to $q = -\frac{1}{4}a^*$ at
the first-order $38\,K$ transition.

Physically, these features can be understood in terms of
interchain Coulomb coupling between the charge density waves.
The Coulomb interaction between two chains with charge density
waves, $\rho_i = \rho_0 \cos\varphi_i$, on each may be written[57, 76]

$$U_{interchain} = \frac{1}{4\,\epsilon_\perp}\,\rho_0^2\,[2K_0(2k_F d)]\,\cos\,(\varphi_i - \varphi_j) \qquad (35)$$

where $\epsilon_\perp$ is the perpendicular static dielectric constant, $K_0(x)$ is
the complete Elliptic Integral of the First Kind, and d is the inter-
chain spacing (for $x \gg 1$, $2K_0(x) \simeq \sqrt{\frac{\pi}{2x}}\,e^{-x}$). For $49\,K < T < 54\,K$,
where only the TCNQ chains are ordered this Coulomb energy is
minimized if $(\varphi_i - \varphi_j) = \pi$; i. e. , the chains are anti-phase ordered.
In this case, the net interaction on a TTF chain located symmet-
rically between is zero. However, if order sets in on the TTF
chain the total energy is lowered by breaking the symmetry and
allowing $(\varphi_i - \varphi_j) = \pi + \epsilon$ so that the total energy is minimized. [77]

Bak and Emery[72] have included these features in their two-
chain coupled order parameter theory. They write the free
energy as

$$F(\Psi_{1q}, \Psi_{2q}, q) = f(\Psi_{1q}^2, q) + A(\Psi_{1q}^2, q)\,\Psi_{1q}\Psi_{2q} + B(\Psi_{1q}^2, q)\Psi_{2q}^2 \qquad (36)$$

where Ψ_{1q} represents the Fourier transform of the order parameter on the TCNQ chain and Ψ_{2q} that on the TTF chain. Minimizing in terms of the phase difference as described physically
above and taking advantage of the symmetry of the structure

$$F(\Psi_{1q}, \Psi_{2q}, q) = f(\Psi_{1q}^2, q=0) + (aq\Psi_{2q} + b\Psi_{2q}^2 + cq^2) + \cdots \qquad (37)$$

where Ψ_{1q} is assumed to take on a finite average value, a, b, and
c are function of Ψ_{1q} and T but not of q, and a vanishes when
$\Psi_{1q} = 0$. Near $54\,K$, the coefficients b and c are positive as

evidenced by the lack of order on the TTF chains and the experimental fact that $q = 0$. Minimizing with respect to q, one finds

$$q = -(a/2c) \Psi_{2q} + O(\Psi_{2q}^3) \tag{38}$$

$$F = f(\Psi_{1q}^{-2}, o) + (b - a^2/4c) \Psi_{2q}^2 + D\Psi_{2q}^4 \tag{39}$$

to fourth order in Ψ_{2q}. In absence of the interchain coupling,
order on the TTF chain would occur when $b < 0$; interchain coupling increases the ordering temperature while at the same time
causing $q \neq 0$. The induced transition occurs at the temperature
T_2 when $(b - a^2/4c) = o$. Assuming mean-field behavior at temperatures just below T_2, i.e., $\Psi_{2q}^2 = \alpha(T_2 - T)$, Bak and Emery
find[73]

$$q^2 = \begin{cases} 0, & T_1 \geq T \geq T_2 \\[2ex] \dfrac{a^2\alpha}{4c^2} (T_2 - T), & T < T_2 \end{cases} \tag{40}$$

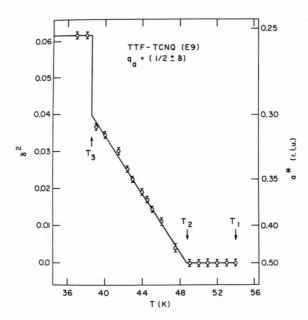

Fig. 11. Plot of observed peak positions along a^* (right scale)
and derived values of δ^2 (left scale). The three transition temperatures are also indicated (ref. 78).

The experimental results of Ellenson et al. [78] are plotted in Fig. 11 as q^2 vs T and are in complete agreement with the Bak-Emery theory.

Thus the observed phase transitions are consistent with the accumulating evidence of a complex order parameter coupled chain system with a scale temperature well above the actual transition temperature. Some evidence of 1 D to 3 D crossover has also been obtained from detailed analysis of the temperature dependent conductivity near 54 K. [79]

Early evidence of the 54 K transition came from the observation of a specific heat anomaly[49] associated with the ordering. However, no excess specific heat (or latent heat) has been observed at either the 38 K transition or the 49 K transition. The absence of an observable maximum near 49 K is somewhat surprising in view of the susceptibility results shown in Fig. 10. The apparent change in density of states on the TTF chain at 49 K is comparable to that on the TCNQ chain at 54 K.

The Pinned Regime at Low Temperatures

The experimental features which most clearly characterize the temperature range below 38 K in TTF-TCNQ are the charge density wave ground state observed in the structural studies and the unusually large low frequency dielectric constant. [16, 31] Lee, Rice and Anderson[61] have calculated the dielectric constant associated with the complex order parameter CDW state in which the phase is pinned. They find

$$\epsilon_1 - 1 = \frac{2}{3} \frac{\omega_p^2}{\omega_G^2} + \frac{\Omega_p^2}{\omega_F^2} \qquad (41)$$

where $\omega_p^2 = 4\pi Ne^2/m^*$ describes the single particle oscillator strength ($\hbar\omega_p = 1.2$ eV) in the presence of an energy gap, ω_G. The single particle oscillator strength is determined by the total electron density, N, and the electron effective band mass, m^*. In addition to this usual contribution to ϵ_1 expected for a simple semiconductor, the second term represents the collective mode oscillator strength due to optically active phase oscillations of pinned CDW where ω_F is the characteristic pinning frequency of the Fröhlich mode. The phase mode oscillator strength is given by $\Omega_p^2 = 4\pi N_s e^2/M^*$ where N_s is the density of condensed electrons ($N_s = N$ in the pure case where the CDW state is not gapless) and M^* is the Fröhlich mass[2, 60, 62]

$$\frac{M^*}{m^*} = 1 + \frac{1}{\lambda} \frac{\omega_G^2}{\omega_o^2} \tag{42}$$

where λ is the dimensionless electron-phonon coupling constant introduced in eq. 18, ω_G is the gap frequency, and ω_o is the bare phonon frequency at $2k_F$.

Extensive microwave measurements have been carried out on TTF-TCNQ to determine the low temperature dielectric constant along the chain b-axis (ϵ_1^b) and the transverse a-axis (ϵ_1^a).[16,31,32 80] The principle results include the unusually large value for $\epsilon_1^b \simeq$ 3500 and an anisotropy, $\epsilon_1^b/\epsilon_1^a$, of several hundred consistent with the 1 D nature of the electronic system. The large value for ϵ_1^b implies the presence of oscillator strength at relatively low frequencies.

Although it is natural to associate the large b-axis dielectric constant with the pinned CDW, let us examine the alternative possibility. If ϵ_1^b were entirely the result of single particle effects, the observed values for ϵ_1^b and $\hbar\omega_p$ would require that the single particle oscillator strength be exhausted at frequencies of order 0.02 eV. The number of effective electrons obtained from the oscillator strength sum rule

$$\frac{m}{m^*} N_{eff}(\omega) = \frac{1}{8}\int_o^\omega \sigma_1(\omega^1) d\omega^1 \Big/ \omega_p^2 \tag{43}$$

is shown[12] in Figure 12. The single particle oscillator strength does not begin to take on appreciable values until well above 1000 cm^{-1}. Figure 12 is obtained from analysis of the room temperature reflectance. However, the data at low temperatures[13, 69] lead to the same conclusion. Direct analysis of the 4.2 K far IR reflectance data[69] yields $\epsilon_1^b \simeq 90$ in the frequency range 20 cm$^{-1} <$ $\overline{\nu} < 80$ cm^{-1}. This value is consistent with the single particle contribution expected from the first term of eq. 42 using the known plasma frequency ($\hbar\omega_p = 1.2$ eV) and the Peierls' gap[12] ($\hbar\omega_G \simeq$ 0.14 eV).

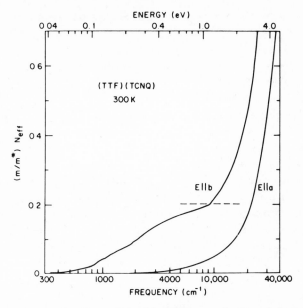

Fig. 12. Number of effective electrons, obtained from the oscillator strength sum rule, for TTF-TCNQ for two polarizations, vs frequency from 300 to 37,000 cm^{-1}. Notice the logarithmic frequency scale. The gentle plateau for E∥b implies the exhaustion of the single tight-binding band oscillator strength just below the onset of interband transitions (ref. 12).

The large microwave value therefore is associated with a relatively small oscillator strength at frequencies well below 20 cm^{-1}. To obtain more detailed information on the distribution of low frequency oscillator strength in the insulating dielectric regime, an attempt was made[69] to analyze the reflectance data using a single Lorentzian oscillator with strength Ω_p^2 pinned at ω_F at 4.2K. The corresponding dielectric function is therefore

$$\epsilon(\omega) = \epsilon_1^{sp} + \frac{\Omega_p^2}{\omega_F^2 - \omega^2 \, i\frac{\omega}{\tau}} \qquad (44)$$

where ϵ_1^{sp}, defined above is the temperature independent single particle contribution ($\epsilon_1^{sp} \simeq 90$). The analysis of the 4.2K reflectance data is shown in Fig. 13. Assuming $\epsilon_{sp} = 90$, eq. 44 contains three adjustable parameters, the oscillator strength Ω_p^2 (4.2 K), the pinning frequency, ω_F, and the pinned mode lifetime.

The dashed curve shown in the figure was obtained with $\Omega_p(4.2\,K)$ = 120 cm^{-1}, ω_F = 2 cm^{-1} and τ_{pin} = 0.35 cm although the shape of the curve is relatively insensitive to the value of the lifetime. The existence of oscillator strength below 20 cm^{-1} causes $\epsilon_1(\omega)$ to decrease as the frequency is decreased leading to the decreased reflectance with a minimum at $\omega = \Omega_p / \sqrt{\epsilon_1^{sp}}$. The above values for Ω_p and ω_F should be regarded as lower limits. The principal complication arises from finite conductivity in the gap region. If there is a small residual conductivity, σ_R, in the gap region, the corresponding $\epsilon_2 = 4\pi\sigma_R/\omega$ becomes progressively more important as the frequency is lowered. The effect would be to hold up the reflectance and distort the low frequency behavior. However, the existence of oscillator strength between 2 cm^{-1} and 10 cm^{-1} is clearly demonstrated by the combined microwave and far IR data. Direct absorbance measurements carried out by Eldridge[81] on crystals of TTF-TCNQ provide additional evidence of pinned mode oscillator strength in this range.

The pinning frequency obtained from these pure single crystals differs substantially from the 80 cm^{-1} value inferred from thin films studies.[13] This difference may be an indication of additional surface pinning contributions associated with the relatively large surface to volume ratio of the crystallites in the sublimed films.

The total oscillator strength associated with the collective mode is small indicating a relatively large Fröhlich effective mass. Using the standard relation, $\Omega_p^2 = 4\pi N_s e^2/M^*$ with $\Omega_p(100\,K)$ = 250 cm^{-1} one obtains $M^* \sim 1500\ m^*$. Using (M^*/m^*) ~ 1500 and $\omega_c \simeq 1100$ cm^{-1}, we find $\lambda^{\frac{1}{2}}\omega_0 = 40$ cm^{-1}. Neutron inelastic scattering studies[39, 40] have shown that the bare acoustic phonon frequencies are of order 50 cm^{-1} for TTF-TCNQ. Moreover, the intramolecular mode stabilization discussed by Rice et al.[82] reduces the effective value of ω_0^2. Thus, although somewhat large, the value obtained for M^* can be understood in the context of the Fröhlich mass with not unreasonable parameters.

Systematic studies of the microwave dielectric constant[80] have included measurements of the isostructural compounds (TTF)(TCNQ), TMTTF-TCNQ, DSeDTF-TCNQ, and TSeF-TCNQ. The results suggest that a common mechanism is involved. The temperature dependence of ϵ_1^b is shown for all of these systems in Fig. 14. For TTF-TCNQ, ϵ_1^b increases slightly from the 3.5×10^3 value at 4.2 K, leveling off at about 30 K, then drops

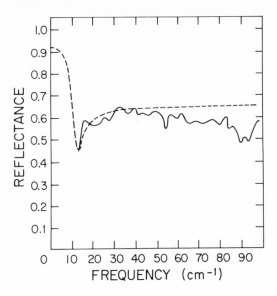

Fig. 13. Reflectance $(E \parallel \underline{b})$ at 4.2 K. The dashed curve represents the single Lorentzian oscillator analysis with $\Omega_p = 120 \text{ cm}^{-1}$ and $\omega_F = 2 \text{ cm}^{-1}$ (ref. 69).

toward zero at 38 K. Above 38 K there is strong indication that ϵ_1^b goes negative; however the experiment is insensitive to a negative ϵ_1 in the presence of a large σ_1. The derivative systems exhibit a qualitatively similar temperature dependence. The TSeF-TCNQ system is exceptional; the 4.2 K dielectric constant is 16,000 and drops monotonically to zero at about 20 K.

IR data are not yet available for the derivative systems so that a decomposition into single particle and collective phase mode contributions is not possible. However, the similar crystal structures, the similar values for ω_p,[10,84] the similar temperature dependence of the magnetic susceptibilities,[33,85] and the observation of 1 D diffuse scattering above the transition temperature[86] all suggest that the energy gap is a common feature and that ϵ_1^b is dominated by the collective phase mode at low temperatures.

The temperature dependence of ϵ_1^b has not yet been analyzed in detail. The success of mean-field theory[72] in explaining the details of the transition region in TTF-TCNQ suggests a simple renormalization of the pinning frequency with a corresponding

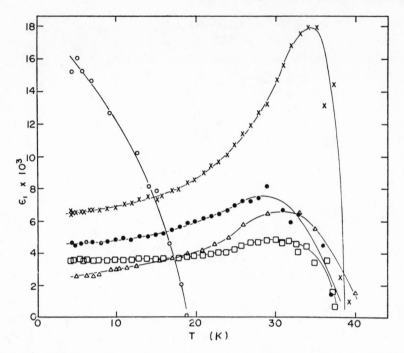

Fig. 14. Temperature dependence of the dielectric constant for TTF-TCNQ and related derivative salts (ref. 80). TTF-TCNQ □□□; TSeF-TCNQ ○○○; TMTTF-TCNQ △△△; DSeDTF-TCNQ ●●●; TTF$_{.97}$TSeF$_{.03}$-TCNQ xxx

increase in the microwave dielectric constant and conductivity. When the pinning frequency approaches zero, ϵ_1^b is driven negative by the low frequency oscillator strength as can be seen from the Kramers-Kronig transform,

$$\epsilon_1(\omega) = 1 + 8P \int_0^\infty \frac{\sigma_1(\omega_1)\,d\omega_1}{\omega_1^2 - \omega^2} \tag{45}$$

$\epsilon_1(\omega) < 0$ requires that $(d\sigma_1/d\omega) < 0$ in the vicinity of ω. Alternatively, fluctuation effects may be important even in the low temperature regime leading to thermally depinned regions which increase in fractional volume with increasing temperature. Such an effective medium approach[87] has been applied to TTF-TCNQ with results qualitatively similar to the experimental temperature dependence.

Nonlinear Transport in TTF-TCNQ at Low Temperatures

The unusually large low temperature dielectric constant and
the related low resonance frequency for the coherent pinned phase
mode imply relatively weak pinning forces and suggest the possibility that an applied electric field at low temperatures can cause
the weakly pinned CDW condensate to partially depin and thereby
become conducting. The recent experimental observation[88] of
nonlinear I-V characteristics for TTF-TCNQ at low temperatures
provides initial evidence of such CDW phenomena.

Figure 15 shows the I-V curves of a crystal of TTF-TCNQ
measured along the principal conducting b-axis. The I-V curves
are nonlinear with the dynamic conductance, dI/dV, increasing
with increasing voltage. As the temperature is lowered, the low
field conductance decreases, but the degree of nonlinearity increases dramatically. The I-V curves qualitatively appear to be
approaching an off-on situation at T = 0 K where current is not
generated until a "critical" electric field is reached. The degree
of non-linearity is illustrated in detail in the full logarithmic plot.
The results shown in Figure 15 represent current levels as low as
10^{-10} A and power inputs as low as 10^{-11} Watts. The solid line
represents ohmic behavior and is shown for comparison. At the
lowest temperatures, the I-V characteristics of TTF-TCNQ are
non-linear over the entire range measured.

To clarify any possible role of crystalline defects or imperfections in these nonlinear transport phenomena, a study of the
effects of radiation damage on the electrical properties of TTF-
TCNQ was initiated.[88] Samples were bombarded with a total flux
of 5 x 10^{14} cm^{-2} of 8 MeV deuterons from the University of
Pennsylvania tandem accelerator. Crystals of TTF-TCNQ of
typical thickness (50 μm) are essentially transparent to 8 MeV
deuterons which lose approximately 100 KeV during passage. The
induced damage was monitored by observing the room temperature
resistance during irradiation; initial studies were on samples with
room temperature resistance increased by 20%. The electron
spin resonance of the damaged samples at 4.2 K indicated an
induced concentration of spin $\frac{1}{2}$ impurities of order one-tenth
percent. The actual defect concentration was probably somewhat
higher as judged by the smearing of the phase transitions.[89]

The low temperature I-V curves were measured before and
after irradiation with several important results. The low field

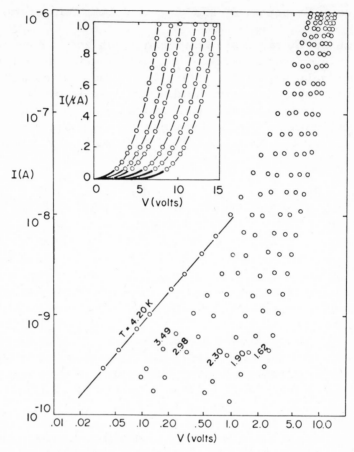

Fig. 15.   Logarithmic plot of I vs V for TTF-TCNQ showing the nonlinear transport at low temperatures.   The data are plotted as I vs V in the inset (ref. 88).

ohmic conductance at 4. 2 K was increased by more than two orders of magnitude and was sample independent.   This defect contribution to the low temperature transport is <u>ohmic</u> and described by $\sigma_d = \sigma_d^0 \exp(-E_d/T)$ with $E_d = 20$ K.   The previously observed nonlinearity is masked at 4. 2 K by the larger defect contribution.   At 1. 6 K where σ_d is small and unimportant, the nonlinearity shows up clearly with identical curvature as a function of applied voltage. The absolute current density was reduced by a constant factor of three due to the irradiation induced defects.

A trivial source of nonlinearity in a semiconductor sample is self-heating.   The ohmic behavior of the irradiated samples at

4. 2 K when combined with the measured temperature dependence of σ_d allows an upper limit for self-heating of order 0. 1 K at an input power level of 10^{-3} W. Since the data of Fig. 15 are restricted to power levels below 10^{-5} W, the maximum sample heating is less than 10^{-3} K, or completely negligible.

The I-V characteristics of a semiconductor can be nonlinear if the number of carriers is determined by the applied electric field through injection, impact ionization by accelerated carriers, or field ionization of neutral impurities. Each of these mechanisms can be ruled out on theoretical grounds. To achieve the measured current densities by injection would require a mobility in excess of 10^3 cm^2/V-sec.[90] Similarly, impact ionization at the observed electric field of order 10^2 volts/cm of carriers with activation energy of order 20 K would require $\mu > 10^3$ cm^2/V-sec.[91] These values are three to four orders of magnitude greater than found in even the purest single crystal organic semiconductors. Moreover, defects and chain breaks are especially important in quasi-one-dimensional systems and will limit the mobility. Field ionization is a tunneling process and the available energy is determined by the applied electric field integrated over the length of the impurity wave function. To ionize carriers with binding energy of order 20 K with E $\simeq 10^2$ V/cm would require that the impurity wave function extend over lengths of order 10^3 A. Direct experimental evidence that traditional semiconductor mechanisms are not involved in the present observations comes from the relative insensitivity of the low temperature nonlinearity to induced defects at the fraction of a percent level.

The data from Fig. 15 are replotted in Fig. 16 as log I vs 1/T at fixed electric field (fixed voltage). The general features are those of a thermally activated conductivity with the activation energy, Δ/k_B, decreasing with increasing electric field. The field dependent activation energy as obtained from the slopes of the curves is plotted in the inset to Figure 16 which shows Δ decreasing approximately linearly with increasing field and extrapolating to zero at $E_0 \simeq 125$ V/cm. Thus for E < E_0, the data are adequately described by an expression of the form $(j/E) = \sigma_0 \exp\left[-\frac{\Delta}{T}(1 - E/E_0)\right]$. The value for E_0 as obtained from several samples if 150 ± 50 V/cm with the variation from sample to sample consistent with the errors involved in determining the distance between voltage contacts on the small samples. The activation energy is approximately $\Delta/k_B = 14$ K with little variation (± 1 K) from sample to sample. The prefactor, σ_0, is somewhat sample

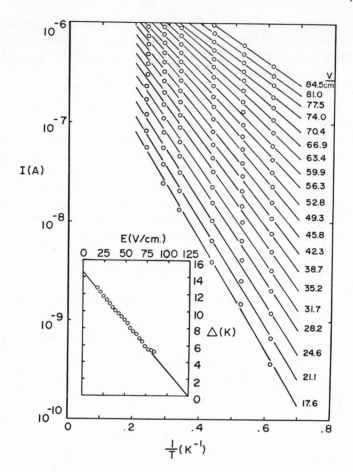

Fig. 16. log I vs T^{-1} at various values of the applied electric field, the solid lines represent least-square fits to the data at each field. The field-dependent activation energy is shown in the inset (ref. 88).

dependent ranging from 5×10^{-5} to 3×10^{-4} $(\Omega\text{-cm})^{-1}$ with a typical value of 10^{-4} $(\Omega\text{-cm})^{-1}$. The irradiation experiments suggest that σ_o is dependent on sample perfection as described above. For $E < E_o$, the prefactor is only weakly field dependent with σ_o increasing by less than a factor of two whereas (j/E) changes by three orders of magnitude at the lowest temperatures. Preliminary evidence suggests that at higher fields $(E \geq E_o)$ the prefactor increases more rapidly, approximately exponentially, with increasing electric field.

The critical field for depinning the CDW at $T = 0\,K$ can be estimated from the following argument. We assume a pinning potential $V(\varphi) = V_o f(\varphi)$ where $f(\varphi)$ is periodic in φ. The real space periodicity would depend on the pinning mechanism. [64] If the pinning is due to high order commensurability for example, the potential would be periodic in the equilibrium lattice constant, $\varphi = (2\pi/b)x$. If the pinning arises from interchain Coulomb interactions, the potential would be periodic in the superlattice constant i.e., $\varphi = (2\pi/\lambda_s)x$ where, for $(TTF)(TCNQ)$ $\lambda_s = (3.4)b$. In order to depin, the electric field energy in moving the CDW a periodicity length (ℓ) must be greater than V_o; thus $E_c \sim V_o/\ell$. In the simplest case, $f(\varphi) = (1 - \cos\varphi)$ and V_o can be related to the fundamental pinning frequency, ω_F, and the Fröhlich effective mass M^* by expanding about the pinned equilibrium and retaining only the harmonic term. [62]

$$eV_o = M^* \omega_F^2 \left[\frac{d\varphi}{dx} \Big|_o \right]^{-2} \tag{46}$$

$$E_c = \frac{M^* \omega_F^2}{e\,\ell} \left[\frac{d\varphi}{dx} \Big|_o \right]^{-2} \tag{47}$$

Using values of M^* and ω_F^2 from infrared studies, [69] E_c is estimated to be of order several hundred volts/cm; a rough estimate due to uncertainties in the values of the parameters and, more importantly, the actual form of $f(\varphi)$.

For a rigid CDW, the nonlinear current density should be zero at $T = 0\,K$ for $E < E_c$. Inhomogeneous pinning or phase-soliton[64] excitations will allow thermally activated precursor effects to show up below the critical field. E_c is therefore identified with the field needed to reduce the activation energy to zero; $E_c \simeq 1.5 \times 10^2$ V/cm. From this point of view, the CDW conductivity in the regime $(E \gtrsim E_c)$ is itself non-ohmic, increasing approximately exponentially with applied field.

Although the identification of the observed nonlinear transport with phase-soliton excitations and CDW depinning is speculative, the possibility of experimentally studying such charged anharmonic domain-wall type excitations[92] is exciting and of broad current interest. [93] These 1 D charge density wave systems may provide a unique "laboratory" for the study of such phenomena.

CONCLUSION

There has been remarkable progress in the study of organic one-dimensional conductors in the two years since the previous NATO school on this subject. The Peierls' instability and CDW ground state have been established as experimental fact as a result of the structural measurements. The detailed role played by the TCNQ and TTF chains has been clarified through measurements of the local susceptibility using NMR techniques and g-value labeling. The important theoretical contributions to the developing understanding of the phase transitions have provided confidence in an approach based on a complex order parameter description of the coupled chain system.

The observation of $4k_F$ diffuse scattering,[37] it origin and implications, remains to be understood in detail. Emery's[55] analysis of these results represents one of the first direct applications of the exact solution of 1 D models to a particular experimental problem.

The strength and importance of Coulomb electron-electron interactions on the individual chains remains to be settled. Nuclear magnetic resonance relaxation studies[53,54] in combination with the local chain susceptibilities[33,54] provide the basic necessary experimental information. However, to date the analysis relies on random-phase-approximation results[94] carried over from the three dimensional problem. The observation of frequency dependence[95] in the nuclear relaxation rate suggests that specific 1 D features may be playing a role.[96,97] A more detailed analysis of the effect of electron-electron interactions on the nuclear relaxation rates may be required.

Perhaps the most important question remaining to be settled is whether the Fröhlich collective mode is the dominant transport mechanism in the conducting state. The large conductivity in the presence of well-defined $2k_F$ scattering which gives rise to the pseudo-gap as observed in the optical and magnetic properties provides the primary evidence of collective behavior. The presence of oscillator strength at very low frequencies in the dielectric regime as indicated by the microwave and far IR studies provides similar evidence in the low temperature regime. On the other hand it has been argued that the thermal conductivity,[50] optical scattering time,[98] and temperature dependence[99] of the conductivity are all consistent with simple metallic behavior with electron single particle transport limited by mean-free-path scattering.[100]

The existence of high metal-like conductivities in quasi-1 D systems as evidence for many-body effects remains under continued active study. For example, Bychkov,[101] Berezinsky,[102] and Gogolin, Mel'nikov, and Rashba[103] have shown that for independent electrons in a strictly 1 D system containing impurities or defects, the conductivity should decrease as the temperature is lowered, approaching zero with a temperature dependence determined by the ratio of the impurity and phonon scattering rates, $\tau_{imp}/\tau_{ph}(T)$. Thus experiments involving controlled amounts of impurities and/or defects should be of increasing importance. The role of defects or impurities as potential pinning centers for the collective mode has been discussed[104] with the suggestion that in the presence of such centers, the fluctuations will be resistive. This important theoretical question has not yet been settled. Experimentally,[15,89,105] however, impurities and defects do give rise to resistive fluctuations leading to a lower peak conductivity occurring at a higher temperature.

As a result of the lack of long-range order and the presence of intrinsic dissipative processes, the Peierls'-Fröhlich condensate is not a superconductor. However, a broken symmetry conduction mechanism arising from a (nearly) phase independent charge-density wave would represent the only example yet found in nature of collective electron transport which is different from pairing superconductivity. The experimental evidence for collective transport in these 1 D materials is accumulating and is a principal reason for their interest.

ACKNOWLEDGEMENT

The work of many people is summarized in this brief review. I particularly wish to acknowledge the contributions of Prof. A. F. Garito. His care and insistence on the highest quality materials, together with a constant attention to detail on many phases of the experimental studies, have been of clear importance. I thank C. K. Chiang, Marshall J. Cohen, L. B. Coleman, C. R. Fincher, W. J. Gunning, S. K. Khanna, P. R. Newman, P. Nigrey, E. F. Rybaczewski, J. C. Scott, D. B. Tanner, and T. S. Wei for recent crucial contributions. I am especially grateful to Dr. R. Comès and Dr. G. Shirane and their colleagues for their beautiful experimental studies of the structural aspects of the Peierls instability in TTF-TCNQ and for many valuable discussions.

REFERENCES

1. R. E. Peierls, Quantum Theory of Solids (Oxford U. P., London, England, 1955), p. 108.

2. H. Fröhlich, Proc. R. Soc. A 223, 296 (1954).

3. A. W. Overhauser, Phys. Rev. Lett. 4, 62 (1960).

4. For a series of articles, see Low Dimensional Cooperative Phenomena edited by H. J. Keller (Plenum, New York, 1975); and One Dimensional Conductors edited by H. G. Schuster (Springer, Berlin 1975).

5. R. P. Shibaeva and L. O. Atovmyan, J. Struct. Chem. 13, 546 (1972); F. H. Herbstein, in Perspectives in Structural Chemistry, edited by J. D. Dunitz and J. A. Ibers (Wiley, New York, 1971), Vol. IV, pp. 166-395.

6. For a description of the crystal structure of TTF-TCNQ, see T. J. Kistenmacher, T. E. Phillips and D. O. Cowan, Acta Cryst. B30, 763 (1974).

7. D. Kuse and H. R. Zeller, Phys. Rev. Lett. 27, 1060 (1971).

8. P. Bruesch, S. Strässler, and H. R. Zeller, Phys. Rev. B 12, 219 (1975).

9. H. R. Zeller, in Low-Dimensional Cooperative Phenomena, edited by H. J. Keller (Plenum, New York, 1975), p. 215.

10. A. A. Bright, A. F. Garito, and A. J. Heeger, Solid State Commun. 13, 943 (1973); Phys. Rev. B 10 1328 (1974).

11. P. M. Grant, R. L. Greene, G. C. Wrighton, and G. Castro, Phys. Rev. Lett. 31, 1311 (1973).

12. D. B. Tanner, C. S. Jacobsen, A. F. Garito, and A. J. Heeger, Phys. Rev. Lett. 32, 1301 (1974); Phys. Rev. B13, 3381 (1976).

13. C. S. Jacobsen, D. B. Tanner, A. F. Garito, and A. J. Heeger, Phys. Rev. Lett. 33, 1559 (1974).

14. M. J. Minot and J. H. Perlstein, Phys. Rev. Lett. 26, 371 (1971); H. R. Zeller and A. Beck, J. Phys. Chem. Solids 35, 77 (1974); D. Kuse and H. R. Zeller, Solid State Commun. 11, 355 (1972).

15. M. J. Cohen, L. B. Coleman, A. F. Garito, and A. J. Heeger, Phys. Rev. B 10, 1298 (1974).

16. S. K. Khanna, E. Ehrenfreund, A. F. Garito, and A. J. Heeger, Phys. Rev. B 10, 2205 (1974).

17. A. J. Heeger and A. F. Garito, in Low Dimensional Cooperative Phenomena, edited by H. J. Keller (Plenum, New York, 1975), p. 89.

18. Morrel Cohen, Lake Arrowhead Conference on One-Dimensional Conductors, Lake Arrowhead, Calif., 1974 (unpublished).

19. M. Boudeulle, Ph. D. Thesis, U. Claude-Bernard de Lyon (1972); M. Boudeulle, Cryst. Struct. Commun. 4, 9 (1975).

20. Marshall J. Cohen, A. F. Garito, A. J. Heeger, A. G. MacDiarmid, C. M. Mikulski, M. S. Saran, and J. Kleppinger, J. Amer. Chem. Soc. 98:13, 3844 (1976).

21. A. A. Bright, M. J. Cohen, A. F. Garito, A. J. Heeger, C. M. Mikulski, P. J. Russo, A. G. MacDiarmid, Phys. Rev. Lett. 34, 206 (1975).

22. A. A. Bright, M. S. Cohen, A. F. Garito, A. J. Heeger, C. M. Mikulski, A. G. MacDiarmid, Appl. Phys. Lett. 26, 612 (1975); see also Marshall J. Cohen, Thesis, University of Pennsylvania (1975).

23. H. Kamimura, A. J. Grant, F. Levy, A. D. Yoffe and G. D. Pitt, Solid State Commun. 17, 49 (1975).

24. L. Pinstchovius, H. P. Geserich and W. Moller, Solid State Commun. 17, 477 (1975).

25. J. Bordas, A. J. Grant, H. P. Hughes, A. Jakobson, H. Kamimura, F. A. Levy, K. Nakao, Y. Natsume, and A. D. Yoffe, J. Phys. C 9, L277 (1976).

26. C. H. Chen, J. Silcox, A. F. Garito, A. J. Heeger, and A. G. MacDiarmid, Phys. Rev. Letters 36, 525 (1976).

27. V. V. Walatka, Jr., M. M. Labes and J. H. Perlstein, Phys. Rev. Lett. 31, 1139 (1973); C. H. Hsu and M. M. Labes, J. Chem. Phys. 61, 4640 (1974).

28. R. L. Greene, P. M. Grant, and G. B. Street, Phys. Rev. Lett. 34 89 (1975).

29. R. L. Greene, G. B. Street and L. J. Suter, Phys. Rev. Lett. 34, 577 (1975).

30. J. H. Perlstein, J. P. Ferraris, V. V. Walatka, and D. O. Cowan, AIP Conf. Proc. 10, 1494 (1972); J. P. Ferraris, D. O. Cowan, V. V. Walatka, and J. H. Perlstein, J. Amer. Chem. Soc. 95, 948 (1973)

31. S. K. Khanna, A. F. Garito, A. J. Heeger, and R. C. Jaklevic, Solid State Commun. 16, 667 (1975).

32. A. N. Bloch, J. P. Ferraris, D. O. Cowan, and T. O. Peohler, Solid State Commun. 13, 753 (1973). The value of ϵ_1 (4.2 K) = 50 reported has been demonstrated to result from a lack of sample purity (Ref. 16). With increased sample quality, their result has come into agreement with the value reported in Ref. 16 (A. N. Bloch, APS Meeting, Denver, Colo., 1975 (unpublished).

33. J. C. Scott, A. F. Garito, and A. J. Heeger, Phys. Rev. B 10, 3131 (1974).

34. F. Denoyer, R. Comès, A. F. Garito, and A. J. Heeger, Phys. Rev. Lett. 35, 445 (1975).

35. S. Kagoshima, H. Anzai, K. Kajimura, and T. Ishiguro, J. Phys. Soc. Japan 39, 1143 (1975).

36. R. Comès, S. M. Shapiro, G. Shirane, A. F. Garito, and A. J. Heeger, Phys. Rev. Lett. 35, 1518 (1975).

37. J. P. Pouget, S. K. Khanna, F. Denoyer, R. Comès, A. F. Garito, and A. J. Heeger, Phys. Rev. Lett. (August, 1976).

38. S. Kagoshima, T. Ishiguro, and H. Anzai, J. Phys. Soc. Japan (in press).

39. H. A. Mook and C. R. Watson, Phys. Rev. Lett. 36, 801 (1976).

40. G. Shirane, S. M. Shapiro, R. Comès, A. F. Garito, and A. J. Heeger, Phys. Rev. B (in press).

41. A. M. Afanas'ev and Yu. Kagan, Zh. Eksp. Teor. Fiz. 43, 1456 (1963) [Sov. Phys.-JETP 16, 1030 (1963)].

42. C. G. Kuper, Proc. Royal Soc. (London), A227, 214 (1955).

43. M. J. Rice, and S. Strässler, Solid State Commun. 13, 697 (1973); 13, 1389 (1973).

44. J. R. Schrieffer, in Collective Properties of Physical Systems, edited by B. Lundquist and S. Lundquist (Nobel Foundation, Stockholm and Academic, New York, 1973), p. 142.

45. P. A. Lee, T. M. Rice, and P. W. Anderson, Solid State Commun. 31, 462 (1973).

46. P. Nielsen, D. J. Sandman, and A. J. Epstein, Solid State Commun. 17, 1067 (1976); 15, 53 (1974); W. D. Grobman, R. A. Pollak, D. E. Eastman, E. T. Maas, and B. A. Scott, Phys. Rev. Lett. 32, 534 (1974); P. Coppens, Phys. Rev. Lett. 35, 98 (1975); W. D. Grobman and B. D. Silverman, Solid State Commun. 19, 319 (1976).

47. D. Jérome, W. Müller, and M. Weger, J. Phys. (Paris) 35, L77 (1974); J. R. Cooper, D. Jérome, M. Weger, and S. Etemad, ibid. 36, L219 (1975); A. J. Berlinsky, T. Tiedje, J. F. Carolan, L. Weiler, and W. Friesen, Bull. Am. Phys. Soc. 20, 465 (1975); S. Etemad, T. Penney, and E. M. Engler, ibid. 20, 496 (1975); see also C. W. Chu, J. M. E. Harper, T. H. Geballe, and R. L. Greene [Phys. Rev. Lett. 31, 1491 (1973)] who saw indications of a second transition at a somewhat higher temperature and first conjectured that the TTF and TCNQ chains might be undergoing separate transitions. The second peak in the anisotropy (Ref. 15) results from the 38 K transition.

48. S. Etemad, Phys. Rev. B13, 2254 (1976).

49. R. A. Craven, M. B. Salomon, G. DePasquali, R. M. Herman, G. Stucky, and A. Schultz, Phys. Rev. Lett. 32, 769 (1974).

50. M. B. Salomon, J. W. Bray, G. DePasquali, R. A. Craven, G. Stucky, and A. Schultz, Phys. Rev. B11, 619 (1975).

51. P. M. Chaikin, J. F. Kwak, T. E. Jones, A. F. Garito, and A. J. Heeger, Phys. Rev. Lett. 31, 601 (1973).

52. J. F. Kwak, P. M. Chaikin, A. A. Russel, A. F. Garito, and A. J. Heeger, Solid State Commun. 16, 729 (1975).

53. E. F. Rybaczewski, A. F. Garito, A. J. Heeger, and E. Ehrenfreund, Phys. Rev. Lett. 34, 524 (1975).

54. E. F. Rybaczewski, L. S. Smith, A. F. Garito, A. J. Heeger, and B. Silbernagel, Phys. Rev. B (in press).

55. V. J. Emery, Phys. Rev. Lett. 37, 107 (1976).

56. J. B. Torrance, to be published.

57. P. A. Lee, Aspen Workshop on 1 D Conductors, Aspen, Colo., 1975 (unpublished); M. J. Rice, Aspen Workshop on 1 D Conductors, Aspen, Colo., 1975 (unpublished).

58. D. J. Scalapino, Y. Imry, and P. Pincus, Phys. Rev. B11, 2042 (1975).

59. S. Barisic and K. Uzelac, J. Physique 36, 1267 (1975).

60. J. Bardeen, Solid State Commun. 13, 357 (1973); D. Allender, J. W. Bray, and J. Bardeen, Phys. Rev. B9, 119 (1974).

61. P. A. Lee, T. M. Rice, and P. W. Anderson, Phys. Rev. Lett. 31, 462 (1973); see also M. J. Rice and S. Strässler, Solid State Commun. 13, 1389 (1973); and A. Bjelis and S. Barisic, J. Physique Lett. 36, L169 (1975).

62. M. J. Rice, in Low-Dimensional Cooperative Phenomena, edited by H. J. Keller (Plenum, New York, 1975), p. 23.

63. D. J. Scalapino, M. Sears, and R. A. Farrell, Phys. Rev.
B6, 3409 (1972); R. Balian and G. Toulouse, Ann. Phys. 83,
28 (1974).

64. M. J. Rice, A. R. Bishop, J. A. Krumhansl, and S. E.
Trullinger, Phys. Rev. Lett. 36, 432 (1976).

65. P. Manneville, J. Physique 36, 701 (1975).

66. A. Karpfen, J. Ladik, G. Stollhoff and P. Fulde, Chem.
Phys. 8, 215 (1975); A. J. Berlinsky, James F. Carolan
and Larry Weiler, Solid State Commun. 15, 795 (1974);
F. Herman (to be published) finds W = 0. 3 - 0. 4 eV from
calculations of the TCNQ dimer.

67. P. M. Chaikin, R. L. Greene, S. Etemad and E. M. Engler,
Phys. Rev. B13, 1627 (1976).

68. C. S. Jacobsen, Thesis, The Technical University of
Denmark, Lyngby, Denmark (1975).

69. L. B. Coleman, C. R. Fincher, Jr., A. F. Garito, and
A. J. Heeger, Physica Status Solidi(b) 75, 239 (1976).

70. M. Cohen, S. K. Khanna, W. J. Gunning, A. F. Garito,
and A. J. Heeger, Solid State Commun. 17, 367 (1975).

71. David Penn, Phys. Rev. 128, 2093 (1962).

72. Per Bak and V. J. Emery, Phys. Rev. Lett. 36, 978 (1976).

73. T. Ishiguro, S. Kagoshima, H. Anzai, J. Phys. Soc. Japan
(in press).

74. Y. Tomkiewicz, B. A. Scott, L. J. Tao and R. S. Title,
Phys. Rev. Lett. 32, 1363 (1974).

75. Y. Tomkiewicz, A. R. Taranko, and J. B. Torrance, Phys.
Rev. Lett. 36, 751 (1976).

76. K. Saub, S. Barisic, and J. Friedel, Phys. Lett. 56A, 302
(1976).

77. T. D. Schultz and S. Etemad, Phys. Rev. B13, 4928 (1976);
see also Guy Deutscher, Phys. Rev. B13, 2714 (1976).

78. W. D. Ellenson, R. Comès, S. M. Shapiro, G. Shirane,
     A. F. Garito, and A. J. Heeger, Solid State Commun. (in
     press).

79. P. Horn and D. Rimai, Phys. Rev. Lett. 36, 809 (1976).

80. W. J. Gunning, S. K. Khanna, A. F. Garito, and A. J.
     Heeger, Solid State Commun. (submitted).

81. J. E. Eldridge, Bull. Amer. Phys. Soc. 20, 495 (1975);
     and to be published.

82. M. J. Rice, C. B. Duke and N. O. Lipari, Solid State
     Commun. 17, 1089 (1975).

$$\omega^{-2}_{o} = \sum_n g^2_n \omega^{-3}_n / \sum_n g^2_n \omega^{-1}_n$$

     where ω_n are the various mode frequencies and g_n are the
     respective coupling constants.

83. S. Etemad, T. Penney, E. M. Engler, B. A. Scott, and
     P. E. Seiden, Phys. Rev. Lett. 34, 741 (1975).

84. P. M. Grant, P. Mengel, E. M. Engler, G. Castro, and
     G. B. Street, Bull. Am. Phys. Soc. 21(3), 254 (1976).

85. Studies carried out on samples prepared independently at
     IBM (S. Etemad, E. M. Engler, and J. C. Scott, unpub-
     lished) and Penn (T. Wei and A. F. Garito, unpublished)
     show behavior qualitatively similar to that observed in
     TTF-TCNQ.

86. C. Weyl, E. M. Engler, S. Etemad, K. Bechgaard, and
     J. Jehanno, Solid State Commun. (to be published).

87. F. P. Pan, D. Stroud, and D. B. Tanner, Solid State
     Commun. (in press).

88. Marshall J. Cohen, P. R. Newman, and A. J. Heeger,
     Phys. Rev. Lett. (submitted).

89. C. K. Chiang, M. J. Cohen, P. R. Newman, and A. J.
     Heeger (to be published).

90. See, for example, F. Gutmann and L. E. Lyons, Organic
     Semiconductors (Wiley, New York, 1967).

91. N. Sklar and E. Burstein, J. Phys. Chem. Solids $\underline{2}$, 1 (1957).

92. J. A. Krumhansl and J. R. Schrieffer, Phys. Rev. B$\underline{11}$, 3535 (1975); T. R. Koehler, A. R. Bishop, J. A. Krumhansl and J. R. Schrieffer, Solid State Commun. $\underline{17}$, 1515 (1976).

93. R. F. Dashen, Phys. Rev. D$\underline{11}$, 3424 (1975).

94. E. Ehrenfreund, E. F. Rybaczewski, A. F. Garito, and A. J. Heeger, Phys. Rev. Lett. $\underline{28}$, 873 (1972).

95. G. Soda, D. Jérome, M. Weger, J. M. Fabre, and L. Giral, Solid State Commun. (in press).

96. J. Villain, J. Phys. Lett. $\underline{36}$, L173 (1975).

97. F. Devreux, J. Phys. C $\underline{8}$, L132 (1975); Phys. Rev. B$\underline{13}$, 4651 (1976).

98. P. E. Seiden and D. Cabib, Phys. Rev. B$\underline{13}$, 1846 (1976).

99. R. P. Groff, A. Suna, and R. E. Merrifield, Phys. Rev. Lett. $\underline{33}$, 418 (1974).

100. G. A. Thomas et al., Phys. Rev. B$\underline{13}$, 5105 (1976).

101. Y. A. Bychkov, Zh. Eksp. Teor. Fiz. $\underline{65}$, 427 (1973) [Sov. Phys. -JETP $\underline{38}$, 209 (1974)].

102. V. L. Berezinskii, Zh. Eksp. Teor. Fiz. $\underline{65}$, 1251 (1973) [Sov. Phys. -JETP $\underline{38}$, 620 (1974)].

103. A. A. Gogolin, V. I. Mel'nikov, and E. I. Rashba, Zh. Eksp. Theor. Fiz. $\underline{69}$, 327 (1975).

104. H. Fukuyama, T. M. Rice, and C. M. Varma, Phys. Rev. Lett. $\underline{33}$, 305 (1974).

105. L. B. Coleman, Thesis, Univ. of Pennsylvania, 1975; Marshall J. Cohen, Thesis, Univ. of Pennsylvania, 1975.

THE ROLE OF COULOMB INTERACTIONS IN TTF-TCNQ

Jerry B. Torrance

IBM Research, Yorktown Heights, New York 10598 and

IBM Research, San Jose, California 95193 USA[†]

In this article we demonstrate that Coulomb interactions play a major role in the solid state properties of TTF-TCNQ. We start by considering a model of non-interacting electrons and then examine the new features caused by adding the strong Coulomb repulsion between these electrons. These new features include:

(1) The appearance of a new optical excitation whose energy is a measure of the strength of the effective Coulomb interaction;

(2) An enhancement in the magnitude of the magnetic susceptibility and effects on the NMR relaxation rate;

(3) The decoupling of spin and translational degrees of freedom and hence the appearance of spin waves; and

(4) The doubling of the characteristic wavevector of the charge excitations, from $2k_F$ to "$4k_F$".

In the first section of this article, we discuss a theoretical model which describes the effects of strong Coulomb interactions. The remaining sections are devoted to presenting the evidence for each of these new features contained in measurements of optical absorption, magnetic susceptibility, NMR relaxation, inelastic neutron scattering, and diffuse X-ray scattering.

I. MODEL OF STRONG COULOMB INTERACTIONS

In order to construct a theoretical framework for the discussion of the experiments which will follow, we first consider the simplest model for TTF-TCNQ: a one-dimensional stack of organic molecules, with an average of ρ conducting electrons per molecule, which form a tight binding band of width 4t, where t is the transfer (or resonance) integral. We would now like to examine the effect of the Coulomb repulsion between these electrons on the properties of such a stack. Since the electrons are constrained to be on the molecules (and not inbetween), the Coulomb energy, e^2/r, between a pair of electrons at a distance r takes on discrete values, U, V_1, V_2, V_3,..., corresponding to the cases where the two electrons are on the same molecule (U), on nearest neighboring molecules (V_1), on next nearest neighboring molecules down the stack (V_2), etc. Unfortunately, the full effects of all of these terms are generally not yet known and we must consider some approximation to this interaction. The simplest and most extensively studied such approximation is that of the Hubbard model [1,2,3], where only the repulsion, U, between electrons on the same molecule is retained; the other, longer range interactions (V_i) are neglected. For some properties, it is possible [4] to include both U and V_1 in an extended Hubbard model with an effective $U_{eff}=U-V_1$. In most cases, particularly those where $\rho \neq 1$, however, it is not known if the effects of V_1 are correctly included by such a single U_{eff}. Even within the Hubbard model, most of the properties are not known for all values of $U/4t$ and ρ, particularly for $\rho \neq 1$, which is very different from the familiar case of $\rho=1$. Since the few results which are known are often for the case of large or infinite U, we shall have to restrict ourselves to the $U \gg 4t$ limit of the Hubbard model. In a few instances, we will include both U and $V_1 \gg 4t$ (which we will call the Wigner crystal limit) for physically describing the limit of strong Coulomb interactions.

Thus, we compare a model of non-interacting ($U=0$) electrons with a large U Hubbard model and look for the new features associated with the presence of Coulomb interactions. Since it is not possible to give a rigorous comparison, we will present a simple, qualitative picture, which emphasizes the difference in the physics of these two limits. This comparison is certainly an extreme one, but with a number of advantages: more is known about the properties at these limits and a clearer physical description can be given, enabling us to more easily and dramatically see the effects of Coulomb interactions. In addition, these limits represent the two extreme views which have been expressed about the importance of Coulomb interactions in TTF-TCNQ: (1) they play only a minor role and can be neglected [5-7]; or (2) they play a major role and must be included to describe the most basic properties [8-11]. Since we will be concerned with this major question involving potentially large quantities (U and $4t$), we will make only brief contact with the phase transitions and relatively weak interactions, such as interstack coupling, electron-phonon interactions, etc. For simplicity and because of evidence from g-value measurements

[12,13], we will assume that both stacks behave equivalently and hence have comparable values of U and 4t.

We first consider a one-dimensional tight-binding band of non-interacting electrons, for which the dispersion relation is shown in Fig. 1(a):

$$E^{U=0}(k) = -2t \cos kb \qquad (2N_0 \text{ states}) \qquad (1)$$

Since each momentum state can accommodate two electrons because of spin degeneracy, this band contains $2N_0$ states, where N_0 is the number of molecules. At T=0 the band is filled by ρN_0 electrons up to the Fermi wavevector k_F, where

$$2k_F^{U=0} \equiv 2k_F^0 = \rho b^*/2 = \pi\rho/b \ . \qquad (2)$$

Here $2k_F^0$ is in the b-direction, where b is the spacing between molecules along the stack, and b*=2π/b.

We now want to examine the effects on this band caused by large Coulomb interactions between the electrons. In the Hubbard model with U>>4t, the energy of those states with two electrons on the same molecule will be raised by an energy ~U. Since U>>4t, these states will be excluded from the one-electron band in Fig. 1(a); they will then form a band of anti-bound states at an energy ~U above the remaining states, as shown schematically in Fig. 1(b). For ρ≤1, these excluded states will not be occupied, since kT is generally much less than U, but they can be observed by optical transitions from the occupied states [9,14,15], shown schematically in Fig. 1(b) by the transition labeled B. This transition corresponds to exciting an electron to an already occupied molecule. Thus, the presence of strong Coulomb interactions will give rise to a new peak in the optical absorption, at an energy ~U.

There are other consequences of the exclusion of these states with two electrons on a molecule. Since these excluded states have paired spins, they are non-magnetic. The remaining occupied states will then have a larger fraction of magnetic states, and hence a larger susceptibility. Thus, the magnitude of the magnetic susceptibility will be enhanced [16].

Another feature caused by the presence of Coulomb interactions is the decoupling of the spin and translational degrees of freedom, or equivalently, the spin and charge excitations. In the absence of these interactions (U=0), the energy E(q) required to excite an electron from a momentum state k to a state above the Fermi level at k+q is shown in Fig. 2(a). The degeneracy at $q=2k_F^0$ corresponds to the fact that it takes zero energy to excite

an electron from k_F^0 across the Fermi surface to $-k_F^0$. Since all
the states are occupied by two, spin-paired electrons at T=0, in
order to reverse an electronic spin the electron must be excited
to another unoccupied k-state. Thus, the spectrum for excitations
involving a reversal of the electron spin is the same as that of
the spin conserving spectrum (Fig. 2(a)). The equivalent
excitation spectrum E(q) for the large U Hubbard model has been
calculated by Coll [17] and is shown in Fig. 2(b). The spectrum
involving spin reversals is now not the same as the spectrum of
exciting electrons into higher states in the band. For large U,
these spin excitations appear <u>below</u> the electronic continuum and
are spin waves. The dispersion relation calculated for these
spin waves is given [17] by:

$$E(q) = \pi J \sin (qb/\rho) \, , \qquad\qquad\qquad (3)$$

where J is related to U, 4t, and ρ by

$$J = \frac{2t^2}{U} \rho \, (1 - \frac{\sin 2\pi\rho}{2\pi\rho}) \, . \qquad\qquad\qquad (4)$$

For $\rho=1$, Eq. (3) reduces to the familiar des Cloizeaux-Pearson
[18] dispersion relation, with [2,3] $J=2t^2/U$.

This decoupling of the spin and translational degrees of
freedom can also be seen from the following physical picture:
consider a case where U and V_i are so large compared to 4t that
the ground state is a Wigner crystal, i.e., the charges form an
ordered lattice, as in Fig. 3(a). Because of the bandwidth,
there will be an antiferromagnetic coupling between the spins,
so that the T=0 ground state may be schematically viewed as in
the figure. In this model, we can imagine excitations in which
the charges are fixed, but one of the spins has been reversed.
Similarly, we could keep the spin orientations fixed and move
one of the charges, thus illustrating the decoupling of spin and
translational or charge excitations. This decoupling is also
clear from a mathematical point of view in the infinite U limit
[19-24] of the Hubbard model. This effect has been observed in
two ways: by measuring the spin waves directly with inelastic
neutron scattering [10]; and by observing that the activation
energies at low temperature for the conductivity (charge
excitations) and the magnetic susceptibility (spin excitations)
are quite different [25].

More precise information about states in the lower band may
be obtained in the U$\to\infty$ limit [19-24]. In this limit the problem
of ρN_0 interacting electrons can be transformed, exactly, into
the problem of ρN_0 non-interacting, spinless Fermions and the
energies of these particles are given [21,22] by

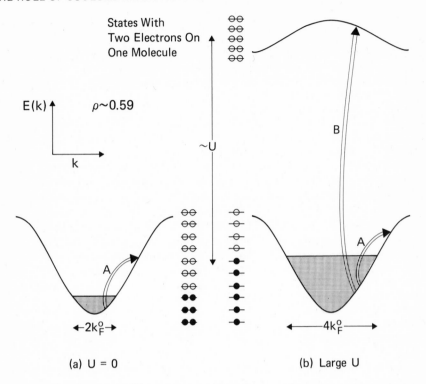

Fig. 1.   Effect of a large intramolecular Coulomb interaction on
           a U=0 band, showing new band at energy U, new optical
transition, and new Fermi wavevector.   (Schematic, for ρ=0.59).

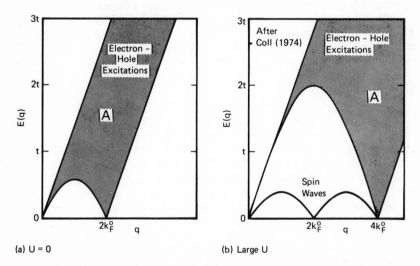

Fig. 2.   E(q) for the spin and charge excitations of the electrons
           in Fig. 1.   (Schematic, for ρ=0.5, without B transitions).

$$E^{U \to \infty}(k) = -2t \cos kb \ . \qquad (N_o \text{ states}) \qquad\qquad (5)$$

Since these Fermions are spinless, each state can contain only one quasiparticle and so the band holds only N_o states. Thus, ρN_o electrons fill the band up to the Fermi momentum [24]:

$$2k_F^{U \to \infty} = \rho b^* = 2\pi\rho/b = 4k_F^o \ , \qquad\qquad (6)$$

by comparison with Eq. (2) for U=0. This factor of two can be seen (less rigorously) from Fig. 1(b), where the doubly occupied momentum states have been excluded by the presence of U. Hence, the electrons fill up twice as many of the new momentum states, up to twice the Fermi wavevector of the U=0 case. This is the same wavevector as that where the energy of the electron–hole excitation spectrum of Coll [17] (Fig. 2(b)) goes to zero, while the first zero in E(q) for the spin waves occurs [17] at $2k_F^o$. Similar conclusions are obtained by examining the schematic picture of the Wigner crystal in Fig. 3(a). There the period of the charges is b/ρ, while that of the spins is $2b/\rho$, corresponding to wavevectors $4k_F^o$ and $2k_F^o$, respectively [10].

Unfortunately, the theory has not progressed far enough to make the physics of the large U limit clear. It is true that the energy spectrum for $U \to \infty$ is the same as for U=0, but the wavefunctions are clearly not the same [19]. For $U \to \infty$ there is no clear physical picture of how the quasi–particles (which are non–interacting spinless Fermions) should be viewed in terms of electrons. Without knowing the wavefunctions, the matrix elements, or the response functions, it is not possible to draw firm conclusions about the effect of weaker interactions, such as the electron–phonon interaction. Thus, it is uncertain how, or if, a possible Peierls instability at $q=4k_F^o$ would differ from the familiar one, which occurs at $2k_F^o$ when U=0. We also do not know what the differences are in the d–c or infrared conductivity (the excitations labeled A in Figs. 1(a) and (b)) between the two limits.

One feature is clear, however: in the presence of strong Coulomb interactions, the characteristic wavevector of the charge excitations changes from $2k_F^o$ to $4k_F^o$, while the wavevector for the spin excitations remains at $q=2k_F^o$. It is less clear, however, exactly what might happen at these wavevectors. Another discussion of the origin and meaning of $4k_F^o$ in the weak-to-intermediate U limit is presented in the article by Emery [26,27] in this book.

One concern of many workers in the field is their feeling that strong Coulomb interactions will tend to somehow limit the conductivity. Probably this feeling is based on an implicit assumption of $\rho=1$ (the half-filled band). In that case, the lower band in Fig. 1(b) is filled. There is a gap [28] in the electron-hole excitation spectrum for all values of $U/4t$, and the conductivity will certainly be much less than in the $U=0$ case. For the general case of $\rho\neq1$, however the lower band in Fig. 1(b) is partially filled and there is no gap [17] in the charge excitations, even for large U. For the extreme case of infinite U, one can show that the velocity operator commutes with the Hamiltonian and hence the mobility is infinite [19,22]. Thus, the conductivity is not limited by the Coulomb interactions, at least in the Hubbard model.

In this section, we have sought to emphasize the difference in the physics between the properties of the $U=0$ and $U>>4t$ limits. The large differences in the properties between these extremes are due to correspondingly large differences in the wavefunctions. An attempt to schematically show the wavefunctions is given in Fig. 3(b) and (c). For $U=0$, the electrons can interpenetrate and move through each other without scattering. For the case of a large U (Fig. 3(c)), the electrons cannot overlap and are thus confined (or localized) to the space inbetween neighboring electrons. The situation is crudely similar to a string of beads. Note that the momentum of one of the electrons (beads) would not be dissipated by collisions between electrons; rather, the momentum would be passed down the stack from one electron to another. There is no agreement yet on which extreme more closely describes the electronic wavefunctions in TTF-TCNQ.

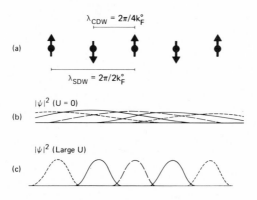

Fig. 3.   (a) A schematic representation of the charge and spin
distribution along a stack in the case of strong Coulomb interactions, showing the difference in their periodicity.   (b) and (c) the large differences in the electronic wavefunctions in the two extreme limits.

II. OPTICAL ABSORPTION SPECTRUM

The complete optical spectrum of TTF-TCNQ from the infrared to the ultraviolet is complex; the correct interpretation is not easily determined and hence not yet agreed upon. In order to understand this spectrum, one must also self-consistently understand the closely related spectra of a large number of other TCNQ salts as well as the spectra of the molecules in solution. On the basis of such a lengthy and detailed comparison, two models have been proposed [9,29] for the intepretation of the TTF-TCNQ absorption spectrum. A complete discussion of these two models and how they interpret all of these different spectra would be too large a diversion for this article, but is described in a very recent review [30], which presents the evidence for and against these two models. Although they differ in some important assignments, both groups [9,29] conclude that Coulomb interactions on the TCNQ stack are large, i.e., $(U-V_1) \sim 1-1\frac{1}{2}$ eV, and are not significantly screened by the excitonic polarizability [31] of the TTF stack [32,33]. In this article, we shall choose one of these models [9] to concentrate on, partly because this model can be easily discussed in terms of the concepts introduced in Section I, and partly because we believe it is the correct one.

The absorption spectra of TTF- and K-TCNQ (as powders dispersed in KBr) are shown [9] in Fig. 4. Starting with the simpler case of K-TCNQ, it has been shown [30] that peak C is an intramolecular transition (not included [2] in our Hubbard model), while peak B is an intermolecular excitation. In addition, it is known that K-TCNQ is an insulator with $\rho=1$. In terms of the U=0 model of Fig. 1(a), K-TCNQ would be highly conducting, without any such high energy peak as that observed at ~1 eV. In terms of the large U picture of Fig. 1(b), on the other hand, the lower band is filled (for $\rho=1$) and K-TCNQ is expected to be an insulator with a gap [28] of order ~U, across which an optical transition is allowed--peak B. For TTF-TCNQ, $\rho<1$ ($\rho=0.59$) and the lower band of Fig. 1(b) is only partially filled, allowing for potentially high conductivity and new, intraband absorption labeled A, in addition to the transition, B, which is still present. This picture thus accounts [9] for both peaks A and B in the spectrum of TTF-TCNQ in Fig. 4. (The intramolecular excitation, peak C, is shifted to somewhat higher energies.) The U=0 model, Fig. 1(a), cannot account for the large energy excitation, peak B, in TTF-TCNQ. In the large U model this peak is interpreted [9] as due to the Coulomb interactions and gives a measure of their strength: $(U-V_1) \sim 1-1\frac{1}{2}$ eV for both K- and TTF-TCNQ, as well as all other TCNQ salts measured [9,29,30].

This comparison illustrates the important difference between the case of $\rho=1$ and $\rho<1$, which in practice is precisely the

principal difference [9,34] between K- and TTF-TCNQ and the origin
of the fact that K-TCNQ has a conductivity of 10^6 times smaller
at room temperature, i.e., this difference is not due to excitonic
screening by the TTF-stack, as had been widely supposed [7,33].
The remaining fact to be understood is then why TTF-TCNQ has $\rho<1$,
while K-TCNQ has $\rho=1$. Most of this difference is believed [34,35]
to be contained in the differences in the electrostatic Coulomb
binding energy of the respective crystals. Briefly and crudely,
TTF is a larger cation with a much larger ionization potential
and, for these reasons, is unable to form a salt with TCNQ which
is as strongly ionic as K-TCNQ.

III.   ENCHANCEMENT OF χ and T_1^{-1}

The experimental magnetic susceptibility, χ, of TTF-TCNQ is
shown by the solid points [5,12] in Fig. 5. The anomalous
decrease in $\chi(T)$ observed below ~280K has been attributed to the
effect of gaps or pseudo-gaps associated with either the phase
transition or interstack coupling (e.g., hybridization). In this
discussion we shall concentrate on the region above ~280K, where
kT is large compared to these gaps and hence the magnitude of χ
should be governed by the largest interactions, which are what
we want to examine. The magnitude of χ in this region is, in
principle, quite sensitive to the presence of significant Coulomb
correlations, since these interactions interactions tend to make
states with paired spins (on the same molecule) energetically
unfavorable, thus increasing the fraction of magnetic states.

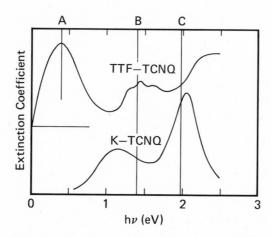

Fig. 4.   A comparison of absorption spectra of K-TCNQ and TTF-
TCNQ, (powdered samples dispersed in KBr). (after Ref. 9)

In order to ascertain whether or not χ is enhanced, the predictions of a number of simple models have been compared [11] to the magnitude of χ at high temperatures. These results will be briefly summarized in this section. Relative to other properties, the magnitude of χ for the one-dimensional Hubbard model is well understood and can be calculated in three limits:

(a) U=0. This is the familiar Pauli susceptibility, which at T=0 is given [16] by:

$$\chi(0) = \chi_p = \frac{N_o \mu_B^2}{\pi t \, \sin\pi\rho/2} \, . \tag{7}$$

For the small bandwidths and Fermi energies required to fit the data in this limit, the zero temperature solution, Eq. (7), is not at all sufficient and the full temperature dependence of χ must be calculated, as in Ref. [11];

(b) U<<4t. In this limit χ can be viewed [11,16] simply as an enhanced Pauli susceptibility:

$$\chi = \frac{\chi_p(t,T)}{1-U/4\pi t \, \sin\pi\rho/2} \, ; \tag{8}$$

and (c) U>>4t. In this atomic limit, χ is given [36] by the Bonner-Fisher susceptibility [37], which at T-0 is:

$$\chi_{BF}(0) = \frac{2 \, N_o \mu_B^2}{\pi^2 J} \, \rho \, , \tag{9}$$

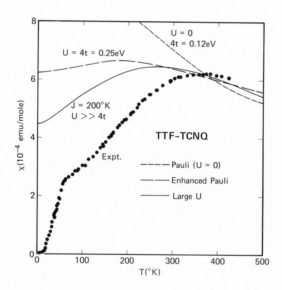

Fig. 5.   The measured χ of TTF-TCNQ and three attempts to fit its magnitude at 350K using three different limits. (Ref. 11)

where the one parameter is the exchange interactions, J, which is related [17] to U, 4t, and ρ by Eq. (4). For $\rho=0.59$ (as in TTF-TCNQ), Eq. (4) gives $J=1.3 \, t^2/U$.

Thus, the observed magnitude of $\chi=6\times10^{-4}$ emu/mole at, say, ~350K can be fitted by using these three expressions, with the appropriate choice of the parameters U and 4t or J (assuming both TTF and TCNQ stacks behave equivalently [12,13]). Examples of the predicted $\chi(T)$ for each of the three limits are shown [11] in Fig. 5. Again, the most important point is not the lack of agreement with the temperature dependence of χ at low temperatures. (This agreement could presumably be improved by including other, weaker interactions into the calculation.) Rather, we want to concentrate on obtaining agreement with the magnitude of χ at high temperatures. For U=0, this agreement is obtained by choosing 4t=0.12 eV, while for the U>>4t case, we choose J=200K $(=1.3t^2/U)$. For U<<4t, fitting to the magnitude of χ yields a relationship between U and 4t, or, equivalently, between U/4t and 4t. The latter relationship is shown in Fig. 6, along with a similar relationship obtained from the large U result that $J=1.3t^2/U=200K$. These results are plotted as solid lines only in the region of U/4t where they are valid. It is clear that one can connect a smooth curve from the solution valid for low U to the large U solution, thus providing a relationship between U/4t and 4t which is valid for all U/4t.

Measurements of the plasma edge [38] and energy loss [39] as well as calculations of the molecular overlap [40] all indicate [41] that 4t~0.5 eV for TTF-TCNQ. Using this value and the relationship in Fig. 6 derived from the magnitude of χ, we obtain U/4t~2-3 or U~1-1½ eV. Some of the initial work [5] incorrectly

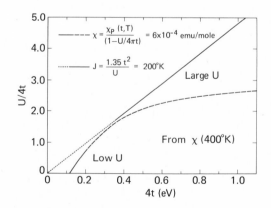

Fig. 6. The relationships between U/4t and 4t determined by the constraint of fitting the magnitude of χ at 400K. (Ref. 11)

concluded that the magnitude of χ was not appreciably enhanced, whereas we find that it is, in fact, enhanced by a factor of ~3. This enhancement of χ has also been noted by Walsh et al. [13] and more recently by Lee, Rice, and Klemm [42], the former pointing out that $\chi(300K)$ is ~50% of the Curie susceptibility.

While the complete interpretation of the NMR relaxation rate, T_1^{-1}, data is still controversial [6,42-45], both the theory and experiment have been recently extended by Jerome and coworkers [43-45]. As discussed in another article of this book [45], they conclude that T_1^{-1} measurements also indicate strong Coulomb interactions; $U/4t\sim1-2$.

IV. SPIN WAVES

As discussed in Section I, another consequence of large Coulomb interactions is the appearance of well defined spin wave excitations, as shown [17] in Fig. 2. In simpler, one-dimensional magnetic systems with $\rho=1$, spin waves have been observed [46,47] by inelastic neutron scattering and were found to obey the dispersion relation given by Eq. (3). By analogy, we would hope that spin waves could be similarly observed in TTF-TCNQ. Since the value of $\rho=0.59$ is known (see Section V) and J should be near ~200K (from fitting χ in Section III), the $E(q)$ predicted for the spin waves from Eq. (3) is largely determined. In Fig. 7 we show only the very low energy portion of $E(q)$, Fig. 2, in order to get it on the same energy scale as the acoustic phonons, whose energies (~60K) are much less than those ($\pi J\sim600K$) of the highest spin waves. Actually, the predicted $E(q)$ shown in Fig. 7 is for a slightly different value of $J(=150K)$, in order to better fit the neutron data [10,48,49]. Also, the dispersion relation near $0.59b^*$ is folded back across the Brillouin zone to near $q=0.41b^*$. On the same figure, we show the original inelastic neutron scattering results of Mook and Watson [48] as well as additional scattering recently reported [10].

In addition to the acoustic phonons, there is some anomalous scattering near $q=0.295b^*$. The original scattering data in this region (open circles) were initially attributed [48] to a giant Kohn anomaly in the longitudinal acoustic (LA) phonon branch. This assignment is, however, not readily consistent [50] with the temperature independence of the experimentally observed $E(q)$, with diffuse X-ray scattering measurements [51-54], and with other measurements in TTF-TCNQ. For example, such a giant Kohn anomaly at $2k_F^0$ would strongly decrease the magnitude of χ at 300K. The observed magnitude of χ is sufficiently large [11], however, as to rule out any such decrease. These apparent inconsistencies are largely removed if the neutron scattering is associated with spin waves. In fact, the additional scattering

which is predicted for the spin waves at energies <u>above</u> the
acoustic phonons has been recently observed [10]. As seen in
Fig. 7 these new scattering peaks (solid circles) are in
remarkable agreement with the simple spin wave prediction.

It must be recognized, however, that the evidence for spin
waves is not complete. In fact, with the relatively small size
of the crystals of TTF-TCNQ presently available, the experiments
[10,48-50] are <u>extremely</u> difficult. Spin wave scattering is
generally extremely weak compared to phonon scattering and as a
result it requires approximately 20-30 hours to make one scan at
constant energy in the region of $q \sim 0.295 b^*$ in order to obtain
two of the data points shown in Fig. 7. Even with such long
scans, the amplitude of the intensity at the peak of the
scattering is only ~3 times the statistical noise $\sqrt{N}$ of the
incoherent background, where N is the background intensity. These
uncertainties and difficulties with the experiments are comparable
with those of the theory: nothing is known about the temperature
dependence of spin waves or their intensities in delocalized,
metallic Hubbard system with $\rho \neq 1$, let alone one which may be
undergoing a distortion (a spin-Peierls distortion, as we shall
suggest in Section VII). For this reason, any comparisons to
the simple case of the $\rho = 1$, insulating, Heisenberg chain must be
made with caution. Nevertheless, some observations can be made:
it would be somewhat surprising to observe a well defined spin
wave at 300K in a system where J is only ~150-200K. Indeed, the
scattering observed in TTF-TCNQ at 300K is not well defined or

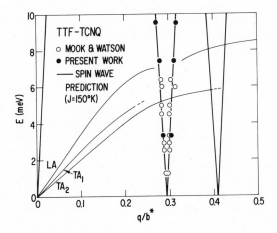

Fig. 7. The results of inelastic neutron scattering, showing the
         three acoustic phonon branches and the scattering near
q=0.295b* attributed to spin waves, including the data <u>above</u> the
phonons. (after Refs. 10 and 49)

sharp at all--it is highly damped, with an energy width comparable
to the energy itself. At lower temperatures the scattering peaks
do narrow and become more well defined. The measured integrated
intensity (with a factor of ~2 uncertainty) is only weakly
dependent on temperature, energy (near 0.295b*) and total momentum
transfer (Q). Some of these crude observations are not exactly
what we might expect, but neither are they wildly far off. The
fact that no spin wave scattering has been observed near q=0 or
q=0.41b* is consistent with the simplest expectations [46].

Thus, it is not possible to make a final judgement on spin
waves in TTF-TCNQ, but since no more evidence (for or against)
is expected in the near future (until much larger crystals become
available), some tentative evaluation should be made based on
the existing data. Probably the strongest and most significant
evidence is the observation of scattering <u>above</u> the acoustic
phonon energies near q=0.295b*. Since the scattering data shown
in Fig. 7 are quite weak, an additional scan has very recently
been made [10] at 9.5 meV with much lower resolution, but with
correspondingly larger scattering intensity. Although the
two-peak structure was not resolved, the peak scattering intensity
was ~5-6 times the noise, thus clearly demonstrating the existence
of some kind of scattering above the acoustic phonons near
q=0.295b*. It has been suggested [55,56] that this scattering
might involve optical phonons; however, a number of points favor
the spin wave interpretation. First, the agreement with the E(q)
of the simple Hubbard model, and with a value of J (~150K)
comparable to that (~200K) estimated independently. Also, the
fact that the scattering intensity is appreciable near (0, 0, 0)
favors [57] the spin wave assignment--a technical point. Finally,
a very special optical phonon is required [55] to remove the
apparent inconsistencies [50] with the diffuse X-ray results:
for example, the large scattering seen at 300K in the neutron
scattering should, in general, be observed as strong diffuse
X-ray scattering, but the latter [51-54] is very weak at 300K
near q=0.295b*. This is consistent with the spin wave
interpretation, since X-rays are not scattered by spin waves.
The diffuse X-ray scattering at $2k_F^o$ observed at lower temperatures
is attributed, not to spin waves, but to the phonon anomaly
associated with a spin-Peierls distortion (Section VII). Thus,
the available evidence at this stage firmly supports the
suggestion [10] that spin waves are observed in TTF-TCNQ, but
the details of these spin waves must await further experiments
(i.e., larger crystals) and some theoretical calculations of
their properties.

Another consequence of the decoupling of spin and charge
excitations is found below the phase transition: In general,
one should expect that the gap (and hence the activation energy)

in the spin excitation spectrum should be different from that of
the charge spectrum. In fact, it has been shown [25] that $T^{\frac{1}{2}}\chi$
and σ, have very different temperature dependences. Below the
phase transition at 38K, for example, the apparent activation
energy for σ is ~2½ times larger. This fact does not indicate
unambiguously that there are strong Coulomb interactions, since
the difference between $T^{\frac{1}{2}}\chi$ and σ might be due to an activated
mobility [25].

<h2 style="text-align:center">V. DIFFUSE X-RAY SCATTERING AT "$4k_F^o$"</h2>

Our understanding of the properties of TTF-TCNQ has made
dramatic progress in the last year and a half, largely because
of the X-ray experiments conducted at Orsay [51,53] and Tokyo
[52,54] and the neutron experiments performed at Brookhaven
[50,58] and Oak Ridge [10,48,49]. These experiments have provided
much more powerful, definitive, and useful information on TTF-TCNQ
than previous bulk measurements. Since these experiments will
be described in detail by Dr. Comes in another article [59] in
this book, they will only be briefly discussed here. Similarly,
part of Dr. Emery's article [27] will overlap some of the
interpretations presented in this section. In general, the two
interpretations involve some of the same broad conclusions, but
by complimentary approaches.

The X-ray diffuse scattering measurements have been most
useful. The data of Kagoshima, Ishiguro, and Anzai [54], shown
in Fig. 8, are taken using photon counting techniques and are
similar to the earlier results of Pouget et al. [53], who used
photographic methods. The most significant feature of the data
of both groups is that there is diffuse X-ray scattering at two
inequivalent wavevectors: $q/b*=0.295$ and 0.41 (also at the
equivalent wavevectors 0.705 and 0.59). It was immediately
recognized [53,54] that there is a simple relationship between
these wavevectors: one of them is twice the other-- $2\times0.295=0.59$,
which is equivalent to 0.41, after reflection through the zone
boundary at $q=0.5b*$.

It is important to examine the temperature dependence of the
observed intensities of the scattering at these two wavevectors,
evident in Fig. 8 and shown in Fig. 9 for the results of Khanna,
et al. [53]. The scattering at $q=0.295b*$ decreases rapidly as
the temperature is increased from 60 to 120K and is difficult to
observe above ~150K. Although the scattering intensity at
$q=0.59b*$ also decreases between 60 and 120K, it remains quite
strong all the way up to 300K. It might be suspected that the
scattering at $q=0.59b*$ is simply a harmonic of that at $q=0.295b*$,
caused by some non-linearity. The evidence against such a simple
possibility is that (at high temperatures) the scattering at the

(supposed) harmonic is stronger than that at the fundamental
(Figs. 8 and 9). A more complicated harmonic mechanism has
recently been suggested by Sham [60], according to which there
is no electronic anomaly at $q=4k_F^0$--only a phonon anomaly. The
phonons at $4k_F^0$ soften by coupling to two, soft $2k_F^0$. phonons. In
fact, Sham's calculations show that the $4k_F^0$ phonon can be softer
than the phonons at $2k_F^0$. The relatively temperature independent
intensity of the $4k_F^0$ scattering at high temperatures (Fig. 9)
would tend to cast doubt on this mechanism. Although we cannot
rule this mechanism out, we will concentrate on the other
mechanisms. Since we are interested in the strongest
interactions, we must concentrate on the scattering which persists
up to the highest temperature--that at $q=0.59b^*$.

The fact that the scattering is observed at incommensurate
wavevectors strongly suggests that these wavevectors are somehow
related to that of the Fermi surface. This relationship will
depend both on the degree of charge transfer, ρ, and on the model
used to interpret the results. Using a U=0 model, one expects
scattering only at $2k_F^0$ and hence one cannot explain the
observation of scattering at two different wavevectors.
Historically, X-ray [51,52] and neutron scattering [48] were
first discovered at $q=0.295b^*$, which was interpreted as $2k_F^0$ (U=0),
giving $\rho=0.59$ via Eq. (2). On the other hand, in a large U model,
scattering at two wavevectors might be expected and was in fact
predicted [61] for TTF-TCNQ: $4k_F^0$ is the wavevector characteristic
[10] of the charge excitations and the charge density, while $2k_F^0$
is the principal wavevector of the electronic spins. Also, a
model with intermediate U might involve [26,27] wavevectors of
the charge density at both $4k_F^0$ and $2k_F^0$. In the large U and
intermediate U models, these two wavevectors would match the
experimental observations if $\rho=0.59$. Thus, the major theoretical
approaches have some common conclusions: $\rho=0.59$; $2k_F^0=0.295b^*$;
and $4k_F^0=0.59b^*$ (=0.41b^*$), even though the physics and
interpretation of these approaches are very different. This
value of ρ is very consistent with independent estimates from
X-ray intensity [62] and ESCA measurements [63].

In the large U model, the occurrence of the wavevector $4k_F^0$
can be viewed in a number of different (but related) ways, as
discussed in Section I. For example, $4k_F^0$ is the characteristic
wavevector of the electron-hole excitations [17]. It is also
the width in momentum of the Fermi surface of the spinless
Fermions in the $U \to \infty$ limit. Perhaps the most physical description
of this wavevector is from the picture of a Wigner crystal (Fig.
3(a)). There the period is readily calculated to be $b/\rho=2\pi/4k_F^0$
(via Eq. (2)), in agreement with Eq. (6). In the familiar case
of a Peierls distortion, there is a pairing of electrons and the
characteristic period is the <u>spacing between the pairs</u>, in

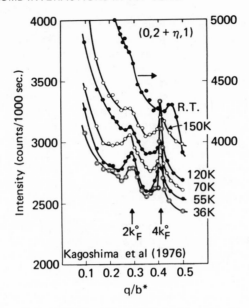

Fig. 8. The diffuse X-ray scattering intensity vs. wavevector, showing the scattering at both $2k_F^o$ and $4k_F^o$.

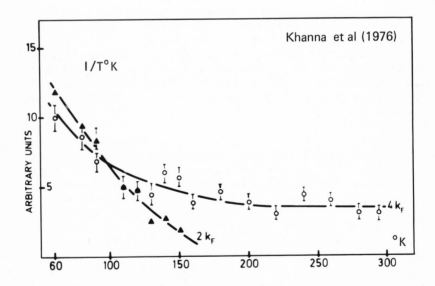

Fig. 9. The temperature dependence of the diffuse X-ray scattering intensity at $2k_F^o$ and $4k_F^o$, showing the dominance of $4k_F^o$ scattering at high temperatures.

contrast to the spacing between the electrons in the Wigner
crystal (Fig. 3(a)). This difference is crudely, but physically,
the origin of the factor of two difference between the
characteristic wavevector for U=0 and large U limits. In an
approach very similar to ours, Sumi [65] has very recently
extended this type of discussion to describe the $2k_F^O$ and $4k_F^O$
distortions present in several complex TCNQ salts. From earlier
studies of the crystalline electrostatic energy, a Wigner crystal
has been predicted for TTF-TCNQ [64,35]. In addition, it was
shown [35] that there is a direct relationship between the
occurrence of modulated charge density (Wigner crystal) and $\rho \neq 1$,
i.e., incomplete charge transfer is stabilized by the formation
of a Wigner crystal--otherwise, $\rho = 0$ or 1.

The significance of the wavevector $4k_F^O$ has been recently
emphasized by Emery [26,27] from a different point of view.
Starting with a band of non-interacting electrons, as in Fig.
1(a), he considers the effect of including weak-to-intermediate
Coulomb interactions and thus treats the important regime
inbetween the two extreme limits of Section I. For sufficiently
strong interactions, he finds that the electronic system is
unstable at the wavevector $q=4k_F^O$ to the formation of a new kind
of correlated state of charge density waves (CDW). Some physical
insight into this picture can be gained by a comparison with the
usual Peierls instability at $2k_F^O$, which can occur because it
requires no energy to transfer an electron across the Fermi
surface. In the other case, the ($4k_F^O$) instability can occur
because it requires no energy to transfer two electrons across
the Fermi surface, involving a change in momentum of $4k_F^O$. More
quantitatively, his calculations [26] show that the $2k_F^O$
instability (usual Peierls) is divergent for all values of U
considered, but the new instability at $4k_F^O$ is divergent only for
sufficiently strong U, such that $U/4t \geq 1.5$. In order to account
for the experimental fact [53,54] that the $4k_F^O$ scattering
dominates that at $2k_F^O$, an even larger value of U is required:
$U/4t \geq 2$. It is uncertain exactly what is the maximum strength of
Coulomb interaction that can be treated by this approach (which
starts from U=0); perhaps one indication is that the exponent of
the divergence becomes imaginary for $U/4t \geq 2.5$.

Although ours and Emery's models both stress the importance
of Coulomb interactions as the origin of the scattering observed
at $q=0.59b^*$, a discussion of some of the differences may help us
better understand each of these models. Since Emery is basically
extending the U=0 picture (Fig. 1(a)) toward large U, the
magnitude of $2k_F$ is $0.295b^*$ and $q=0.59b^*$ occurs at $4k_F$. Since we
are starting from a large U model (Fig. 1(b)), the magnitude of
$2k_F$ is $0.59b^*$, i.e., the characteristic wavevector. In our case
the scattering at $q=0.295b^*$ occurs at "k_F" and is tentatively

attributed to a spin–Peierls anomaly (see Section VII). The
ground state in our picture (Fig. 3(a)) is similar to a Wigner
crystal, i.e., an ordering of electronic charges (see Section
VII), while Emery's ground state is a new kind of correlated
state of CDW's. At this stage it appears difficult to understand
and visualize how these two approaches blend together and match
in the regime of intermediate-to-high U.

The simple experimental situation outlined at the beginning
of this section is complicated by the following observation
[53,54]: As the temperature is increased between ~150 and 200K,
the observed wavevector of the diffuse X-ray scattering decreases
from q=0.59b* to 0.55b*, and appears to remain at q=0.55b* until
300K, as seen in Fig. 8. On the other hand, the inelastic neutron
data at 300K remain centered at q=0.295b*. This remains a major
unsolved question, although there are a number of possible
explanations. For example, if the effective dimensionality were
somewhat larger than one, the electrons in the large U case would
still fill up twice as many momentum states as in the U=0 case,
but the Fermi wavevector would be less than twice $2k_F^o$.

The observation of the wavevector $4k_F^o$ is most probably related
to Coulomb interactions. The fact that any scattering is observed
at $4k_F^o$ is an indication that these interactions play a role. The
fact that the scattering at $4k_F^o$ is the dominant scattering at
high temperature qualitatively suggests that Coulomb interactions
play a dominant role in TTF-TCNQ. This conclusion has been
substantiated quantitatively Emery's calculations [26,27], which
indicate $U/4t \geq 2$.

VI. SCREENING

The magnitude of the Coulomb repulsion between two electrons
is given by

$$\frac{e^2}{r} = \frac{14.4 \text{ eV}}{r(\text{Å})} \tag{10}$$

where r is the distance between electrons in Å. Since
characteristic distances within and between molecules are
typically of the order 3–7Å, this energy is ~2–4 eV. Since this
energy is clearly much larger than the largest estimates of 4t
(~0.5 eV), why was the importance of the Coulomb interaction
neglected in TTF-TCNQ? There are two reasons: (1) The relevant
energies might involve small differences between large Coulomb
energies, e.g., $U_{eff}=U-V_1$, where U_{eff} may be 1 eV, even though U
and V_1 may be 3 and 2 eV, respectively; and (2) The Coulomb
interactions might have been screened by a number of other
interactions. In order to consider the screening of the repulsion
between two electrons which are ~4A apart, one cannot simply

divide the interaction by the <u>macroscopic</u> dielectric constant--one must consider each microscopic mechanism in detail:

(a) <u>Excitonic screening</u> [32,33,66]. This is similar to the mechanism first proposed by Little [31] for possible superconductivity in organic materials. It was applied to the TCNQ salts by LeBlanc [32] and subsequently by Garito and Heeger [33]. The idea is that the excitonic polarizability of the molecules in the TTF-stack, for example, might screen the Coulomb interaction between electrons on the TCNQ-stack. Optical measurements show [9,29] that U on the TCNQ-stack is independent of the cation for a wide variety of cations, with widely varying polarizabilities. This screening would also markedly narrow the bandwidth [66], but there is no evidence from plasma frequency measurements [38,39] for any major reduction in this bandwidth. Thus, the Little-LeBlanc mechanism is apparently not effective in this class of materials. This conclusion is quite consistent with recent work by Davis, Gutfreund, and Little [67], who show that in order for the Coulomb interaction to be appreciably screened, a large density of very highly polarizable molecules must be extremely close (even bonding) to the electrons which are to be screened. The TTF molecules are neither highly polarizable nor extremely close in TTF-TCNQ.

(b) <u>Optical phonon screening</u>. Optical phonons are expected to act just like the excitons in the Little mechanism to screen the Coulomb interaction. The screening of a Hubbard U may be described in terms of a polaron binding energy, E_B, which can be related [68] to the dimensionless electron-phonon coupling constant λ by:

$$U_{eff} - U = -2E_B \simeq -4t\lambda \lesssim 0.1 \text{ eV} . \tag{11}$$

By analysis [69,70] of the relaxation time of the infrared plasma edge, an upper bound of $\lambda \sim 0.2$ is obtained for the optical-phonon (Holstein) contribution. Assigning all of this relaxation to one stack, the maximum possible reduction of U is expected to be of order 10% due to this mechanism.

(c) <u>Metallic screening</u>. Very little is known about metallic screening in such systems as TTF-TCNQ. The experimental fact that U is only weakly screened indicates that this mechanism is also ineffective. This may be due to the one-dimensionality and/or the low electron density.

VII.   DISCUSSION AND SPECULATION

It would be important to compare the magnitude of these Coulomb interactions with that of the electron-phonon

interactions. These electron-phonon interactions have been used
to describe many of the properties of TTF-TCNQ. Initially, the
peak in the d-c conductivity near 60K of some crystals was
attributed [71] to BCS-like superconducting fluctuations.
Subsequently, electron-phonon interactions, described in terms
of Peierls-Fröhlich giant charge density waves, have been used
to describe [7] the conductivity, the temperature dependence of
χ, the "energy gap" in the infrared, the Kohn anomaly, and the
metal-non-metal transition (Peierls transition). While some of
these interpretations remain controversial, recent X-ray and
elastic neutron scattering measurements [59] have demonstrated
that the electron-phonon interactions do play a crucial role in
TTF-TCNQ at low temperatures.

The problem is how to quantitatively compare the
electron-phonon interaction to the strength of the Coulomb
interaction. One way is to recognize that the electron-phonon
interaction gives rise to an attractive electron-electron
interaction, which can be compared with the Coulomb repulsion
between electrons. We may view the electron-phonon interaction
as either a BCS-like interaction, a Peierls-Fröhlich mechanism
or as an attractive Hubbard model (U<0). In all three cases,
the attractive interaction gives rise to an energy gap: the BCS
energy gap, the Peierls gap, or the gap of magnitude $|U|$. The
maximum estimate [7] of this gap, and hence the attractive
electron-electron interaction, is ~0.12 eV (from the infrared
"gap" [72]). Since this magnitude is considerably smaller than
the most conservative estimate of U (Table I), we conclude that
Coulomb interactions cannot be neglected from any discussion
involving electron-phonon interactions. In fact, an improved
approach would be to start with a model of strong-to-intermediate
Coulomb interactions and add the electron-phonon interacting as
a perturbation.

A detailed description of the properties of TTF-TCNQ does
not immediately emerge from the knowledge that the Coulomb
interactions are larger than the bandwidth and the electron-phonon
interaction. This fact merely guides us to the correct starting
point, from which there are a large number of directions in which
one may proceed. In this section, we shall discuss the model
which we believe best represents the way in which the electrons
are affected by the Coulomb interactions in TTF-TCNQ. While
there is little doubt that these interactions are strong, it is
not yet clear to what extent our specific model (discussed below)
correctly describes the way in which the Coulomb interactions
manifest themselves.

Let us imagine a stack of N_0 molecules in which there are
ρN_0 electrons. The bandwidth of these electrons causes them to

move rapidly back and forth (over short distances) along the
stack, but the Coulomb repulsion tends to keep them away from
each other. On a low frequency time scale, this rapid motion of
the electrons is averaged and we imagine that the resulting
average electron density is somewhat similar to that shown
schematically in Fig. 3(a). There is a periodic modulation of
the charge density at the wavevector $q=4k_F^o=2\pi\rho/b$, which is
incommensurate with the underlying lattice of molecules (not
shown for simplicity). The overlap between these electrons gives
rise to an effective antiferromagnetic exchange interaction
between the spins of these electrons of magnitude
$J\sim1.3t^2/U\sim150-200K$. The characteristic wavevector of the spin
system occurs at $q=2k_F^o$, as seen from Fig. 3(a).

Before describing this model further, we want to emphasize
the large difference between our model and the commonly discussed
CDW and SDW, or charge and spin density waves. A metallic
one-dimensional band of electrons for U=0 is unstable to a variety
of perturbations which can potentially lower the kinetic energy
of the electrons. The essential feature of this perturbation is
that it has an amplitude at $q=2k_F^o$ and some coupling to the
electrons. The most familiar example of such a perturbation is
a distortion of the lattice, i.e., a Peierls distortion, but
there are examples of perturbations where there is no distortion.
Such a perturbation is a deviation from the normally uniform
charge density in the form of a CDW with $q=2k_F^o$. Similarly, a
periodic modulation of the spin density, or SDW, can also take
advantage of this instability. This situation is very different
from our model: the charge density modulation imagined in Fig.
3(a) is not formed because of an instability; the energy gained
is Coulombic (not kinetic); and the electron interactions are
repulsive (while they are often attractive for CDW formation).
For the usual SDW, the charge density remains uniform, while an
important feature of Fig. 3(a) is the strong modulation of the
charge density. Also, the usual SDW instability occurs in order
to gain kinetic energy, without a distortion, whereas we shall
discuss the spin-Peierls "instability" in which magnetic energy
is gained by a distortion of the lattice at $q=2k_F^o$. Because of
these large differences, we have attempted to avoid the use of
the terms CDW and SDW since their usual meaning does not describe
our picture of the electronic system. A name which attempts to
contain the physics of this picture would be an antiferromagnetic
Wigner wave (AWW). Of course, this wave is not a static, standing
wave, but is dynamic and able to slide its phase along the stack.
Similarly, the antiferromagnetic correlations do not correspond
to a static Néel ground state, but are the dynamic ones of a
linear Heisenberg chain of spins. One further note about CDW,
SDW, and AWW: neither of these electronic effects have yet been
observed directly; the X-rays and neutrons are scattered by

distortions of the lattice, presumably related to these waves
via the electron-phonon interaction.

In order to understand the temperature dependence of this
antiferromagnetic Wigner wave (AWW), we must recognize several
features of the model. First, the magnitude of the Coulomb
interactions is large compared to kT, so there should not be a
major temperature dependence to the modulation of the charge
density. J~150-200K, so that over the range 60-300K large changes
can be expected in the degree of antiferromagnetic correlation
between spins. The above effects are without a coupling to the
lattice. There are two effects of this phonon coupling
considered: (1) the charge density modulation with $q=4k_F^O$ causes
a phonon modulation (possibly involving optical phonons) at that
wavevector, off which the X-rays are scattered; and (2) the
phonons can couple to the spin waves to give rise to a
spin-Peierls anomaly at $q=2k_F^O$, particularly at low temperatures
where the antiferromagnetic correlations are strong. The
spin-Peierls instability has been studied theoretically [73,74]
and experimentally [75] for the case of $\rho=1$, where a linear
Heisenberg chain of spins has been shown to be unstable to a
distortion which dimerizes the chain, in a manner very similar
to the usual Peierls distortion. We suggest that this concept
of a spin-Peierls distortion can be readily generalized to the
case where $\rho\neq1$, in which the spin system can distort at $q=2k_F^O$.
Thus, we shall tentatively interpret the observed anomaly at $2k_F^O$
as associated with a spin-Peierls anomaly, while that at $4k_F^O$ is
related to the amplitude of the charge modulation of the AAW.

Using this picture of an AAW, the interpretation of the
experiments goes something like this: Since the Coulomb
interactions which cause the charge modulation are much stronger
than kT, there is a reasonably well defined charge modulation,
and hence diffuse X-ray scattering at $4k_F^O$, at all temperatures.
The major temperature dependence is then in the spin correlations
and J is the scale temperature. For T~500K, the spins are highly
paramegnetic, but for T~300K there apparently is enough
antiferromagnetic correlation to form highly damped spin waves.
For T~150K, the spin correlations are much stronger, since T~J.
As a consequence, the spin waves are narrower and the spin-Peierls
correlations (and diffuse X-ray scattering at $2k_F^O$) start to build
up. Correspondingly, this distortion appreciably reduces the
magnetic susceptibility. For T~60K, the spin correlation is now
so large that the spin-Peierls fluctuations at $2k_F^O$ become dominant
and may even influence the effects at $4k_F^O$. These distortions
become three-dimensional below the phase transitions.

Although the details and specifics of this model are extremely
vague, the important points are (1) that the major temperature

dependence is related to the spin correlations, so that the
characteristic or scale temperature is T~J; and (2) the diffuse
scattering at $2k_F^o$ is attributed to a spin-Peierls anomaly, as is
the decrease in $\chi(T)$ between 300 and 60K (Fig. 5).

At this stage we see a number of major unsettled questions:

(1) Is the phonon anomaly at $2k_F^o$ observed by diffuse X-rays
     associated with the spin excitations (i.e., spin-Peierls,
     as we have suggested)? or the charge excitations, as in
     the usual Peierls case?

(2) What is the origin of the temperature dependent shift
     of the wavevector $4k_F^o$ from q=0.55b* to 0.59b*?

(3) Why is scattering in TSeF- and HMTSeF-TCNQ observed at
     only one wavevector? Is it at $2k_F^o$ or $4k_F^o$?

(4) What is happening at T_c?

and

(5) How should we view the wavefunctions in the large U
     limit? As a Wigner crystal with a sinusoidal charge
     density, as we have implied? or as a commensurate Wigner
     crystal with defects? In the former case, the
     conductivity would occur via a collective mode, very
     similar to the Peierls-Fröhlich case [7], while in the
     latter it would be via single defect conduction.

VIII. CONCLUSIONS

Do these experiments give consistent values for the magnitude
of U and 4t in TTF-TCNQ? In Table I we summarize the values of
U, 4t, and U/4t obtained from the various measurements discussed
above. It is clear from the Table that the following values are
quite consistent with these experiments:

$$4t \sim 0.5 \text{ eV}$$
$$U \sim 1 \text{ eV}$$
$$U/4t \sim 2\text{-}3 \ .$$

Care must be taken not to attribute too much significance to
these values, since they are based on the Hubbard model, in which
other interactions (V_i) have been neglected. Nevertheless, within
this context, the conclusion is that TTF-TCNQ is inbetween the
two extreme limits discussed (U<<4t), and the estimates given
above are important to quantitatively determine where TTF-TCNO
is in relation to these extremes. It may be more important,
however, for a field at such an early stage of development as
this one is, to consider some more general, qualtiative questions
such as: Which of the two extremes, U<<4t or U>>4t, best
represents the physics and the features of the experiments in
TTF-TCNQ?; or, Do Coulomb interactions play only a minor role,
or a major one, in the properties of this material? As discussed

in the previous sections, there are a number of basic experimental observations which cannot be understood within a U=0 model. On the other hand, the model at the other extreme (with U>>4t) is readily able to account for these experiments. It is within this context that we draw our qualitative conclusion (the quantitative conclusion is Table I): Coulomb interactions play a major role and are the dominant interactions in TTF-TCNQ and can no longer be neglected.

This conclusion, however, does not solve all the questions and problems of TTF-TCNQ--it merely gives us the correct starting point from which to proceed. We must still discover the way in which other, weaker interactions (e.g., electron-phonon, interstack coupling) act on this starting point to give the anomalous temperature dependence of χ, the phase transitions, the mechanism which limits the conductivity, and the other anomalous properties of TTF-TCNQ, which still remain to be understood.

From the very beginning, interest in conducting organic materials has been stimulated by the suggestion of Little [31] that in such materials high temperature superconductivy might be stabilized. Hope for this goal was generated by the original interpretation [71] that the rise in the conductivity in TTF-TCNQ was caused by superconducting fluctuations. Since then, a considerable theoretical and experimental effort has been exerted to understand and verify (or disprove) this conjecture. It appears [9,29] now that the Little mechanism of excitonic polarizability is ineffective in these materials. Furthermore, there is strong evidence (summarized in this article) that the interactions between electrons, in fact, are repulsive, and not

TABLE I

Experiment	Estimate	Reference
Plasma frequency, energy loss	$4t \sim 0.5eV$	[38, 39]
Optical absorption at U	$U \sim 1-1\frac{1}{2}eV$	[9, 29]
Enhancement of χ	$U/4t \sim 2-3$	[11]
NMR T_1^{-1}	$U/4t \sim 1-2$	[45]
Spin Waves	$U/4t \gtrsim 1$ (?)	[10]
"$4k_F^0$" Scattering	$U/4t \gtrsim 2$	[26, 27]

attractive. In this case, it appears that superconducting phenomena are unlikely in this class of materials.

ACKNOWLEDGMENTS

The understanding of the physics of TTF-TCNQ presented in this article is _not_ the result of thinking quietly and working alone in isolation. Many of the ideas described were stimulated by discussions (and often arguments) with other workers in the field. This aspect has been a particularly enjoyable, as well as educational, one. I have enjoyed working and interacting with my collaborators and colleagues at IBM, particularly Sam La Placa, Erling Pytte, Ted Schultz, Bruce Scott, Dave Silverman, and Yaffa Tomkiewicz, as well as many others outside, including especially Robert Comes, Vic Emery, Patrick Lee, Herb Mook, Gen Shirane, and Zoltan Soos. Finally, I would like to thank Rick Greene and Hans Morawitz for carefully reading the manuscript.

REFERENCES

† Present address.

1. J. Hubbard, Proc. Roy. Soc. (London) A276, 238 (1963); A277, 237 (1963); A281, 401 (1964).

2. Z. G. Soos, Ann. Rev. Phys. Chem. 25, 121 (1974).

3. P. Pincus, in Low-Dimensional Cooperative Phenomena, ed. H. J. Keller (Plenum Press, N. Y. 1974).

4. G. Beni and P. Pincus, Phys. Rev. B 9, 2963 (1974).

5. J. C. Scott, A. F. Garito, and A. J. Heeger, Phys. Rev. B 10, 3131 (1974).

6. E. F. Rybaczewski, A. F. Garito, A. J. Heeger, and E. Ehrenfreund, Phys. Rev. Letters 34, 524 (1975).

7. A. J. Heeger and A. F. Garito, in Low-Dimensional Cooperative Phenomena, ed. H. J. Keller (Plenum Press, N. Y. 1974), p. 89.

8. J. B. Torrance, B. A. Scott, D. C. Green, P. Chaudhari, and D. F. Nicoli, Solid State Comm. 14, 100 (1974).

9. J. B. Torrance, B. A. Scott, and F. B. Kaufman, Solid State Comm. 17, 1369 (1975).

10. J. B. Torrance, H. A. Mook, and C. R. Watson, submitted for publication.

11. J. B. Torrance, Y. Tomkiewicz, and B. D. Silverman, submitted to Phys. Rev. B.

12. Y. Tomkiewicz, B. A. Scott, L. J. Tao, and R. S. Title, Phys. Rev. Letters 32, 1363 (1974).

13. W. M. Walsh, Jr., L. W. Rupp, Jr., D. E. Schafer, and
    G. A. Thomas, Bull. Am. Phys. Soc. 19, 296 (1974), and preprint.

14. I. Sadakata and E. Hanamura, J. Phys. Soc. Japan 34, 882
    (1973); K. A. Kikoin and V. N. Flerov, Sov. Phys. Solid State
    16, 237 (1974).

15. S. K. Lyo and J-P. Gallinar, preprint (1976).

16. H. Shiba, Phys. Rev. B 6, 930 (1972).

17. C. F. Coll, Phys. Rev. B 9, 2150 (1974).

18. J. des Cloizeaux and J. J. Pearson, Phys. Rev. 128, 2131
    (1962).

19. W. F. Brinkman and T. M. Rice, Phys. Rev. B 2, 1324 (1970).

20. J. B. Sokoloff, Phys. Rev. B 2, 779 (1970).

21. A. A. Ovchinnikov, Sov. Phys. JETP 37, 176 (1973).

22. G. Beni, T. Holstein, and P. Pincus, Phys. Rev. B 8, 312
    (1973).

23. D. J. Klein, Phys. Rev. B 8, 3452 (1973).

24. J. Bernasconi, M. J. Rice, W. R. Schneider, and S. Strässler,
    Phys. Rev. B 12, 1090 (1975).

25. Y. Tomkiewicz, A. R. Taranko, and J. B. Torrance, Phys. Rev.
    B (in press).

26. V. J. Emery, Phys. Rev. Letters, 37, 107 (1976).

27. V. J. Emery, Proc. NATO Adv. Study Institute on "Chemistry
    and Physics of One-Dimensional Metals," Bolzano, Italy, Aug.
    1976, to be published, H. J. Keller, ed. (Plenum Press, 1977).

28. A. A. Ovchinnikov, Sov. Phys. JETP 30, 1160 (1970).

29. J. Tanaka, M. Tanaka, T. Kawai, T. Takabe, and O. Maki, Bull.
    Chem. Soc. Japan 49, 2358 (1976).

30. J. B. Torrance, Proc. Conf. on Organic Metals and
    Semiconductors, Siofok, Hungary, Sept. 1976 (to be published).

31. W. A. Little, Phys. Rev. A134, 1416 (1964).

32. O. H. Le Blanc, J. Chem. Phys. 42, 4307 (1965).

33. A. F. Garito and A. J. Heeger, Accts. of Chem. Res. 7, 232
    (1974).

34. J. B. Torrance, to be submitted.

35. J. B. Torrance and B. D. Silverman, Bull. Am. Phys. Soc. 20,
    498 (1975); and Phys. Rev. B (in press).

36. D. J. Klein and W. A. Seitz, Phys. Rev. B 10, 3217 (1974).

37. J. C. Bonner and M. E. Fisher, Phys. Rev. $\underline{135}$, A640 (1964).

38. A. A. Bright, A. F. Garito, and A. J. Heeger, Phys. Rev. B
    $\underline{10}$, 1328 (1974); P. M. Grant, R. L. Greene, G. C. Wrighton,
    and G. Castro, Phys. Rev. Letters $\underline{31}$, 1311 (1973).

39. J. J. Ritsko, D. J. Sandman, A. J. Epstein, P. C. Gibbons,
    S. E. Schnatterly, and J. Fields, Phys. Rev. Letters $\underline{34}$, 1330
    (1975).

40. A. J. Berlinsky, J. F. Carolan, and L. Weiler, Solid State
    Comm. $\underline{15}$, 795 (1974); D. R. Salahub, R. P. Messmer, and F.
    Herman, Phys. Rev. B $\underline{13}$, 4252 (1976); F. Herman, D. R.
    Salahub, and R. P. Messmer, submitted to Phys. Rev. B.

41. This estimate of 4t is a lower bound, since it is obtained
    from a U=0 tight-binding band.

42. P. A. Lee, T. M. Rice, and R. A. Klemm, preprint (1976).

43. D. Jerome and L. Giral, Proc. Conf. on Organic Metals and
    Semiconductors, Siofok, Hungary, Sept. 1976 (to be published).

44. G. Soda, D. Jerome, M. Weger, K. Bechgaard, and E. Pedersen,
    Solid State Comm. (in press).

45. D. Jerome and M. Weger, Proc. NATO Adv. Study Institute on
    "Chemistry and Physics of One-Dimensional Metals," Bolzano,
    Italy, Aug. 1976, to be published, H. J. Keller, ed. (Plenum
    Press, 1977).

46. M. T. Hutchings, G. Shirane, R. J. Birgeneau, and S. L. Holt,
    Phys. Rev. B $\underline{5}$, 1999 (1972).

47. Y. Endoh, G. Shirane, R. J. Birgeneau, P. M. Richards, and
    S. L. Holt, Phys. Rev. Letters $\underline{32}$, 170 (1974).

48. H. A. Mook and C. R. Watson, Jr., Phys. Rev. Letters $\underline{36}$, 801
    (1976).

49. H. A. Mook, submitted to Phys. Rev. B.

50. G. Shirane, S. M. Shapiro, R. Comes, A. F. Garito, and
    A. J. Heeger, Phys. Rev. B $\underline{14}$, 2325 (1976).

51. F. Denoyer, R. Comes, A. F. Garito, and A. J. Heeger, Phys.
    Rev. Letters $\underline{35}$, 445 (1975).

52. S. Kagoshima, H. Anzai, K. Kajimura, and T. Ishiguro, J.
    Phys. Soc. Japan $\underline{39}$, 1143 (1975).

53. J. P. Pouget, S. K. Khanna, R. Comes, A. F. Garito and
    A. J. Heeger, Phys. Rev. Letters $\underline{35}$, 445 (1975); S. K. Khanna,
    J. P. Pouget, R. Comes, A. F. Garito, and A. J. Heeger, to
    be submitted.

54. S. Kagoshima, T. Ishiguro, and H. Anzai, submitted to J.
    Phys. Soc. Japan.

55. K. Carneiro, Phys. Rev. Letters 37, 1227 (1976).

56. H. Morawitz, private communication; and Phys. Rev. Letters 34, 1096 (1976).

57. G. E. Bacon, Neutron Diffraction (Clarendon Press, Oxford, 1975).

58. R. Comes, S. M. Shapiro, G. Shirane, A. F. Garito, and A. J. Heeger, Phys. Rev. Letters 35, 1518 (1975); Phys. Rev. B 14, 2376 (1976).

59. R. Comes, Proc. NATO Adv. Study Institute on "Chemistry and Physics of One-Dimensional Metals," Bolzano, Italy, Aug. 1976, to be published, H. J. Keller, ed. (Plenum Press, 1977).

60. L. J. Sham, preprint (1976).

61. Both the "4k$_F$" scattering and the spin waves were first predicted for TTF-TCNQ by J. B. Torrance, private communication, January 9, 1976, to Drs. Comes, Kagoshima, Mook, Shirane, and Weyl.

62. P. Coppens, Phys. Rev. Letters 35, 98 (1975).

63. W. D. Grobman and B. D. Silverman, Solid State Comm. 19, 319 (1976); A. J. Epstein, N. O. Lipari, P. Nielsen, and D. J. Sandman, Phys. Rev. Letters 34, 914 (1975).

64. V. E. Klymenko, V. Ya. Krivnov, A. A. Ovchinnikov, I. I. Ukrainsky, and A. F. Shvets, Soviet Phys. JETP 42, 123 (1976).

65. H. Sumi, preprint (1976).

66. P. M. Chaikin, A. F. Garito, and A. J. Heeger, Phys. Rev. B 5, 4966 (1972).

67. D. Davis, H. Gutfreund, and W. A. Little, Phys. Rev. B 13, 4766 (1976).

68. G. Beni, P. Pincus, and J. Kanamori, Phys. Rev. B 10, 1896 (1974).

69. P. E. Seiden and D. Cabib, Phys. Rev. B 13, 1846 (1976).

70. J. B. Torrance, E. E. Simonyi, and A. N. Bloch, Bull. Am. Phys. Soc. 20, 497 (1975), and (submitted).

71 L. B. Coleman, M. J. Cohen, D. J. Sandman, F. G. Yamagishi, A. F. Garito, and A. J. Heeger, Solid State Comm. 12, 1125 (1973).

72. D. B. Tanner, C. S. Jacobsen, A. F. Garito, and A. J. Heeger, Phys. Rev. B 13, 3381 (1976).

73. G. Beni, J. Chem. Phys. 58, 3200 (1973).

74. E. Pytte, Phys. Rev. B 10, 4637 (1974).

75. I. S. Jacobs, J. W. Bray, H. R. Hart, Jr., L. V. Interrante,
    J. S. Kasper, G. D. Watkins, D. E. Prober, and J. C. Bonner,
    Phys. Rev. B 14, 3036 (1976).

ELECTRONIC PROPERTIES OF THE SUPERCONDUCTING POLYMER   (SN)$_x$

R. L. Greene and G. B. Street

IBM Research Laboratory

San Jose, California 95193 (U.S.A.)

I.   INTRODUCTION

This paper will review our present understanding of the physical properties of the inorganic sulfur-nitrogen polymer, (SN)$_x$. This material is rather unique with properties quite different from both conventional polymers and the prototype one-dimensional (1D) materials being discussed extensively at this summer school, e.g., TTF-TCNQ and its derivatives [1]. Studies during the past few years have shown that it is more reasonable to consider (SN)$_x$ as a highly anisotropic metal rather than a quasi 1D metal as originally thought. Perhaps the most suggestive evidence for this is the electrical resistivity along the polymer chain which decreases monotonically with decreasing temperature and eventually vanishes at a superconducting transition temperature (T$_c$) near 0.3°K without exhibiting the Peierls metal-insulator transition characteristic of the quasi 1D metals. This behavior makes it important to understand the properties of (SN)$_x$ and to compare them with those of the quasi 1D metals in the hope of gaining insight into the stabilization of the metallic (and possibly the superconducting) state in other anisotropic materials. This is perhaps even more relevant in light of the suggestions of Little and coworkers [2] that some 1D materials could exhibit superconductivity at temperatures far above that of conventional superconductors (at present limited to T$_c \gtrsim 23$°K).

(SN)$_x$ is not a new material; it was first reported by Burt in 1910 [3]. Although some work was done in the intervening years, it was not until after 1973 that extensive research began

on this material. This increased activity stemmed partly from
the already existing widespread interest in 1D materials and from
three important measurements on $(SN)_x$, namely the determination
of the structure by Boudeulle and coworkers [4], the observation
of high conductivity between 300°K and 4°K by Labes and coworkers
[5] and the discovery of superconductivity below 0.3°K [6]. This
was the first report of superconductivity in a polymer.

The history of $(SN)_x$ research up to 1973 and a discussion of
the chemical properties and preparation procedures has been given
in several recent reviews [7,8]. Therefore, in this paper, we
will restrict ourselves to a summary of the structure, the normal
state physical properties, the superconducting properties and
the electronic band structure calculations. In the conclusion
we discuss the potential for achieving chemical analogs or
modifications of $(SN)_x$ and try to show how our understanding of
$(SN)_x$ may have implications for the materials more appropriately
described as quasi 1D metals e.g., TTF-TCNQ.

II.   CRYSTAL STRUCTURE

$(SN)_x$ is prepared from the solid state polymerization of S_2N_2
which is itself first formed by the thermal decomposition of S_4N_4.
The $(SN)_x$ crystals which result from this process are shiny,
brass-colored, and have typical dimensions of a few millimeters.
Unfortunately even the best crystals produced to date are highly
imperfect. Scanning electron microscope pictures and X-ray
diffraction experiments show that all these crystals are b-axis
oriented bundles of fibers with a diameter of only a few hundred
Å. In poor quality crystals the fibers are very apparent but in
better crystals they are not so obvious because there are no
visible discontinuities between them. This is illustrated in
Fig. 1. Besides their fibrous nature, $(SN)_x$ crystals have other
structural complications. All crystals are twinned to some extent
with (100) as the twin plane [4]. There is a defect site
occupancy of about 10% [9-11] and between two and seven atomic
percent of hydrogen impurity is present [7]. Finally the X-ray
diffraction spots corresponding to perpendicular directions are
rather wide (3-10 degrees). This is caused either by the small
size of the fibers, by a mosaic spread of the fibers about the
b axis or by a combination of both effects. The fibers and many
of the other imperfections apparently result from the solid state
polymerization of S_2N_2 to $(SN)_x$. One possible polymerization
mechanism and the origin of the defect structure has been
discussed by Baughman, Cohen, and coworkers [9-11]. Other
mechanisms are conceivable [7] and it seems clear that more work
is needed to completely understand, and perhaps eliminate, these
crystal imperfections.

(a)

(b)

Fig. 1. Scanning electron micrographs of an (SN)$_x$ crystal. (a) A ruptured region. (b) A smooth area of the surface.

The crystal defects and fibrous morphology lead to difficulties in the measurement of many of the physical properties. This is particularly true of the crystal structure. Two independent structure determinations have been made, the first reported by Boudeulle et al. [4] using electron diffraction and the other more recently reported by Mikulski et al. [12] using X-ray diffraction. A comparision of these two suggested monoclinic structures is given in Fig. 2. Both structures have high residual factors (Boudeulle 19%, Mikulski 11%) but agree on the lattice constants and on the almost planar, cis-trans, nature of the SN chains along an axis coinciding with the crystallographic b direction (the direction of high electrical conductivity). Both works also find nearest neighbor interchain distances that are considerably larger than the intrachain distances. The major difference between the structures is in the exact location of the four S and four N atoms within the unit cell. A preliminary neutron diffraction study [8] agrees with the X-ray structure and it is our opinion that the X-ray structure is the better approximation to the true structure of crystalline $(SN)_x$.

Figure 3 shows a view down the b axis of the $(SN)_x$ structure and illustrates how the centro-symmetrically related, translationally inequivalent chains in Fig. 2 alternate in the ($\bar{1}02$) plane. The shortest interchain bond lengths are indicated in Fig. 2. Comparison of these bond lengths with the corresponding van der Waals distances [7] (3.70Å for S-S, 3.35Å for S-N) suggests that the interaction between the chains is weak. On this basis it was initially thought that this highly anisotropic material was quasi one-dimensional. However, as we shall discuss later, other evidence indicates that the interchain interactions are more significant than suggested by these rather large interchain spacings and thus $(SN)_x$ is not one-dimensional but is a highly anisotropic three-dimensional metal.

The structure shown in Figs. 2 and 3 is the monoclinic β phase of $(SN)_x$. Very recently Baughman et al. [13] have shown that on shearing or grinding the β phase of $(SN)_x$ partial conversion (<75%) to a new orthorhombic phase takes place. The major difference between the β phase and the orthorhombic phase appears to be in the relative arrangement of the chains. A clear evaluation of the properties of the orthorhombic phase must await its preparation in pure form and detailed measurements. In this review we will discuss only the properties of the β monoclinic phase.

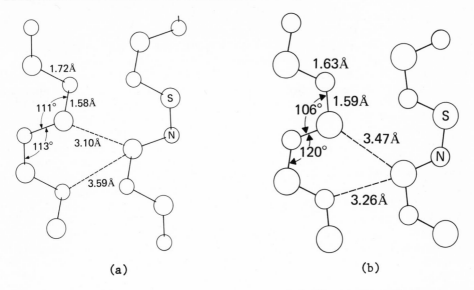

(a)   (b)

Fig. 2.   Crystal structure of β phase of (SN)$_x$ projected onto
(102) plane.   (a) Determined by electron diffraction
[4].   (b) Determined by X-ray [12].

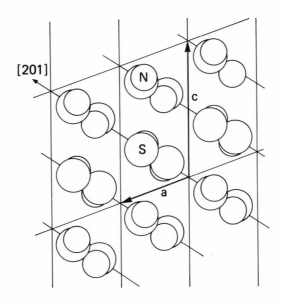

Fig. 3.   View down b-axis of electron diffraction structure of
(SN)$_x$.

III. BAND STRUCTURE

Before discussing the physical properties of $(SN)_x$ it is
useful to understand the electronic band structure that emerges
from the crystal structure. The unusual metallic and
superconducting properties of $(SN)_x$ have stimulated a large number
of band structure calculations [14-26]. These range from simple
1D Extended Huckel (EH) tight binding methods [14] to a complete
3D Orthogonalized-plane-wave (OPW) calculation [20]. The
essential physics of all these calculations can be discussed
quite simply. We first note that each SN molecule contains an
odd number of electrons and therefore the highest occupied energy
level of the molecule will be a half filled π^* antibonding
orbital. If these SN molecules were separated by distance b/2
along a single chain and interacted via a transfer integral $t_\parallel$
this orbital would become a half-filled tight binding band of
the solid. This is illustrated by the cosine-like band labeled
(1) in Fig. 4a. Assuming that the chains are non-interacting a
Peierls distortion will occur (dimerizing the chain) which splits
this band at the Fermi level and produces an insulating state.
Now in the actual crystal there are two chains per unit cell each
with two SN pairs related by a screw axis symmetry. For this
case a 1D model gives the tight binding band shown as curve (2)
in Fig. 4a. Note that there is a degeneracy at the new Brillouin
zone boundary $k_F=\pi/b$ (the Z point) which will be split by a
Peierls distortion. In order to explain the absence of the
Peierls insulating state in $(SN)_x$ the more rigorous 1D
calculations [14,16-18,21,24] all suggest that there are
overlapping bands at the Fermi level which somehow do not distort.
Recently, however, Berlinsky [27] has shown that any strictly 1D
band structure of $(SN)_x$ is unstable and will exhibit a Peierls
distortion to give an insulating state. Thus, we must consider
the role of 3D interactions. Even with interchain interactions
we might naively expect $(SN)_x$ to be an insulator since there are
an even number of electrons per unit cell. In reality our
discussion thus far suggests that metallic behavior will result
from the screw symmetry degeneracy which is stabilized against
the Peierls distortion by interchain coupling. This is in fact
the picture which emerges from the 3D EH calculations [15,22,25]
and the more complete 3D OPW [20], LCAO [26], and pseudopotential
[19] methods.

With these simple ideas in mind we can now discuss the details
of the 3D calculations. The basic physics for why $(SN)_x$ remains
metallic can be seen in the early EH results of Friesen et al.
[15]. However, a much more complete understanding of the 3D band
structure comes from the work of Rudge and coworkers [20,28,29].
In Fig. 4b we show the important bands near the Fermi level which
result from the OPW calculation of Rudge and Grant [20] based on
the X-ray crystal structure. Several features are immediately

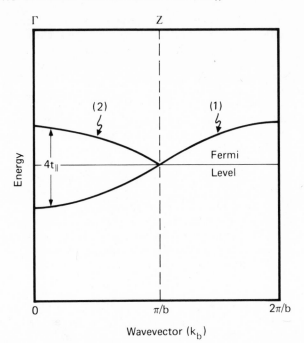

Fig. 4a. Tight binding conduction band for single (SN)$_x$ chain with (1) one SN molecule per unit cell and (2) two SN molecules per unit cell related by a screw symmetry.

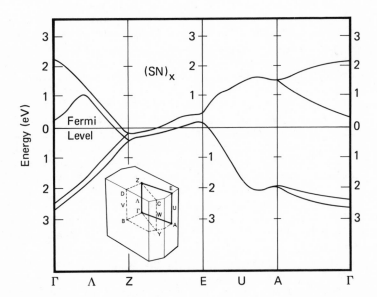

Fig. 4b. Energy bands near the Fermi level of (SN)$_x$ from a three dimensional OPW calculation based on the X-ray crystal structure [20]. The Brillouin zone is shown in the inset.

apparent and are qualitatively the same in the OPW calculation
based on the Boudeulle structure. First we see that the Fermi
level crosses overlapping bands corresponding to the polymer
chain direction (Γ to Z) and perpendicular directions (Z to E).
This results in electron (at Z) and hole (at E) pockets which
give $(SN)_x$ semimetallic character. The degeneracy at Z which
emerges from the simple considerations discussed above is also
seen. The splitting of the bands at Z and the perpendicular
dispersion of the bands are both rough measures of the strength
of the interchain coupling. The interactions between chains in
both the ($\overline{1}02$) plane and the (100) plane are important [25].
Note that the parallel bandwith is $W_{||}\approx 2.5$ eV and the perpendicular
band width is $W_{\perp}\approx 0.5$ eV i.e., $W_{\perp}$ is as wide as the parallel
bandwidth in TTF-TCNQ [1]. This band structure clearly shows
that $(SN)_x$ must be considered as a 3D semimetal and not as a
quasi 1D metal. The complete band structure for all the states
is shown in Fig. 5a along with the resulting density of states.

Though the stability of $(SN)_x$ against a Peierls distortion
is suggested by these results, the point is made even more
apparent from the shape of the Fermi surface. This has been
calculated by Rudge et al [28] from a pseudopotential interpolation
of the OPW results. The Fermi surface is shown in Fig. 5b. Here
it is seen that $(SN)_x$ has electron and hole pockets which close
in one direction. The pockets are roughly ellipsoidal tubes;
however the actual situation is more complicated since there are
holes within the tubes. These points are discussed in detail in
Ref. 28. The essential point is that there is no Peierls
distortion which would place a gap at every point on a Fermi
surface like that shown in Fig. 5b and hence one does not expect
a Peierls metal-insulator transition in $(SN)_x$. This work shows
that the stability of metallic conduction in $(SN)_x$ arises from
interchain coupling which leads to a closed Fermi surface. All
the other 3D calculations [15,19,22,25,26] are in qualitative
agreement with the OPW results except for the work of Kamimura
et al. [23]. Since that work is an EH calculation it is not clear
why the results differ from the other EH calculations [15,22].
This discrepancy remains to be resolved [30]. It is our opinion
that the OPW results are the best approximation yet to the correct
Fermi surface and band structure of·crystalline $(SN)_x$. This view
is supported by the recent critical review of the $(SN)_x$ band
structure calculations by Salahub and Messmer [25]. They report
their own molecular cluster and EH calculations on $(SN)_x$ which
agree with the OPW work. In addition they find several other
interesting results:

   1) The atomic 3d state contribution to the SN molecular
       orbitals is negligible.
   2) The theoretical charge transfer from S to N is
       0.5 electron.

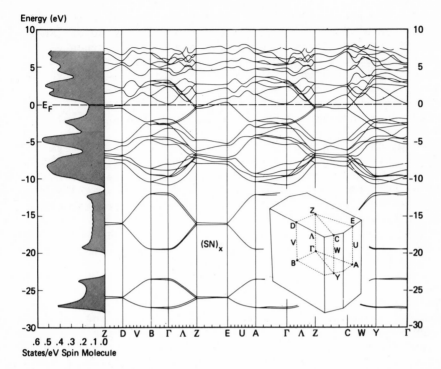

Fig. 5a. Complete OPW band structure of (SN)$_x$ with corresponding
density of states [20].

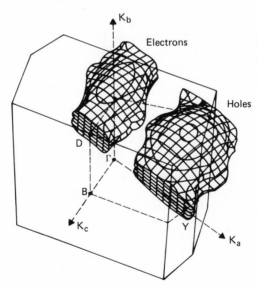

Fig. 5b. Outermost electron and hole Fermi surfaces of (SN)$_x$
derived from OPW band structure [28,29]. Not shown is
a second set of surfaces nested under those given above.

3)   There are significant S-N interactions between chains
      in both the $(\bar{1}02)$ plane and the (100) plane of $(SN)_x$.
4)   The conduction band is only due to π electrons.
This latter point, which has also been found by others, is
interesting because the π orbitals are directed between the chains
and may also contribute to interchain bonding. This is in
contrast to the 1D materials, like TTF-TCNQ and KCP, where the
conduction band orbitals are directed along the chain axis.

As we shall see later, the OPW band structure and Fermi
surface calculations agree very well with many experimental
results. In order to compare with experiment, the band structure
must be used to calculate useful quantities such as the dielectric
response function $[\varepsilon_2(\omega)]$, the photoemission response function,
and the plasma tensor, i.e.

$$(\omega_p^2)_{\mu\nu} = 4\pi e^2 \int_{S_F} dS \; \frac{v_\mu v_\nu}{|v|} \; , \qquad v_\mu = \frac{1}{\hbar} \frac{\partial E}{\partial k_\mu} \; .$$

These calculations have been done by Rudge et al. [28,29] using
the OPW results. Many of the transport, optical and
superconductivity properties of a metal may be related to the
plasma tensor. For metals with complicated band structures like
$(SN)_x$, the plasma tensor occupies the role played by n/m^* in the
theory of simple metals. Rudge et al. find the principal axes of
the combined hole-electron plasma tensor to be nearly contained
within and orthogonal to the $(\bar{1}02)$ plane. The magnitudes of the
principal axes are in the ratios 1:0.13:0.09 with respect to the
directions $b:(\bar{1}02)_{\parallel}:(\bar{1}02)_{\perp}$. This important result will be compared
with the anisotropy found experimentally in the next section.

IV.   PHYSICAL PROPERTIES

In this section we discuss the physical properties of $(SN)_x$.
We shall show that the results of the experimental measurements
are strongly affected by the fibrous morphology and chain
imperfections found in all known crystals. Although these
extrinsic material features cause difficulties, it has been
possible to combine the results of many different experiments to
determine the intrinsic properties of $(SN)_x$.

a)   Normal State Transport and Optical Properties

For many years $(SN)_x$ was though to be a semiconductor; however
in 1973 Walatka et al. [5] published dc electrical conductivity

and thermoelectric power data which strongly suggested that (SN)$_x$ was metallic down to 4.2°K. This conjecture was confirmed by the subsequent discovery of superconducitivity below ~0.3°K [6]. Since that time several groups have measured the temperature dependence of the dc conductivity (σ) of (SN)$_x$ [31-36]. The results show that the conductivity is quite dependent on the crystal preparation method. This is illustrated by our own data [32-34] in Fig. 6, which shows the b axis or parallel resistivity ($\rho_{\parallel} = 1/\sigma_{\parallel}$) of (SN)$_x$ normalized to the room temperature value as a function of temperature. For comparison, curve (a) shows the semiconducting behavior observed for compressed pellets, curve (b) shows some of the original data of Walatka et al. [5] and curve (c) shows data for a crystal grown in our laboratory using the preparative method of Walatka et al. Curves, d, e, and f result from a slightly different growth technique [33] and room temperature annealing for increasing lengths of time. As annealing time increases the resistivity minimum becomes less pronounced and shifts to lower temperature, the resistivity ratio [$\rho(300°K)/\rho(4°K)$] increases to ~250, the room temperature conductivity increases from 1000 to 4000 ohm^{-1}cm^{-1} and the superconducting transition temperature (T_c) increases from 0.26 to 0.35°K. These results indicate that extrinsic contributions, such as chain breaks and defects, can mask the intrinsic conductivity and cause the low resistivity ratio, the resistivity minimum and a depression of T_c. The annealing process apparently allows more complete polymerization of the (SN)$_x$ chains and minimizes the chain imperfections. The curves d)-f) which presumably comes nearest to the intrinsic temperature dependence show that between 30°K and 250°K $\rho_{\parallel}(T)=AT^B$ with an exponent $B\approx2$. Chiang et al. [35] have suggested that the typical electron-phonon scattering of usual metals is not the dominating process in (SN)$_x$ but that electron-electron Umklapp scattering is responsible for the unusual T^2 behavior of ρ. For the Fermi surface shown in Fig. 5b with the electron and hole pockets separated by half a reciprocal lattice vector (G/2) this scattering mechanism seems plausible, however more experiments are needed to prove the matter conclusively. Similarities may exist with the layered material TiS$_2$ [37] and with TTF-TCNQ [38] where T^2 dependences of ρ have also been observed and analyzed in terms of electron-electron scattering.

Although this recent work has achieved considerable improvement in the dc electrical properties of (SN)$_x$ crystals we expect further progress as developments in the crystal growth process lead to crystals of even greater purity and perfection. Some support of this view comes from the fact that the room temperature optical conductivity (σ_{opt}) is about ten times greater than the highest value reported for the dc conductivity. The optical reflectivity, from which σ_{opt} is determined, has been

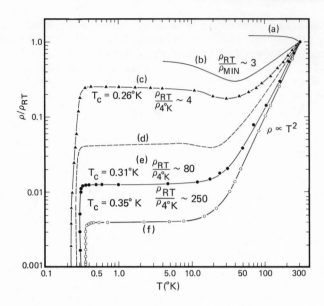

Fig. 6.   Temperature dependence of the dc resistivity of $(SN)_x$
crystals along the polymer chains (b-axis).   Different
curves are discussed in text.

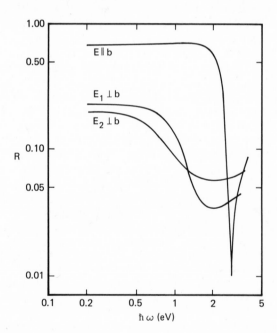

Fig. 7.   Polarized reflectance (R) of $(SN)_x$ crystals at room
temperature [43].

measured by several groups [17,39-44]. Typical results [43] are
shown in Fig. 7 for normally incident light with E vector polarized
parallel and perpendicular to the b axis. For E∥b polarization a
steep, metallic reflectance edge is observed. These data have
been analyzed using the Drude model where the complex dielectric
constant $\varepsilon(\omega)=\varepsilon^b-\omega_p^2/(\omega^2+i\omega/\tau)$ and ω_p is the plasma frequency, τ
the scattering time and ε^b the background dielectric constant.
The results of this analysis give $h\omega_p$=6.9 eV, τ=2.6×10^{-15} sec and
ε^b=6.5, values similar to those found by the other groups. The
contribution of interband transitions is included in ε^b and causes
a shift of the plasma energy from the free electron value of
$h\omega_p$=6.9 eV to $h\omega_p/(\varepsilon_b)^{1/2}$=2.8 eV. From these parameters one finds
$\sigma_{opt}=\omega_p^2\tau/4\pi\approx2.5\times10^4\Omega^{-1}cm^{-1}$ compared with the current maximum of
$\sigma_{dc}\approx4\times10^3\Omega^{-1}cm^{-1}$ at room temperature.

The perpendicular conductivity [31,34,44] is more difficult
to interpret. The dc conductivity has an almost temperature
independent value of $\sigma_{dc\perp}\approx10\Omega^{-1}cm^{-1}$, a behavior very much like that
found in some granular metals [45]. There is a small decrease
of $\sigma_\perp$ below ~50°K until superconductivity occurs at the same T_c as
for $\sigma_\|$. The perpendicular reflectivity (Fig. 7) shows structure
below 1 eV which has been interpreted as a strongly damped plasma
edge. A Drude analysis of $R_{2\perp}$ gives parameters $h\omega_{p\perp}\approx2.4$ eV,
$\tau_\perp\approx3.3\times10^{-16}$ sec, $\varepsilon_\perp^b$=3.6, and $\sigma_{opt\perp}$=380$\Omega^{-1}cm^{-1}$. Note that $\sigma_{opt\perp}$ is
40 times greater than $\sigma_{dc\perp}$ but is still apparently limited by the
fiber boundaries which cause the very short $\tau_\perp$. The $R_\perp$ and the
derived parameters depend somewhat on the direction of E with
respect to the plane of the chains [8]. A measure of the
experimental anisotropy is determined from $\omega_{p\|}^2/\omega_{p\perp}^2$=8.3, $\tau_\|/\tau_\perp$=7.9,
and $\sigma_{opt\|}/\sigma_{opt\perp}$=65. The experimental anisotropy in ω_p^2 falls within
the intrinsic theoretical estimates of 8-11 found from the plasma
tensor. Most authors have used the simple tight binding
approximation to the band structure to obtain values for the
anisotropy in terms of the effective mass (m*) or the transfer
integral (t). This analysis [8] applied to the data in Fig. 7
gives m^*=0.9 m_e, $m_\perp^*$=7 m_e, $t_\|/t_\perp$=8, assuming a carrier concentration
which corresponds to one conduction electron per SN molecule,
i.e. n=3×10^{22}cm^{-3}. The anisotropy ratio obtained from this
analysis agrees with that found from ω_p as expected, however,
the absolute values of m* obtained are not meaningful because
the actual band structure is complex. As we pointed out earlier
the only quantity of significance is n_{eff}/m^*. The effective
number of carriers, n_{eff}, can be roughly estimated from
the volume of the electron and hole pockets relative to the
Brillouin zone volume. From Fig. 5 we find n_{eff}=0.07n=2.1×10^{21}cm^{-3}.
This is in remarkably good agreement with the value of
n_{eff}=3.2×10^{21}cm-3 found by Kahlert [46] from Hall effect measurements.

It does appear that the intrinsic perpendicular conductivity
is metallic. This conclusion is supported by the following
experimental results:

1) Thermoelectric power measurements which show that between
    4.2-300°K $(SN)_x$ has a small (of order a few uV/°K)
    negative thermopower, typical of metals, both parallel
    [5,36,42,47] and perpendicular [47] to the b axis.

2) Energy loss experiments [48] which show a well defined
    plasmon, with metallic angular dependence, in ∥ and ⊥
    directions. An instrument limited $\tau_\perp$ is found that is
    three times longer than the $\tau_\perp$ found earlier from the
    optical reflectivity, which implies that the intrinsic
    $\sigma_\perp$ of $(SN)_x$ crystals is greater than $1000\Omega^{-1}cm^{-1}$.

3) A Maxwell-Garnett analysis [49] of the optical data in
    terms of metallic fibers embedded in a dielectric matrix
    which gives good agreement with experiment.

4) The magnetoresistance measurements of Gill and Beyer
    [50].

These latter results at 4.2°K are shown in Fig. 8a. The current
flow (J) was chosen ⊥ to the b-axis so that the transverse
magnetoresistance could be measured for H∥b and H⊥b. At low fields
a negative effect is observed with a positive quadratic effect
dominant at higher fields. The origin of the negative effect is
not known but it probably is caused by the fibers or other
perpendicular disorder. This view is supported by the work of
Kahlert and Seeger [36] who measured the dc conductivity and
magnetoresistance of $(SN)_x$ with current parallel to the fibers
before and after mechanical deformation. It seems likely that
deformation introduces defects since it causes the resistivity
ratio to significantly decrease and a negative, non-quadratic
magnetoresistance to appear. In the work of Gill et al. [50] the
negative magnetoresistance was subtracted out and the positive
effect analyzed for its angular dependence. This is shown in
Fig. 8b. The observed anisotropy ratio is ≈3 for H⊥b over H∥b.
Using the plasma tensor calculation of Grant et al. [29], the
galvanomagnetic tensor for $(SN)_x$ can be calculated in a
two-carrier conductivity model assuming a single isotropic
scattering time. The angular dependence of the magnetoresistance
predicted from this calculation is shown as the solid curve in
Fig. 8b. The calculated anisotropy is 5.7 in reasonable agreement
with the measured anisotropy. With J parallel to the fibers, a
transverse magnetoresistance of the same magnitude as for J⊥b
was found. Since the magnitude of $\Delta\rho/\rho B^2$ is the same even though
$\rho_\parallel$ and $\rho_\perp$ differ by about four orders of magnitude (at 4.2°K) this
implies that $\sigma_\perp$ should be expressed as $\sigma_\perp = P\sigma_{11}$ where P is a
transmission probability between fibers and σ_{11} is the intrinsic
conductivity of a single fiber. The magnetoresistance then
appears to be a direct and reasonably accurate measure of the
intrinsic anisotropy of the transport properties of $(SN)_x$.

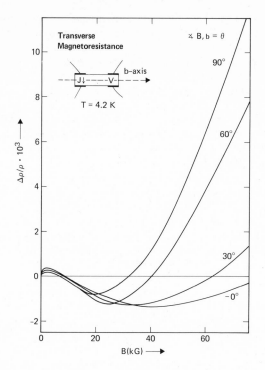

Fig. 8a. Magnetoresistance of (SN)$_x$ crystal at 4.2°K [50].

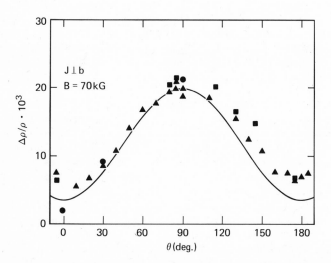

Fig. 8b. Angular dependence of magnetoresistance for several
samples at 4.2°K. Solid curve shows predicted anisotropy
from calculated galvanomagnetic tensor [50].

So what can we conclude about the normal state conductivity of $(SN)_x$? In our opinion the appropriate model for $(SN)_x$, based on the present data, is to consider the crystals as intrinsically anisotropic granular metals, with the crystalline fibers acting as the metal grains. The intrinsic anisotropy in conductivity is ≤ 10. The measured perpendicular dc and optical conductivity is limited by tunnelling between single crystal fibers. The measured parallel conductivity also seems to be influenced by defect scattering but not nearly to the extent of $\sigma_\perp$. An unusual scattering mechanism is present, perhaps e-e scattering, which clearly warrants future investigation.

b)   Other Normal State Properties

The lattice properties of $(SN)_x$ have been investigated by a variety of experiments. The gross features are found from the low temperature specific heat measurements of Harper et al. [51] shown in Fig. 9. The electronic contribution to the specific heat $(C_e=\gamma T)$, when corrected for the electron-phonon coupling constant $(\lambda \approx 0.35)$ determined from T_c, gives a Fermi level density of states of 0.14 states/eV-spin-molecule in excellent agreement with the OPW result of 0.13 states/eV-spin-molecule. The lattice contribution to the specific heat (C_L) has the temperature dependence expected for a solid with quasi-one-dimensional binding forces. The $T^{2.7}$ region is somewhat unusual and its origin has been attributed to the importance of bond-bending forces along the $(SN)_x$ chain as found in the lattice dynamical model of Genensky and Newell [52]. This theoretical model gives for the fit to the data at very low temperature $C_L=A(T/\theta)^3$, with $\theta=148°K$ (102 cm^{-1}) a measure of the <u>interchain</u> lattice coupling energy. The intrachain modes are at high energy and do not contribute to the low temperature specific heat.

More detailed information on the lattice dynamics of $(SN)_x$ comes from inelastic neutron scattering [8], infrared reflectance (IR) and Raman scattering experiments [54-56]. The neutron experiments are difficult because of the small size and poor quality of the crystals. Nevertheless, Pintschovious [8,53] has been able to measure the dispersion of the acoustic phonons traveling in the chain direction. The preliminary results are shown in Fig. 10. The interesting feature of these data is the large difference in the slopes of the longitudinal and transverse branches and the unusual curvature of the lower transverse branch. The intrachain optical phonon modes have been observed by polarized IR and Raman measurements on $(SN)_x$ crystals [54-56]. They appear at energies between 400 and 1000 cm^{-1}. No interchain modes have been found in these experiments, however a measure of the interchain coupling is obtained by the 30-50 cm^{-1} energy

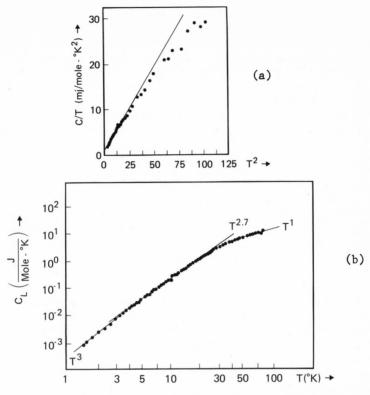

Fig. 9.   (a) Specific heat of (SN)$_x$ from 1.5 to 10°K [32].   Solid
          line shows C=γT+βT^3 fit to data below 4°K.   (b) Lattice
          specific heat, C-γT, up to 80°K [51].

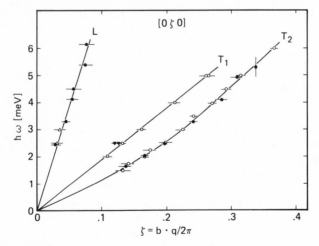

Fig. 10. Dispersion curve of phonons traveling in b direction [8].

difference of certain intrachain IR and Raman modes [54]. This splitting is a result of the presence of two $(SN)_x$ chains per unit cell and is in reasonable agreement with the interchain coupling estimate from the specific heat data. Combining the optical and neutron data Stolz et al. [55] estimate that the interchain force constants are about ten times smaller than the dominant intrachain force constants. This lattice anisotropy is of the same order as the electronic anisotropy found from the transport and optical measurements. It is a bit larger than the lattice anisotropy found in trigonal selenium [57], a rather similar chain-like solid.

Recently, Pintschovius [53] has found preliminary evidence for a weak Kohn anomaly in the LA phonon propagating along the chain direction from inelastic neutron scattering experiments on $(SN)_x$ crystals. The anomaly occurs at a reciprocal lattice wave vector of $q=0.45\ 2\pi/b$. This implies that the Fermi surface must have some planar regions which are spanned by the wave vector $q=2k_F$. As seen in Fig. 5b the Fermi surface has some flat regions which are connected by a wave vector $(2k_F)$ in the chain direction. The theoretical magnitude of $2k_F$ differs from the experimental result. This is not really surprising given the uncertainty as to the exact crystal structure of $(SN)_x$ and the inherent limitations of the band structure calculations. The important point is that the calculated Fermi surface is consistent with the observation of a Kohn anomaly and is in reasonable agreement with all other experimental results reported to date. More work is clearly needed to get quantitative agreement between theory and experiment.

Several independent X-ray photoemission (XPS) experiments have been reported [58-60]. They all give a valence band density of states which agrees well with the band structure predictions shown in Fig. 5a. An analysis of the core level XPS spectrum [59,60] of $S(2p)$ and $N(1s)$ in the $(SN)_x$ crystal give a charge transfer of near 0.5e from S to N, consistent with the theoretical prediction of Messmer and Salahub [25]. Ultraviolet photoemission experiments (UPS) [61,62] are in accord with the XPS results and the OPW band structure except for the region within 0.2 eV of the Fermi level. Here the UPS gives an extremely low density of states which has been theoretically attributed to be the result of a small photoemission transition matrix [62]. Optical experiments which measure interband transitions out to 25 eV have been reported [44,63] and compared with the calculation of $\varepsilon_2(\omega)$ [28,29,63]. The overall agreement is good but more work is needed to clarify some of the details of the structure observed experimentally.

Considerable work has been done on the preparation and properties of films of (SN)$_x$. Films have be prepared by the same procedure used to makes (SN)$_x$ crystals [3] or by direct sublimation of (SN)$_x$ crystals [39]. Typical films are polycrystalline with a grain size of order 1μ. Each crystalline grain has its b axis lying in the plane of the film and the grains are randomly oriented with the film plane. Completely oriented films have been obtained on polymeric substrates such as teflon and mylar [41]. The films have the same optical reflectivity spectrum and X-ray powder pattern as (SN)$_x$ crystals implying that the crystallites are metallic. However, the dc conductivity has the temperature dependence of a semiconductor (with room temperature $\sigma_{dc} \approx 50\Omega^{-1}cm^{-1}$) [41,64,65] showing that there is considerable interparticle resistance even in the parallel direction of the oriented films. One expects the crystalline grains to be superconducting, however four probe conductivity measurements show no definitive evidence of superconductivity in the films prepared to date [64,65]. This may be due to the high resistance of the films at low temperature or to some more fundamental difference between the films and the crystals. This is clearly a problem that warrants more study. One interesting and potentially useful feature of (SN)$_x$ films is their ability to form highly electronegative metallic contacts to some semiconductors. The work of Scranton et al. [66] has shown that (SN)$_x$ is more electronegative than gold and can form Schottky barriers which are significantly larger than those produced by gold on n-type semiconductors.

c) Superconducting Properties

The discovery of superconductivity in (SN)$_x$ has generated considerable interest primarily because it occurs in a material which is a polymer and which contains no metallic elements. Also there has been the hope that modifications of (SN)$_x$ might be possible which could lead to higher transition temperatures. In this section we will review our present understanding of the superconducting properties of this unusual material. As we shall see, the superconductivity, like the normal state behavior, is influenced by the crystal inperfections and fibrous morphology of (SN)$_x$.

The superconductiving experiments which have been reported up to the present are:
1) The variation of T_c with crystal quality [33,67].
2) The angular and temperature dependence of the critical magnetic field. [68-70].
3) The hydrostatic pressure dependence of T_c [71].

4)   The observation of superconducting fluctuations [72,73].

5)   The specific heat from 0.11 to 1.5°K [77].

From Fig. 6 we see that T_c depends on crystal imperfections. Crystals with higher resistivity ratios have higher and sharper (ΔT_c) transition temperatures. The lowering of T_c most likely results from the defect structure along the chain since the fibrous morphology is common to all crystals. The mechanism for the depression of T_c is unclear but it may be related to similar effects observed in the A-15 superconductors [75]. Very poor $(SN)_x$ crystals exhibit some resistance below T_c and very low critical currents, properties which presumably result from chain end effects [68]. The effect, if any, of the hydrogen impurity on the superconductivity, or any other property, has not been established.

The temperature dependence of the critical magnetic field is shown in Fig. 11a. The magnitude of $H_{c\parallel}$ and $H_{c\perp}$ at T=0°K is much greater than our estimate of the thermodynamic critical field, $H_c(0) \approx 20$ Gauss, implying that $(SN)_x$ is a type II superconductor. There are several other unusual features which are discussed in detail in Ref. 68:

1)   The value of $H_{c2\parallel}(0)$ exceeds the paramagnetic limit of $H_p = 18.4\ T_c \approx 5.4$ kOe.

2)   The temperature dependence of $H_{c2\perp}$ is anomalous and different from $H_{c2\parallel}$.

3)   The critical field near T_c is only weakly dependent on $\rho_\parallel(4.2°K)$.

4) $H_{c2\parallel}/H_{c2\perp}$ is large and temperature dependent.

The magnitude of the anisotropy in H_{c2} can be explained by a combination of the intrinsic anisotropy of the band structure and an anisotropy in the scattering time caused by tunnelling between the fibers, which leads to an anisotropy in the coherence length, $\xi(T)$. On the other hand, the weak dependence of H_{c2} on $\rho_\parallel(4.2°K)$, the angular dependence of H_{c2}, and the temperature dependence of $H_{c2\parallel}/H_{c2\perp}$ cannot be explained by the intrinsic anisotropy. This suggests that a model of weakly coupled superconducting fibers of diameter d may be more appropriate. By weak coupling [76] we mean $\xi_\perp$ (isolated fiber) >d and $\xi_\perp$ (crystal) <d. For strong coupling $\xi_\perp$ (crystal) is >d. From the magnitude of $\sigma_\perp$ and $H_{c2\parallel}(0)$ we estimate at T=0°K that $\xi_\perp$ (crystal) ≤ 150Å and $\xi_\perp$ (fiber) ≥ 500Å so that the condition for weak coupling may be fulfilled. The weak coupling view is supported by the work of Civiak et al. [72,73] who have observed superconducting fluctuations in the conductivity of $(SN)_x$ well above T_c. Fluctuation effects would not be expected in the case of large fibers [d>$\xi_\perp$ (fiber)] or for coupled fibers of any size. It should also be noted that similar magnetic critical field effects have been observed in granular superconducting films [77,78] where the coupling between grains plays an important role. The

Meissner effect has not yet been observed in (SN)$_x$, a result
which is not too surprising since we estimate H_{c1} to be
considerably less than the earths' field (0.5 Gauss).

Under hydrostatic pressure [71] T_c and $\sigma_\parallel$(300°K) increase
rather dramatically as shown in Fig. 11b. An even larger increase
is found for $\sigma_\perp$ (300°K) [17,71]. T_c has not been measured at
higher pressure but at 100 kbar there is no superconductivity
above 1.2°K [79]. Assuming that (SN)$_x$ remains in the same
metallic phase, the pressure dependence of T_c must saturate or
change direction. Since it is unusual for T_c to increase under
pressure in a non-tranistion metal superconductor, and the large
increase in $\sigma_\parallel$ is anomalous, it was suggested that electronic band
structure changes with pressure must be significant [71].
However, Schlüter et al. [19] theoretically found that the electron
density of states changed very little with different lattice
parameters and they proposed instead that some phonon modes must
soften under pressure. We consider this latter explanation rather
unlikely since the thermal expansion is quite normal in (SN)$_x$
[10] and the Raman spectrum shows no modes that soften under
pressure increase or temperature decrease. Another possible
origin of the increase in T_c is a stronger coupling of the
superconducting fibers under pressure. It is known from the work
on granular metals [77] that T_c can be decreased from its
intrinsic value by weak coupling between grains. Some support
of this explanation comes from the fluctuation experiments, which
suggest that the intrinsic T_c of (SN)$_x$ may be near 1.4°K [73].

The specific heat measurements below 1°K are shown in Fig. 12
[74]. They confirm that the superconductivity is a bulk effect.
A broad hump in specific heat is observed starting below 0.4°K.
The entropy associated with this hump is comparable to that of
a simulated BCS anomaly with T_c=0.26°K as shown in Fig. 13 by the
solid line. The smearing of the superconducting transition may
be caused by sample inhomogeneities or by fluctuation effects.
The specific heat anomaly is quite sensitive to strain and
magnetic field and more work is needed to completely understand
the details of such data in a complicated material like (SN)$_x$.

Based on these experiments our present view of the
superconductivity in (SN)$_x$ is the following. Polysulfur nitride
is a semimetal with bulk type II superconductivity caused by the
usual BCS mechanism. The transition temperature depends strongly
on the polymer chain perfection and other unknown types of
disorder. The temperature and angular dependence of the critical
field, the pressure dependence of T_c, and the transition width
(as determined from specific heat and conductivity measurements),
result partly from intrinsic properties of (SN)$_x$ and partly from
the fibrous morphology and other imperfections present in (SN)$_x$

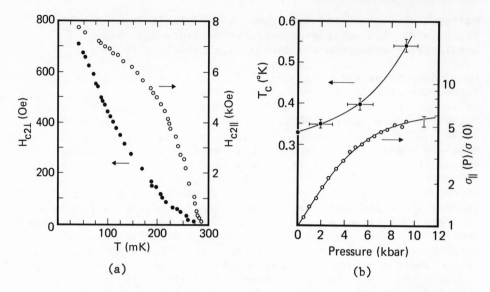

Fig. 11. (a) Upper critical field H_{c2} parallel and perpendicular
to the b axis vs. temperature for crystalline $(SN)_x$ [68].
(b) Pressure dependence of the superconducting transition
temperature (T_c) and b-axis conductivity $(\sigma_\parallel)$ at room
temperature for $(SN)_x$ [71].

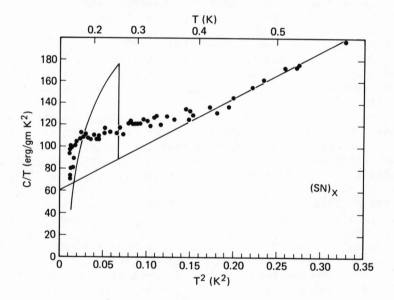

Fig. 12. Specific heat of $(SN)_x$ from 0.1 to 0.6°K [74].

crystals grown to date. The task of quantitatively separating
the intrinsic properties from the extrinsic ones, so as to learn
more about the superconducting behavior of (SN)$_x$, remains an
important problem for future study.

V.   CONCLUSIONS

(SN)$_x$ is the first polymeric solid to exhibit metallic
transport properties and superconductivity. Though its chain
structure is highly anisotropic it does appear that the chains
of sulfur and nitrogen atoms interact sufficiently strongly that
the early description of (SN)$_x$ as quasi one-dimensional is not
appropriate and was in fact partly a reflection of its fibrous
nature. Band structure calculations, along with optical and
transport experiments, show that is more accurate to think of
(SN)$_x$ as an anisotropic three-dimensional semimetal, which
explains why it does not undergo the Peierls metal-insulator
transition found in the quasi 1D metals. The intrinsic
anisotropy, as measured by the lattice binding forces and the
electronic transport properties along the polymer chain axis
relative to these properties perpendicular to the chains, is
between five and ten.

One of the goals of the research on the quasi 1D materials
discussed so extensively at this summer school has been the
stabilization of the metallic state at all temperatures. In our
opinion, one important lesson learned from the research on (SN)$_x$
is that increasing the interchain interaction is the key step to
attaining this goal. This point is supported by the recent
research on HMTSeF-TCNQ [80-82], a quasi 1D metal which apparently
has sufficient interchain interaction to remain semimetallic (but
not superconducting [83]) down to 20 mK. Attempts to vary the
interchain interaction in (SN)$_x$ by intercalating molecules between
the chains have not (to our knowledge) yet been successful. The
newly discovered orthorhomic phase of (SN)$_x$ [13] appears to have
different interchain distances than the β monoclinic phase
discussed in this review. It will be interesting to see if there
are any significant modifications of the physical properties of
(SN)$_x$ in this new phase. Impurities (such as hydrogen), chain
ends, the defect occupied sites, and the fiber boundaries all
may have some influence on the effective interchain interaction.
As we have shown, imperfections such as these certainly affect
the physical properites of (SN)$_x$. One goal of the future chemical
research on (SN)$_x$ should be to gain some control over these
imperfections with the hope of then varying the electronic
properties in a controlled manner.

The remarkable properties of $(SN)_x$ have led to an intensive effort to synthesize analogous compounds. There are several compounds formally analogous to S_4N_4 in which isoelectronic substitutions have been made for nitrogen and sulfur. Messmer and Salahub [84] have theoretically discussed these materials and they conclude that $(SeN)_x$ is the most probable analogue of $(SN)_x$. Unfortunately, polymerization of Se_4N_4 by the methods used for S_4N_4 has not been successful, primarily because Se_4N_4 decomposes in the solid state to the elements before a significant vapor pressure of Se_4N_4 develops [7]. This work and some results on other analogous compounds are discussed in Ref. 7. At this point it seems that in order to produce polymers analogous to $(SN)_x$ new synthetic routes will have to be found. Even though $(SN)_x$ appears to be in a class by itself at the present time there is clearly much more interesting work to be done to understand, and to perhaps alter, the physical and chemical properties of this unique material.

ACKNOWLEDGMENTS

Many of our colleagues at IBM San Jose, UCLA, and Stanford University have made important contributions to the research discussed in this paper. We would particularly like to thank P. M. Grant, W. E. Rudge, W. D. Gill, H. Arnal, P. Mengel, W. Beyer, J. M. E. Harper, W. A. Little, L. J. Suter, L. J. Azevedo, W. G. Clark, G. Deutscher, and P. M. Chaikin for helpful discussions and input throughout the course of this work. In addition, the technical assistance of R. Bingham, J. Vazquez, and D. Miller has been much appreciated. Finally, we thank B. H. Schechtman, P. M. Grant, and J. B. Torrance for their critical reading of this manuscript.

REFERENCES

1. For a review see a) A. J. Berlinsky, Contemp. Phys. <u>17</u>, 331 (1976); b) <u>Low-Dimensional Cooperative Phenomena</u> edited by H. J. Keller, Plenum Press, New York 1975.

2. W. A. Little, Phys. Rev. A <u>134</u>, 1416 (1964); D. Davis, H. Gutfreund, and W. A. Little, Phys. Rev. <u>B13</u>, 4766 (1976).

3. F. B. Burt, J. Chem. Soc., 1171 (1910).

4. M. Boudeulle, Cryst. Struct. Comm. <u>4</u>, 9 (1975) and Thesis, University of Lyon 1974 (unpublished); M. Boudeulle and P. Michel, Acta Cryst. <u>A28</u>, 199 (1972); A. Douillard, Thesis, University of Lyon 1972 (unpublished).

5. V. V. Walatka, M. M. Labes, and J. H. Perlstein, Phys. Rev. Lett. 31, 1139 (1973).

6. R. L. Greene, G. B. Street, and L. J. Suter, Rev. Lett. 34, 577 (1975).

7. G. B. Street and R. L. Greene, IBM Journal of Res. and Dev., in press.

8. H. P. Geserich and L. Pintschovius, in Festkörperprobleme (Advances in Solid State Physics), Vol. XVI, p. 65, J. Treusch (ed), Vieweg, Braunschweig (1976).

9. R. H. Baughman, R. R. Chance, and M. J. Cohen, J. Chem. Phys. 64, 1869 (1976).

10. M. J. Cohen, A. G. Garito, A. J. Heeger, A. G. MacDiarmid, C. M. Mikulski, M. S. Saran, and J. Kleppinger, J. Am. Chem. Soc. 98, 3844 (1976).

11. R. H. Baughman and R. R. Chance, to be published in J. Polymer Science 1976.

12. C. M. Mikulski, P. J. Russo, M. S. Saran, A. G. MacDiarmid, A. F. Garito, and A. J. Heeger, J. Am. Chem. Soc. 97, 6358 (1975).

13. R. H. Baughman, P. A. Apgar, R. R. Chance, A. G. MacDiarmid, and A. F. Garito, preprint.

14. D. E. Parry and J. M. Thomas, J. Phys. C 8, L45 (1975).

15. W. I. Friesen, A. J. Berlinsky, B. Bergensen, L. Weller, and T. M. Rice, J. Phys. C 8, 3549 (1975).

16. V. T. Rajan and L. M. Falicov, Phys. Rev. B12, 1240 (1975).

17. H. Kamimura, A. J. Grant, F. Levy, A. D. Yoffe, and G. D. Pitt, Sol. State Comm. 17, 49 (1975).

18. M. Kertesz, J. Koller, A. Azman, and S. Suhai, Phys. Lett. 55A 107 (1975).

19. M. Schlüter, J. R. Chelikowsky, and Marvin L. Cohen, Phys. Rev. Lett. 35, 869 (1975) and 36, 452 (1976).

20. W. E. Rudge and P. M. Grant, Phys. Rev. Lett. 35, 1799 (1975).

21. A. Zunger, J. Chem. Phys. 63, 4854 (1975).

22. A. A. Bright and P. Soven, Sol. State Comm. 18, 317 (1976).

23. H. Kamimura, A. M. Glazer, A. J. Grant, Y. Natsume, G. Schreiber, and A. D. Yoffe, J. Phys. C9, 291 (1976).

24. C. Merkel and J. Ladik, Phys. Lett. 56A, 395 1976.

25. D. R. Salahub and R. P. Messmer, Chem. Phys. Lett. 41, 73 (1976) and Phys. Rev. B14, 2592 (1976).

26. W. Y. Ching and C. C. Lin, Bull. Am. Phys. Soc. 21, 254 (1976) and to be published.

27. A. J. Berlinsky, J. Phys. C9, L283 (1976).

28. W. E. Rudge, I. B. Ortenburger, and P. M. Grant, Bull. Am. Phys. Soc. 21, 254 (1976) and to be published in Phys. Rev.

29. P. M. Grant, W. E. Rudge, and I. B. Ortenburger, Proceedings of Siofok Conference 1976, to be published.

30. P. M. Grant has found that the overlap integrals computed by Kamimura are incorrect (unpublished results). A discussion of the differences between these calculations is also given by A. D. Yoffe, Chem. Rev. 5, 51 (1976).

31. C. Hsu and M. M. Labes, J. Chem. Phys. 61, 4640 (1974).

32. R. L. Greene, P. M. Grant, and G. B. Street, Phys. Rev. Lett. 34, 89 (1975).

33. G. B. Street, H. Arnal, W. D. Gill, P. M. Grant, and R. L. Greene, Mat. Res. Bull. 10, 877 (1975).

34. P. M. Grant, R. L. Greene, W. D. Gill, W. E. Rudge, and G. B. Street, Mol. Cryst. Liq. Cryst. 32, 171 (1976).

35. C. K. Chiang, M. J. Cohen, A. F. Garito, A. J. Heeger, A. G. MacDiarmid, and C. M. Mikulski, Solid State Comm. 18, 1451 (1976).

36. H. Kahlert and K. Seeger, to be published.

37. A. H. Thompson, Phys. Rev. Lett. 35, 1786 (1975).

38. P. E. Seiden and D. Cabib, Phys. Rev. B13, 1846 (1976).

39. A. A. Bright, M. J. Cohen, A. F. Garito, A. J. Heeger, C. M. Mikulski, P. J. Russo, and A. G. MacDiarmid, Phys. Rev. Lett. 34 206 (1975).

40. L. Pintschovius, H. P. Geserich, and W. Möller, Solid State Comm. 17, 477 (1975).

41. A. A. Bright, M. J. Cohen, A. F. Garito, A. J. Heeger, C. M. Mikulski, and A. G. MacDiarmid, Appl. Phys. Lett. 26, 612 (1975).

42. P. M. Grant, R. L. Greene, and G. B. Street, Phys. Rev. Lett. 35, 1743 (1975).

43. W. Moller, H. P. Geserich, and L. Pintschovius, Solid State Comm. 18, 791 (1976).

44. M J. Cohen, Thesis, University of Pennsylvania 1976 (unpublished).

45. B. Abeles, P. Sheng, M. Coutts, and Y. Aria, Adv. in Phys. 24, 407 (1975).

46. H. Kahlert, contributed results presented at this conference.

47. M. J. Cohen, C. K. Chiang, A. F. Garito, A. J. Heeger, A. G. MacDiarmid, and C. M. Mikulski, Bull. Am. Phys. Soc. 20, 360 (1975).

48. C. H. Chen, J. Silcox, A. F. Garito, A. J. Heeger, and A. G. MacDiarmid, Phys. Rev. Lett. 36, 525 (1976).

49. H. P. Geserich, contributed results presented at this conference.

50. W. D. Gill, W. Beyer, and G. B. Street, to be published in Proceedings of Siofok Conf. 1976 and Solid State Comm.

51. J. M. E. Harper, R. L. Greene, P. M. Grant, and G. B. Street, to appear in Phys. Rev. B; also see Ref. 32.

52. S. M. Genesky and G. F. Newell, J. Chem. Phys. 26, 486 (1957).

53. L. Pintschovius, to be published in Proceedings of Siofok Conf. 1976.

54. H. J. Stolz, A. Otto, and L. Pintschovius, in Light Scattering in Solids, Balkanski, Leite and Porto, eds., (Flammarion, Paris, 1976), p. 737.

55. H. J. Stolz, H. Wendel, A. Otto, L. Pintschovius, and H. Kahlert, preprint.

56. H. Temkin and D. B. Fitchen, Solid State Comm. 19, 1181
    (1976).

57. W. Hamilton, B. Lassier, and M. Kay, J. Phys. Chem. Solids
    35, 1089 (1974).

58. L. Ley, Phys. Rev. Lett. 35, 1976 (1975).

59. P. Mengel, P. M. Grant, W. E. Rudge, B. H. Schechtman, and
    D. W. Rice, Phys. Rev. Lett. 35, 1803 (1975).

60. W. R. Salaneck, J. W. Lin, and A. J. Epstein, Phys. Rev. B13,
    5574 (1976).

61. E. E. Koch and W. D. Grobman, preprint.

62. P. Mengel, W. D. Grobman, I. B. Ortenburger, P. M. Grant, and
    B. H. Schechtman, Bull. Am. Phys. Soc. 21, 254 (1976) and
    P. Mengel, I. B. Ortenburger, W. E. Rudge, and P. M. Grant,
    Proceedings of Siofok Conf. 1976.

63. J. Bordas, A. J. Grant, H. P. Hughes, A. Jakobsson,
    H. Kamimura, F. A. Levy, K. Nakao, Y. Natsume, and A. D. Yoffe,
    J. Phys. C9, L277 (1976).

64. F. de la Cruz and H. J. Stolz, Solid State Comm. 20, 241
    (1976).

65. R. J. Soulen and D. B. Utton, to be published in Solid State
    Comm.

66. R. A. Scranton, J. B. Mooney, J. O. McCaldin, T. C. McGill,
    and C. A. Mead, Appl. Phys. Lett. 29, 47 (1976).

67. P. Civiak, W. Junker, C. Elbaum, H. I. Kao, and M. M. Labes,
    Solid State Comm. 17, 1573 (1975).

68. L. J. Azevedo, W. G. Clark, G. Deutscher, R. L. Greene,
    G. B. Street, and L. J. Suter, Solid State Comm. 19, 197 (1976).

69. R. L. Greene, W. D. Gill, L. J. Azevedo, and W. G. Clark, to
    appear in Ferroelectrics.

70. R. L. Greene, W. D. Gill, L. J. Azevedo, W. G. Clark, and
    G. Deutscher, to be published in Proceedings of Siofok Conf.
    1976.

71. W. D. Gill, R. L. Greene, G. B. Street, and W. A. Little,
    Phys. Rev. Lett. 35, 1732 (1975).

72. R. L. Civiak, C. Elbaum, W. Junker, C. Gough, H. I. Kao,
    L. F. Nichols, and M. M. Labes, Solid State Comm. 18, 1205
    (1976).

73. R. L. Civiak, C. Elbaum, L. F. Nichols, H. I. Kao, and
    M. M. Labes, submitted to Phys. Rev.

74. L. F. Lou and A. F. Garito, preprint.

75. L. R. Testardi, J. M. Poate, and H. J. Levinstein, Phys. Rev.
    Lett. 37, 637 (1976).

76. G. Deutscher, Y. Imry, L. Gunther, Phys. Rev. B10, 4598
    (1974).

77. H. C. Jones, Appl. Phys. Lett. 27, 471 (1975).

78. G. Deutscher and S. A. Dodds, submitted to Physical Review.

79. C. W. Chu, private communication.

80. A. N. Bloch, D. O. Cowen, K. Bechgaard, R. E. Pyle, R. H. Bands,
    and T. O. Poehler, Phys. Rev. Lett. 34, 1561 (1975).

81. J. R. Cooper, M. Weger, D. Jérome, D. Lefur, K. Bechgaard,
    A. N. Bloch, and D. O. Cowan, Solid State Comm. 19, 749 (1976).

82. M. Weger, Solid State Comm. 19, 1149 (1976).

83. D. Jérome and M. Weger, NATO-ASI Proceedings (this volume)
    1976.

84. D. R. Salahub and R. P. Messmer, J. Chem. Phys. 64, 2039,
    (1976).

COLLECTIVE STATES IN SINGLE AND MIXED VALENCE METAL CHAIN COMPOUNDS

P. Day

Oxford University, Inorganic Chemistry Laboratory
South Parks Road, Oxford, OX1 3QR, England

1. INTRODUCTION

The purpose of the present chapter is to survey the structures of inorganic compounds containing chains of interacting metal atoms, to try to see what the molecular requirements are for building up crystals containing such chains, and then to seek correlations between the structures of the chains and the resulting physical properties of the solid. Only a very small fraction of substances containing metal atom chains could in any sense be called one-dimensional metals, and one of the things we wish to do is to see if it is possible to find rules which will help us in looking for potentially metallic systems. Metallic conducting behaviour, as we shall see, is only the limiting case in a sequence of continuously increasing strengths of interaction between adjacent metal ions and evidence for metal-metal interaction stopping short of direct electron exchange can be found in the physical properties of many metal chain compounds. Indeed, in a trivial sence, the electronic states of any assembly of atoms whose positions are governed by translational symmetry have to be called 'collective', because the individual constituent atoms are in principle indistinguishable. However, collective states are not by any means necessarily ones in which electron transfer is freely permitted, and what we shall call 'collective effects' for our present purpose are simply any noticeable departures in the physical properties of the aggregate from what might have been anticipated simply by summing the contributions of the component atoms or ions. A very simple example, which we shall mention again later, is Magnus' Green Salt (MGS), $Pt(NH_3)_4PtCl_4$. Isolated from other ions in solution, or in crystals in which there is no appreciable inter-ionic interaction, the lowest ligand field excited states of

$Pt(NH_3)_4^{2+}$ and $PtCl_4^{2-}$ occur at energies which render the two ions respectively colourless and red.   Clearly MGS does not have a colour anything like the sum of its constituents.   Nevertheless, when stoichiometric it is a pretty good insulator.   Thus we have here a very marked interaction between the ions (at least in their excited states) but no electron exchange in the ground state.

A number of other chapters in this book describe in some detail a convenient theoretical model due to Hubbard (1) which rationalizes the occurrence of metallic or insulating properties in crystals in terms of two adjustable parameters, a transfer integral t which enables the total energy of the system to lower itself by transferring an electron from one site to a neighbour, and an effective on-site Coulomb repulsion U which raises the total energy when a given site becomes doubly occupied.   When $t \gg U$ the electrons delocalize into a band and, if we start with a partly filled shell on each constituent ion, the result is a metal. When $t \ll U$ the electrons will not exchange between ions and we have a magnetic insulator.   The point is that one can find examples both of insulating and conducting metal chain compounds, so we are in a good position to examine the power, and limits, of the Hubbard model.   My own view is that simple one-electron band theory has been applied in much too facile a fashion to the proto-type one-dimensional metals such as KCP $(K_2Pt(CN)_4Br_{0.30}3H_2O)$, and that a wider and more chemical viewpoint on how they relate in structure and properties to other metal chain compounds can only be helpful in enabling theoretical physicists to choose the right starting point for describing these compounds, as well as the purely organic metals.

2.   THE SIMPLEST CONDUCTING METAL CHAIN

As a starting point in building up one-dimensional structures it may be helpful to recall what is probably the simplest inorganic lattice in which chains of closely spaced metal ions are found: that of rutile, TiO_2.   The chains consist of TiO_6 octahedra linked by opposite edges, the remaining octahedral vertices being shared with the next chain (Figure 1). TiO_2 itself, with an electron configuration $3d^0$, is an insulator, but the physical properties of the $3d^1$ analogue VO_2 have been very widely studied.   At high temperature it is a metal but at 341 K it undergoes a first-order phase transition to an insulating state.   It was recently pointed out (2) that no fewer than 18 review articles have been written in the last 6 years on the details of this transition, with no clear unanimity regarding the role of the lattice, the charge carriers, electron correlation or electron-phonon interactions.   I do not wish to add to them here, but a number of points of similarity between VO_2 and the one-dimensional metals are worth stressing. First, in the metallic phase the V-V distances along the chains

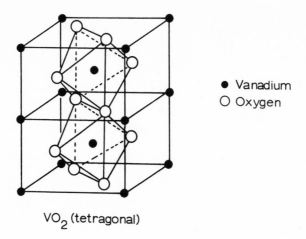

VO$_2$ (tetragonal)

● Vanadium
○ Oxygen

Figure 1.   The structure of rutile.

are all equal at 2.87 Å, but below the 341 K phase transition they
alternate between 2.65 and 3.12 Å.   The electrons have then become
paired into the shorter bonds and the low temperature phase, as
shown by e.p.r. (3), is an antiferromagnetic localized insulator.
The analogy with the simplest possible hypothetical chain containing unpaired electrons, i.e. one consisting of hydrogen atoms,
seems clear.   In this half-filled band (d_{xy} in VO$_2$) an instability
has arisen, a superlattice of two V–V spacings is established and
a gap has thereby opened at the Fermi surface: we have what is
called in other chapters in this book a Peierls distortion.

Of course, a difficulty in this simple-minded view is that
the nearest neighbour V–V spacing between the chains is only about
30% bigger than that within the chains, so the chains must be
coupled quite closely together.   What structural possibilities
are there then for bulking out the chains and so lowering the
dimensionality?   Simplest would be to go from binary compounds
MX$_2$ to ternary, AMX$_3$ or AMX$_3$.2H$_2$O, where A is a large and
electronically 'inert' cation like Cs$^+$ or NR$_4^+$.   Another possibility
would be partly to substitute the coordination sphere of the metal
M with bulky ligands in place of the anions X.   A further solution
would be to replace completely the mono-atomic anions with more
extended organic anions, possibly with large side-chains while

finally one might consider the possibility of using conjugated
organic molecules as bridging groups between the metal ions.
Some examples of each of these types of compound will now be
given, but before doing so, two further subdivisions among the
structure types can be distinguished, which, we shall argue, are
fundamental to the physical properties to be anticipated.

Considering compounds like VO_2 and ternary oxides, which may
be either metallic or insulating depending on the transition metal
ion concerned, Goodenough (4) argued some years ago for the
existence of a critical distance between metal ions, at which a
transition from localized to collective behaviour would take place.
In the context of the Hubbard model (1) this simply means that t
has to be made large enough to overcome U.   In VO_2, containing an
element near the beginning of the 3d—block, whose valence orbitals
are therefore extended, the value of t, which is related to overlap
integrals, is evidently large enough to permit metallic behaviour.
Clearly, therefore, we want to bring our metal ions as close
together as possible in relation to the radii of the valence
orbitals.   Among the metal chain compounds we want to discuss,
we might expect therefore that the properties of those whose
structures permitted direct metal—metal contact would be quite
different from those in which a bridging group joined the nearest
neighbour metal ions.

Another approach to collective electronic behaviour would of
course be to reduce U.   In a divalent transition metal compound,
for instance, U is related to the energy of a process such as
$2M^{2+} \rightarrow M^{3+} + M^{+}$, and consequently, for gas phase ions, to the
difference between the second and third ionization potentials.
For the Group VIII elements, which are going to figure quite
largely in our discussion, the magnitude of this difference is
18.01 and 20.5 eV for Ni and Pd respectively — very big quantities!
On the other hand if we had a mixed rather than single valence
situation (for example a compound containing both Ni^{2+} and Ni^{3+})
the change in intra-site Coulomb repulsion on transferring an
electron from Ni^{2+} to Ni^{3+} would be zero if the two ions had
identical crystal environments.   Since the ligand field site
preferences of different d^n configurations are often quite
different the environments, and hence the site potentials of the
two ions of differing oxidation state also differ quite frequently.
Nevertheless, there can be no doubt that mixed valency is a good
way of reducing U.

With these two points in mind we shall therefore divide our
discussion of metal chain compounds into single and mixed valence
chains, each of these being further divided into bridged and direct
metal—metal contacts (5).

3.   SINGLE VALENCE METAL CHAINS

Anion Bridged Chains

Among anion-bridged single valence chains probably the most ubiquitous are the hexagonal perovskites AMX_3. The structural unit is an MX_6 octahedron and the chains, which are formed from such octahedra by sharing opposite faces, are separated from one another by the much larger A ions (Figure 2). Although there are some examples of mixed cubic-hexagonal perovskites in which X is an oxide ion all the purely hexagonal examples have X as a halide (6). Examples can be found for nearly all divalent 3d ions and halides, though in general the hexagonal chain configuration is favoured by the larger halide ions and larger A.

All known hexagonal chain AMX_3 compounds are magnetic insulators. The magnetic exchange interaction, which can be ferromagnetic or antiferromagnetic for a given M, depending on R and X, is extremely anisotropic, so that in $N(CH_3)_4MnCl_3$ (TMMC) for example, the ratio of the exchange integral within and between the chains is at least 10^3. Probably the most sensitive test of one-dimensionality in magnetism is dispersion of spin-waves propagating perpendicular to the chains. A spin-wave is a collective magnetic excitation whose energy depends on its wave-length (wave-vector). The extent of this dependence (dispersion) is a measure of the near-neighbour coupling strength and can be determined experimentally by inelastic neutron scattering. Two such dispersion curves are shown for comparison in Figure 3. In each, the dispersion perpendicular to the chains is zero within experimental error, while along the chains the nearest neighbour exchange integrals J are -4.6 and -26.7 cm^{-1} for the Mn^{II} (7) and Cr^{II} (8) salts respectively. Both these figures are very much

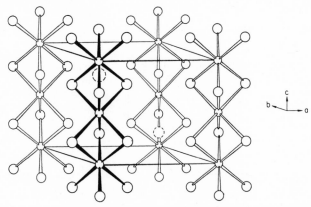

Figure 2. The unit cell of a hexagonal perovskite ABX_3. Small open circles are B, large open circles are X and dotted circles are A.

smaller than the energies of the first electronically excited
states in the two compounds, which lie at 18 730 (9) and
11 700 cm^{-1} (10) respectively.   In fact all the lower excited
states in this type of compound, across the visible and near ultraviolet, are easily classified as ligand field and are therefore
confined to the same Heitler–London configuration as the ground
state transitions, so the energy needed to transfer an electron
from one ion to another is indeed quite large (certainly greater
than 50 000 cm^{-1} in TMMC).   Since one should take account of the

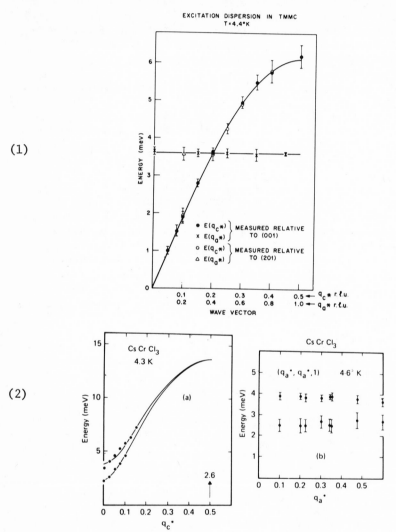

(1)

(2)

Figure 3.   Dispersion of spin-waves propagating along and perpendicular to the metal chains in (1) TMMC (ref. 7) and (2) CsCrCl$_3$
(ref. 8).

translational symmetry along the chain when forming the electronic
as well as the magnetic excitations, the former are excitons (of
the Frenkel, or tight-binding type) while the latter are called
magnons.   Those ligand field excited states which have a different
spin projection from the ground state, and are thus formally
inaccessible via an electric-dipole transition mechanism, can become
accessible if the incoming photon creates a two-particle excitation
involving an exciton $E(k)$ combined with a magnon $M(-k)$, either
through simultaneously creating both $(E(\underline{k}) + M(-\underline{k}))$ or by annihil-
ating a thermally created magnon when the exciton is formed
$(E(\underline{k}) - M(-\underline{k}))$.   When the chain is an antiferromagnetic one the
former process dominates at low temperature because if the exciton
has a lower spin-projection than the ground state (e.g. sextet-
to-quartet, as in TMMC) one can create a magnon on the other
sublattice to increase the spin again.   In a ferromagnetic chain,
on the other hand, there is only one sublattice so at low temperatures the (E+M) process is not available.   At higher temperatures
some magnons which decrease the spin projection of the chain may
be thermally excited, and could be annihilated in an (E-M) process.
Now as we can see from Figure 4 , the magnons have much lower
energies than the exciton states.   Consequently (E+M) combination
bands occur in the optical spectrum quite close to the positions
expected of transitions in the ligand field spectra of related,
but magnetically dilute analogues.   Unless the bands are
particularly sharp, so that E and E+M transitions can be resolved
from one another, the only overt consequence of the one-dimensional
magnetic interaction on the optical spectrum lies in its temperature
dependence.   In the antiferromagnetic case one finds a broad
maximum in the curve of intensity versus temperature whilst for a
one-dimensional ferromagnet the intensity goes down smoothly as
the temperature is lowered.   Two examples from hexagonal perovskite
lattices are shown in Figure 4.   TMMC represents the antiferromagnetic case (9) and $RbFeCl_3$ the ferromagnetic (11).

   Good examples of metal halides in which the MX_6 octahedron is
partly substituted by larger donor molecules are a large set of
compounds $MX_2 \cdot 2Y$, whose structures are shown schematically below:

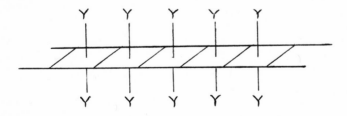

Here M = Cr, Mn, Fe, Co, Ni, Cu and X = Cl, Br, SCN, while Y is
pyridine or H_2O.   The Co^{II} and Fe^{II} examples are ferromagnetic,
the rest antiferromagnetic with exchange integrals of similar

magnitude to the AMX$_3$ compounds.   Particularly interesting is the
effect of interchain separation: the intrachain ferromagnetic
exchange integrals in CoCl$_2$.2H$_2$O and CoCl$_2$.2(pyridine) are very
similar, for instance, but the three-dimensional ordering temperature of the pyridine compound is 3.15 K compared with 17.2 K for
the hydrate.

Organic bridging groups between metal ions can be such bi-
functional molecules as pyrazine(I) or quinoxaline(II), or bidentate chelating agents like oxalate(III) or bipyrimidine(IV).
In all known single valence compounds containing metal atom chains
bridged by these groups the interaction across the ligand   -
system is much too weak to give rise to collective electronic states
and the compounds are magnetic insulators having exchange integrals
of similar magnitude to the partly substituted halides.

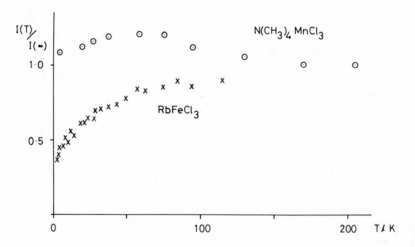

Figure 4.   Temperature variation of the intensity of exciton-
magnon combination bands in the linear antiferromagnet TMMC
(ref. 9) and the linear ferromagnet RbFeCl$_3$ (ref. 11).

Directly Interacting Metal Atoms

Turning to single valence chains in which there is direct 'through space' interaction the first point to note is that in structures like the hexagonal perovskites, or even in the rutile structure as exemplified by VO_2, one may have a combination of 'through space' and 'through bond' interaction as a result of a configuration like

$$M \cdots\cdots M$$
$$\begin{array}{c} X \\ X \end{array}$$

The importance of the balance between these contributions can be seen from the fact that substituting either F for Cl, or $N(CH_3)_4$ for Cs, in $CsNiCl_3$ changes the net intrachain exchange interaction from antiferromagnetic to ferromagnetic (12).   However, steric requirements dictate that the only examples in which there is exclusively a direct interaction are formed by stacking planar complexes one on top of the other:

A tenet of ligand field theory well known to inorganic chemists (13) states that square planar coordination is characteristic of the low spin d^8 configuration so it is there that we must look for single valence unbridged metal chains.   These can be of two kinds, consisting either of identical repeating units or alternating anions and cations.   Chains consisting of identical repeating units can easily be formed by neutral molecules which, since the metal atom is in a +2 oxidation state, must contain anionic ligands or mixed coordination of anions and neutral molecules.   Conjugated organic chelating agents which form anions are such molecules as dimethylglyoximate (dmg) and acetylacetonate (acac).   Ni, Pd and Pt complexes of the former were among the earliest to be investigated for ID interaction properties (14) because of the very obvious colour difference between $Ni(dmg)_2$ as a solid (dark red) and in solution (yellow).   The Ir^I complex $Ir(acac)(CO)_2$ also has quite a different colour in solution and the solid state (15).   Another neat illustration of the effect of metal—metal interaction on visible colours is provided by the two crystalline modifications of $Pt(bipy)_2Cl_2$.   One is red, and contains chains of Pt atoms separated by 3.40 Å, the other, which contains completely isolated $Pt(bipy)_2Cl_2$ planar molecules, with no Pt—Pt contact, is yellow (16).

The most famous examples of metal stacks containing alternating anions and cations are the series of Magnus' Green Salts $(PtA_4)(PtX_4)$

mentioned at the beginning of the chapter (Figure 5).   The extent
to which their colours differ from the sum of their constituent ions
is a function both of the anion X (whether it is Cl or Br) and the
bulkiness of the amine ligands A in the cation, since these two
factors bear directly on the Pt–Pt spacing.   For instance in MGS
itself this spacing is 3.24 Å whilst in the ethylamine analogue,
which is actually pink like K_2PtCl_4 or $PtCl_4^{2-}$ in solution, it has
increased to 3.40 Å.   By looking at the polarised crystal spectra
of several members of the series, including K_2PtCl_4 itself, we
showed many years ago (17) that all the electronic transitions in
these compounds, up to at least 35 000 cm^{-1} were in fact to ligand
field excited states, though their energies had become increasingly
red-shifted with decreasing Pt–Pt spacing.   Their intensities also
become greater as the Pt–Pt spacing diminishes, the component of
electric dipole strength polarised parallel to the chains being
particularly enhanced.   We believe that this is the result of a
very substantial red shift by an intense, electric-dipole-allowed
band, also polarised parallel to the chains.   In K_2PtCl_4 such a
band is found at 42 500 cm^{-1} (18) but in MGS it is at 34 500 cm^{-1}
(19).   Being parity-forbidden the visible transitions gain their
intensity by vibronic mixing with this allowed band.   What we are
called on to explain therefore is the behaviour of the ultraviolet
band.

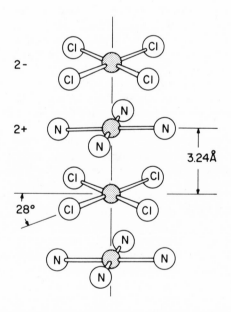

Figure 5.   The Pt atom chains in Magnus' Green Salt.

The starting point for any description of the electronic states in single valence metal chains containing directly interacting square planar d^8 metal ions must be the experimental fact that, like the chains with anion bridges, their ground states are insulating and their band gaps relatively wide. There has been quite a lot of discussion over the years (20, 21, 22) about the precise level of the room temperature conductivity and activation energy of MGS, for example, and values ranging from 10^{-8} to 10^{-2} ohm^{-1}cm^{-1} and 0.6 to 0.2 eV have been quoted. Mehran and Scott's e.p.r. measurements (23) however strongly suggest that such conductivity as there is is determined by trace impurities of Pt^{III}, probably originating from Pt^{IV} complexes introduced during the crystallization. The effect of the Pt^{III} is to give a self-trapped $5d_{z^2}$ hole lying about 0.6 eV above the top of the filled $5d_{z^2}$ valence band. Support is given to this idea by the fact that the conductivity can be further increased by deliberate doping with $K_2Pt(CN)_4Br_2$ (24). Thus the intrinsic band gap is probably much higher, and estimates of up to 4.5 eV have been given (25).

A similar state of affairs obtains in the compounds containing stacks of identical single valence units, whether they be relatively small complex ions like $Pt(CN)_4^{2-}$ or complexes of conjugated ligands such as α-diimines. Thus the Rh^I and Ir^I dicarbonylacetylacetonates have room temperature conductivities of 10^{-11} and 10^{-5} ohm^{-1}cm^{-1} and activation energies of 0.44 and 0.27 eV (26) while Ni, Pd and Pt dimethylglyoximates (22) fall in the same range.

Despite the insulating nature of the ground states very substantial intermolecular interactions are found in the excited states in all these compounds. Most spectacular are probably the tetracyano-complexes in which, merely by changing the cation from a group IA to a IIA element, or even simply the degree of hydration, one can shift the energy of the first excited state in the crystal by anything up to 12 000 cm^{-1}. For instance the $Pt(CN)_4^{2-}$ ion in dilute solution has no absorption bands below 35 000 cm^{-1}, yet all tetracyanoplatinite salts have intense visible colours! The absorption bands in question (e.g. Figure 6) are extremely broad, somewhat asymmetric in shape, very intense (oscillator strengths of the order of 1) and appear only when the incident electric vector is parallel to the metal atom chains. Similar bands are found in the Ir^I and Rh^I compounds mentioned above (15), and also provide the explanation for the red colour of Ni dimethylglyoximate (19).

The two zero-order descriptions of the electronic states of a chain of identical units are those of the simple one-electron band approximation and the excitonic insulator. In the former we would treat the absorption bands we are discussing as transitions from a filled valence band, composed predominantly of $5d_{z^2}$ metal orbitals, to an empty conduction band probably consisting of $6p_z$.

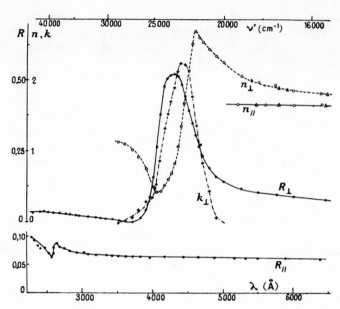

Figure 6.   Reflectivity R, refractive index n and absorption constant k of BaPt(CN)$_4$5H$_2$O (ref. 27).

The shape of the absorption band would then be a convolution of the densities of states of the two energy bands, weighted for transition probabilities.   However, if this description were appropriate we would anticipate that electrons excited to the conduction band would be substantially mobile, and hence that the compounds should be good photoconductors.   In fact, although very weak photoconductivity can be detected, e.g. in barium tetracyanoplatinite (28) its magnitude is no greater than is normally found in molecular crystals containing units with de-localised electrons like the aromatic hydrocarbons.   Thus it appears that we are still firmly within the regime of the Hubbard model in which U dominates t, and the best starting point for a description of the excited states is one based on neglect of intermolecular electron exchange.   The ground state wavefunction of such a crystal is just

$$\Psi_G = \varphi_1 \varphi_2 \cdots \varphi_N \tag{1}$$

with an energy

$$E_G = \sum_{m=1}^{M} w_m + \sum_m \sum_n \langle \varphi_m \varphi_n | v_{mn} | \varphi_m \varphi_n \rangle \tag{2}$$

the φ being molecular wavefunctions, e.g. of $Pt(CN)_4^{2-}$ while V takes account of the interaction between the molecules. An excited state wavefunction of the chain would be

$$\phi_{ip}^r = \varphi_{11}\varphi_{12}\cdots\varphi_{ip}^r\cdots\varphi_{hN} \qquad (3)$$

in which the ith molecule in the pth unit cell has been excited, and $h = M/N$ = the number of molecules per unit cell. The functions ϕ_{ip}^r form an M-fold degenerate set, their energies being simply $(M-1)W + W^r$. However, we must allow for the translational symmetry of the chain by writing

$$\phi_i^r(\underline{k}) = N^{-1/2}\sum_p \exp(i\underline{k}\underline{r}_{ip})\,\phi_{ip}^r \qquad (4)$$

This function permits the excitation to move along the chain, but no transfer of charge takes place. It describes a 'neutral' Frenkel exciton (29). For $\underline{k} = 0$ (to which alone transitions are allowed by light absorption) there is a first-order correction to the transition energy W^r as a result of interaction between the excited and ground state molecules in the chain. This is given by the eigenvalues of a secular determinant

$$\left| \langle\phi_{ip}^r|V|\phi_{ip}^r\rangle - \left\{\langle\Phi_G|V|\Phi_G\rangle + \Delta E\right\}\delta_{ii}\delta_{pp} \right| = 0 \qquad (5)$$

The diagonal elements of the determinant are $D_r + I_{ii}^r - \Delta E$ and the off-diagonal simply I_{ij}^r. D_r measures the difference in electrostatic energy between the ground and excited states, I_{ii}^r the coupling energy between excited molecules on equivalent sites and I_{ij}^r that between excited molecules on inequivalent sites.

When the above theory is applied to molecular crystals such as aromatic hydrocarbons (30) it is usual to expand the interaction operator as a multipole, so that the most important term is the dipole-dipole one:

$$V_{hm} = \left(e^2/R_{hm}^3\right)\sum_k\sum_l\left(x_k x_l + y_k y_l - 2z_k z_l\right) \qquad (6)$$

where z_k ...etc. are measured from the molecules' centres. When the coordinates are transformed to molecular coordinates one finds that the coupling energy terms above become

$$I_{ip,jq}^r = \left(\underline{m}^r\right)^2 R_{ip,jq}^{-3}\left[(x,r_{ip})(x,r_{jq}) + (y,r_{ip})(y,r_{jq}) - 2(z,r_{ip})(z,r_{jq})\right] \qquad (7)$$

where $\underline{M}^r$ is the free molecule transition moment $\langle\varphi|e\underline{r}|\varphi^r\rangle$ and (x,r_{ip}) is the direction cosine between vectors defining the transition moment and the axis of the coordinate system. Eq. (7) is particularly simple to apply to metal chain compounds like the tetracyanoplatinites because, if the planes of the square-planar units are all parallel, and make an angle of $90°$ to the chain axis,

the transition moment vectors will also be parallel.   In the case
that the transition moment vectors are perpendicular to the
molecular planes, the $\underline{k}$ = 0 neutral Frenkel exciton state will be
shifted from the energy of the transition in an isolated molecule
by $D_r + I^r_{ii}$.   Assuming for simplicity that the excitons only
propagate along the stacks, and that coupling between stacks can
be ignored, I^r_{ii} is given by the sum of a series

$$I^r_{ii} = -4(\underline{m}^r)^2\left[(R_{12})^{-3} + (2R_{12})^{-3} + (3R_{12})^{-3} + \cdots\right] \tag{8}$$

Clearly the first term is much the most important.   Its magnitude
is such that the $\underline{k}$ = 0 exciton state originating from a chain of
transition dipoles 1 Å long, separated by 3.25 Å (a reasonable
average spacing for a tetracyanoplatinite salt) would be red-
shifted no less than 24 000 cm^{-1} from the energy of the parent
transition in the isolated molecule.   In a similar fashion the
$\underline{k}$ = 0 exciton arising from a 1 Å transition dipole aligned parallel
to the molecular plane would be blue-shifted by 12 000 cm^{-1}.
Conclusive evidence that the visible absorption bands in the single
valence tetracyanoplatinites do arise from such dipolar shifts
comes from the way in which their energies vary in a series of
salts with varying intermolecular spacings R_{12}.   Figure 7 shows
that a plot of the observed transition energies against R^{-3}_{12} is
indeed a good straight line.   Furthermore, changing the intermolecular spacing in $MgPt(CN)_4 7H_2O$ by compressing shifts the band

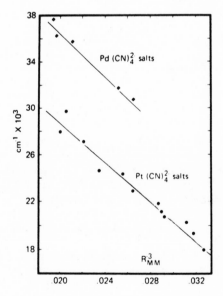

Figure 7.   Energy of the intense transition polarised parallel to
the chains in pallado- and platinocyanide crystals and the M-M
spacing (41).

further to the red along the same line (31).   The limiting value
of the parent transition energy obtained by extrapolating the line
back to R = ∞ is 44 800 cm^{-1} for Pt(CN)$_4^{2-}$, which corresponds nicely
with a $^1A_{2u}$ state identified in the solution spectrum by magnetic
circular dichroism spectroscopy (32).

     Electric dipole allowed intramolecular transitions which would
be polarised perpendicular to the molecular plane are (1) 5d(z^2) →
6p(z), localised on the Pt atom and (2) 5d(z^2) → CN(π*), a metal-
to-ligand charge transfer.   The probability is that at least in
Pt chains, the former is the most important since it is a well known
feature of the electronic structures of the elements towards the
end of the third transition series that the separation between 5d
and 6p becomes small.   Possible allowed transitions polarised
within the planes of the ions are ligand-to-metal charge transfer,
in the present instance ligand (pπ) → 5d(x^2-y^2).   In complexes of
π-acceptor ligands such as CN$^-$ or α-diimines, ligand-to-metal charge
transfer states nearly always occur higher in energy than metal-to-
ligand, but when the ligand is a π-donor like Cl$^-$, transfer of p
electrons to the metal takes place at lower energy.   Instructive
examples are the two ions PtCl$_4^{2-}$ and PdCl$_4^{2-}$.   In the former the
lowest energy allowed transition in the solution spectrum is at
46 300 cm^{-1} and in the latter at 35 500 cm^{-1}.   However, polarised
reflectivity studies of the crystals K$_2$PtCl$_4$ and K$_2$PdCl$_4$, both
containing chains of anions, show that in the Pt salt the first band
is red-shifted to 43 800 cm^{-1} and polarised parallel to the chains
while in the latter there is a small blue-shift and the band is
polarised in the planes of the molecular anions (33).

     So far we have described excited states of single valence metal
atom chains on the assumption that the excited electrons remain
bound exclusively to the atom from which they originate.   The
spectacular shifts in the energies of these states, for example in
square planar Pt^{II}, Ni^{II} and Ir^I salts is simply the result of
coupling between transition dipoles, localised on the individual
centres.   However, if we imagine the overlap between the molecules
in our chains becomes appreciable there is also the possibility that
electrons may be transferred from filled orbitals on one molecule
to empty ones on the next.   Thus instead of neutral Frenkel excitons
we would have ionic ones.   Imagine a transfer from a molecule j in
the chain to another labelled j+β:

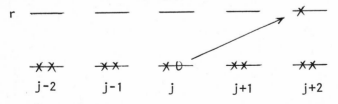

The localized excitation function analogous to eq. (3) is

$$\phi_{j+\beta}^{r} = \varphi_{1}\varphi_{2}\cdots\varphi_{j}^{r(+)}\cdots\varphi_{j+\beta}^{r(-)}\cdots\varphi_{N} \tag{9}$$

and the exciton wavefunction is

$$\phi_{j+\beta}^{r}(\underline{k}) = N^{-\frac{1}{2}}\sum exp\left(i\underline{k}\underline{r}_{j+\beta}\right)\phi_{j+\beta}^{r} \tag{10}$$

Since transfer from j to j+β is completely equivalent to transfer to j−β the ionic exciton wavefunction is a linear combination of the two. For $\underline{k}=0$:

$$\Psi_{\pm\beta}^{r}(0) = 2^{-\frac{1}{2}}\left[\phi_{+\beta}^{r}(0) \pm \phi_{-\beta}^{r}(0)\right] \tag{11}$$

An important feature of $\psi_{\pm\beta}^{r}(0)$ is that it may have the same symmetry as the neutral exciton $\phi_{i}^{r}(0)$ of eq. (4), and so may mix with it. Thus in principle we might have some ionic character, implying charge separation, built into even the first excited state of these crystals.   The existence of ionized states in molecular crystals was first discussed in a comprehensive way by Lyons (34) in connection with the photoconductivity of aromatic hydrocarbons, and applied to one-dimensional crystals by Merrifield (35).   Only in one case, however, do such states appear to have been invoked in order to explain features in the electronic spectra of inorganic chain compounds.   Martin (36) found a band in the polarised crystal spectrum of PtenCl$_{2}$ (en = ethylenediamine) whose temperature dependence shows it to be electric-dipole allowed, but which could not easily be accounted for as an intra-molecular charge transfer transition.   Perhaps an ionized exciton state arises here because the lowest energy intramolecular charge transfer state is of ligand-to-metal type, and consequently shifts to higher energy in the crystal, exposing the ionic exciton to view.   A summary of the properties of single valence compounds with bridging groups and direct interactions appears in Table 1.

4.   MIXED VALENCE METAL CHAINS

Ionic exciton states have energies related to the ionization potential of the donor and the electron affinity of the acceptor, together with the coulomb attraction of the electron-hole pair. When the molecules or ions comprising our one-dimensional chain all have the same ground state charge quite a lot of energy is needed to form an ionic state:

$$M\ M\ M\ M\ \ldots \rightarrow M\ M^{+}\ M\ M^{-}\ M\ \ldots$$

as pointed out already.   On the other hand it costs much less energy simply to move a charge fluctuation along a chain of ions:

$$M \ M^+ \ M \ M \ M \ \ldots \qquad M \ M \ M \ M^+ \ M \ \ldots$$

In fact, only a Franck-Condon barrier is now involved, a function
of the elastic deformation energy of the crystal around the charge
fluctuation.   Metal atom chains which already have charge fluctuations built into them are found in so-called 'mixed valence'
compounds.   (It should be pointed out that the term mixed valence
as used here by no means represents the same phenomenon as when it
is used by physicists to describe such compounds as SmS.   In the
latter, we have on average the same number of electrons on each
atom, and it is really the electron configuration which fluctuates.)
Mixed valency in our sense represents a situation in which different
ions or complexes in the lattice have different oxidation states,
and hence numbers of electrons.   Some 40 elements in the Periodic
Table form compounds of this kind (37), though of course only a
small proportion of them contain metal atom chains.   As with the
single valence chains we can divide them into two groups depending
on whether the atoms of differing oxidation state are connected by
bridging anions or molecules, or whether they are in direct contact.

Bridged Chains

There is a fundamental demarcation among mixed valence compounds between those in which, on a true average, all the metal ion
sites are crystallographically equivalent, and those in which, to
a first approximation, different sites may be ascribed to ions

TABLE 1.   PHYSICAL PROPERTIES OF SINGLE VALENCE
METAL ATOM CHAINS

(1) Bridged

   (a) Local $(d \rightarrow d)$ transitions of metal ions only
        (No low-lying collective transitions)

   (b) Magnetic exchange in one dimension
        (Magnons couple to crystal field Frenkel excitons)

   (c) Insulators

(2) Direct metal-metal contact

   (a) Crystal field transitions (if any) strongly perturbed
        by charge transfer states in the ultraviolet.
        Electric dipole allowed transitions have large shifts
        from isolated molecule energies (dipolar interactions).

   (b) Diamagnetic

   (c) Insulators, becoming conducting under pressure.

carrying the differing oxidation states.   In the former, the
lattice deformation which we expect to accompany the charge fluctuation from site to site is small or zero, whilst in the latter
it is sufficiently large to trap the fluctuation on to a distinct
set of sites, forming a superlattice.   In our classification of
mixed valence compounds some years ago (37) we called these
respectively classes III and II, class I being reserved for
compounds which had such grossly different stereochemistries around
the two sites of differing oxidation state that any interaction
between them could be safely ignored.

With this nomenclature one can then state that all the
examples of mixed valence metal chain compounds known at the
present time which have either anions like halide or neutral
molecules like pyrazine bridging between the metal centres belong
structurally to our class II, and hence behave as semiconductors.
Although there have been a number of very bold attempts by
synthetic inorganic chemists (38, 39) to construct mixed valence
chains containing coordination complexes or organometallics as
units, for example the compounds (V) and (VI) shown below, they
have not yet succeeded in joining together enough monomers to give
chains

(V)

(= acetate

(VI)

on which to test conductivity on a macroscopic scale.   Nevertheless, it is already clear from their optical spectra that chains
like (V) and (VI) have gap energies of at least 1-2 eV and so,
even if extended, would probably prove semiconducting.   By a long
way the most numerous and widely studied mixed valence chains with
bridging groups are those containing alternating square planar and
octahedral units with effective electron configurations d^8 and d^6
respectively.   Work on the conductivities of many of these compounds is summarized by Miller (40).   A typical example would be
Wolfram's Red Salt, whose structure is shown in Figure 8.   We
briefly discussed the optical properties of these compounds (41),
showing that an electric dipole allowed transition could carry an
electron from the filled $d(z^2)$ orbital of the square planar d^8 ion
into the empty $d(z^2)$ on the octahedrally coordinated d^6, thus
generating the intense band, totally polarised parallel to the

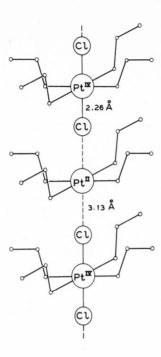

Figure 8.   The structure of Wolfram's Red Salt.

chain axis, which covers most of the visible and gives the
compounds their very deep colours.   Elegant confirmation of the
essential correctness of this assignment has come recently from
resonance Raman spectra, excited within the visible band (42).
Of all the vibrational modes, the only one whose intensity is
resonance enhanced is the axially symmetric metal–halogen stretch
which, if it had a big enough amplitude, would render the two metal
ion sites equivalent in the electronic excited state.   So it
appears that the charge fluctuations built into this type of chain
by the presence of two distinct oxidation states is big enough to
lower the energies of what amount to first–nearest neighbour ionic
exciton states ($\Psi_{\pm 1}^{r}(0)$ of eq. (11)) so that they are now the lowest
excited states in the crystal, well below the neutral excitons
formed from ligand field or intramolecular charge transfer.
However, since the metal ion sites are about 5.5 Å apart in the
chains one might wonder how transitions to the ionic exciton states
come to have such high intensities.   In fact a similar situation
is found in other class II mixed valence compounds containing
bridging ligands, the most famous being Prussian Blue,
$Fe^{III} Fe^{II}(CN)_{6}$ $_3 14H_2O$, where Fe^{II} and Fe^{III} ions, 5.1 Å apart,
are separated by CN^- groups.   The answer, as we showed (43) in a
general theoretical treatment of valence delocalization in such
systems, is that 'local' charge transfer states involving donation

from the low valent ion to the lowest empty orbital of the bridge,
or from the highest filled orbital of the bridge to the higher
valent ion, mix with the direct metal-to-metal transition.   If
the energies of the local charge transfer states are known, for
example by simply measuring the spectra of the individual molecular units in isolation, one may then calculate the valence delocalization coefficient, i.e. the degree of mixing between the
two zero-order distributions of oxidation state in the ground
state, and hence the intensity of the optical transition.   This
we have done for Prussian blue, yielding good agreement between
observed and calculated intensities when the ground state valence
delocalization coefficient has the order of magnitude 0.1.
Another way of saying this is that the ground state of Prussian
blue is 99% low spin Fe^{II} attached to the carbon ends of the CN^-
groups, and high spin Fe^{III} at the nitrogen ends, mixed with 1%
of low spin Fe^{III} at the carbon ends and high spin Fe^{II} at the
nitrogen ends.   A good way of verifying the extent of valence
delocalization in Prussian blue is by polarised neutron diffraction
since it happens to order at low temperature (5.5 K) as a ferromagnet (44).   In this way we have found (45) a value of about
1.5% for the ground state delocalization.

     It is much harder to estimate the ground state delocalization
in such compounds as Wolfram's Red Salt, but it is almost certainly
no larger than in Prussian blue.   This agrees with their being
diamagnetic insulators.   Perhaps the most interesting feature of
their electronic structures though is that at first sight they
appear to be exceptions to the Mott-Hubbard view of the role of
intra-site electron correlation in bringing about electron localization and insulating behaviour.   This comes about because in a
compound like $Pt(NH_3)_2Cl_3$ or Wolfram's Red Salt there is actually
an integral number of valence electrons per metal atom, namely
seven.   However, the compound chooses to distribute these in
doubly filled z^2 orbitals on half the metals, alternating with
empty z^2 on the others.   Hubbard's model (1), on the other hand,
describes a crystal which, for a half-filled band, behaves as a
metal in the limit of zero correlation energy and as an insulator
at zero bandwidth, with each site occupied by one electron.   That
the d^6, d^8 chain compounds present such a dilemma was pointed out
by Ionov, Makarov and their colleagues (46), who suggested that a
possible reason for the failure of the simple Hubbard model was its
neglect of inter-site correlation.   A further one is the electron-
phonon interaction, resulting from the site preferences (octahedral
for low spin d^6 and square planar for low spin d^8) familiar to
coordination chemists.

Directly Interacting Metal Ions

As Clark's resonance Raman spectra, on the d^6, d^8 bridged
mixed valence chains showed so beautifully (42), it is the very
existence of the bridging group between each pair of metal ions
which leads to the electron localization in these compounds,
because the single vibrational degree of freedom represented by
its motion along the chain axis produces the required difference
between the ligand fields at alternate metal ion sites.   One
could say that this electron-phonon coupling, has combined with
the integral number of electrons per metal ion to induce the
simplest possible kind of charge density wave, with a periodicity
of two lattice spacings.   To unpin such a charge density fluctuation from the lattice then, two modifications to the chain are
needed.   First, one should remove the bridging ion, and hence the
mode largely responsible for the electron-phonon interaction.
This of course has the further consequence of increasing direct
overlap between the metal ions, and hence the bandwidth, measured
by the Hubbard t.   Second, one could attempt to ensure that if
there were to be charge density waves, they should be incommensurate with the lattice.   To achieve this one would need to have
on average a genuinely non-integral number of electrons per metal
ion site, i.e. the compound would belong to class III in our earlier
classification.   Both criteria are satisfied by the known mixed
valence chain compounds with directly interacting metal ions, of
which the most famous is KCP.   Some other examples of class III
mixed valence chains from the recent literature are listed in
Table 2, but for fuller details, Miller's review (40) should be
consulted.   Physical properties of KCP are described in much more
detail in other chapters and have been reviewed at an earlier NATO
Advanced Study Institute (56).   Suffice it to say that it is
indeed metallic, though the presence of charge density waves,
detected by neutron diffraction, opens a gap in the conduction band,
so that at low temperatures it behaves as a semiconductor.

Apart from other examples of Pt chains in which the repeating
unit is a square planar molecular complex ion, Table 2 also refers
to some continuous lattice compounds, some having structures
remarkably reminiscent of the class of A-15 β-tungsten compounds
to which belong the high temperature superconductors like Nb_3Sn.
Subhalides (54), like Gd_2Cl_3, containing chains of linked octahedra
of metal ions, will also form a fruitful area for searching out
one-dimensional conductors.

Finally, Table 3 gives a digest of the physical properties of
mixed valence chains comparable to the one in Table 1 for single
valence chains.

TABLE 2.   SOME EXAMPLES OF CLASS III MIXED VALENCE
METAL ATOM CHAINS

(1) Square planar d^{8-x}

	Repeat Unit	R(M–M)	$\delta_{RT}(ohm^{-1}cm^{-1})$	Ref.
$K_2Pt(CN)_4Br_{0.30}3H_2O$	$Pt(CN)_4$	2.88 Å	300 (11)	47
$K_{1.74}Pt(CN)_43H_2O$	$Pt(CN)_4$	2.96	–	48
$K_{1.64}Pt(C_2O_4)_2xH_2O$	$Pt(C_2O_4)_2$	2.81	42	49
$Ir(CO)_3Cl_{1+x}$	$Ir(CO)_3Cl$	2.85	0.2	50

(2) A15 (β-Tungsten)

$Hg_{2.86}AsF_6$	..–Hg–Hg–..	2.64	8,000	51
$Cd_{0.3}Pt_3O_4$	PtO_4	2.82	100	52
$Ni_{0.25}Pt_3O_4$	PtO_4	2.80	3,000	53

(3) Miscellaneous

Gd_2Cl_3	Gd_6		–	54
[Fe Fe](TCNQ)$_2$		–	10	55

TABLE 3.   PHYSICAL PROPERTIES OF MIXED VALENCE METAL CHAINS

(1) Bridged

   (a) Lowest excited state is M → M charge transfer.
        (Local excitations such as d → d transitions only
        weakly perturbed.)

   (b) Diamagnetic semiconductors (become conducting under
        pressure).

   (c) Mixed valence Robin–Day class II.

(2) Direct M–M interaction

   (a) Metallic optical properties (plasma edge in visible).

   (b) Metallic conductivity at room temperature.

   (c) Mixed valence Robin–Day class IIIB.

5.   SUMMARY

In this chapter we have tried to relate the properties of
metal atom chain compounds to their structures. Four classes of
structure are distinguished, each having a characteristic pattern
of physical behaviour. Single valence chains with bridging groups
between the metal ions are magnetic insulators, with near neighbour
exchange integrals in the range 5–100 cm^{-1}. Directly interacting
metal ions in single valence chains are formed exclusively from
molecular units with low spin d^8 electron configurations. They
too are insulators, or wide band gap semiconductors, although they
can be made conducting by extrinsic doping or high pressure. In
their electronic excited states, however, such chains evidence much
stronger intermolecular coupling than the bridged chains, dipolar
shifts of 10 000 cm^{-1} being quite common. Nevertheless, the lowest
excited states can still be described to a good approximation as
neutral Frenkel excitons. If some metal atoms in the chain have
different electron occupation numbers even in the ground state, the
energy required to transfer an electron from site to site is very
much reduced, and in such mixed valence chains the lowest excited
state may then become an ionic exciton. On the other hand,
bridging groups between metal sites in mixed valence chains introd-
uce vibrational modes which may trap the electrons on to alternate
sites in the chain through electron–phonon coupling. So finally,

to be at all confident of achieving a metallic ground state in a metal atom chain the chemical recipe must be to eliminate bridging groups between the units, and try to ensure that the average number of electrons per atom is not an integer or even, if possible, a simple fraction.

The simplest prototype one-dimensional metals like KCP have now been quite thoroughly investigated. The next stage lies with the synthetic inorganic chemist, to turn the general principles we have been describing into more new substances.

REFERENCES

1.   J. Hubbard, Proc. Roy. Soc. A276, 238 (1963).

2.   J.M. Honig and L.L. Van Zandt, Ann. Rev. Mat. Sci. 5, 225 (1975).

3.   V.S. Grunin, V.A. Ioffe and I.B. Patrina, Phys. Stat. Sol. B63, 629 (1974).

4.   J.B. Goodenough, Magnetism and the Chemical Bond, New York, Interscience, 1963.

5.   P. Day, ch. 17 in 'Extended Interactions between Metal Ions in Transition Metal Complexes' ed. L. Interrante, Amer. Chem. Soc. Symposium Series, No. 5, 1974.

6.   G.D. Stucky, ch. in Ref. 5.

7.   M.T. Hutchings, G. Shirane, R.J. Birgeneau and S.L. Holt, Phys. Rev. B5, 1999 (1972).

8.   M.T. Hutchings, P.Day, A.K. Gregson, D.H. Leech and B.D. Rainford, to be published.

9.   P. Day and L. Dubicki, J. Chem. Soc., Faraday Trans. II, 69, 363 (1973).

10. G.L. McPherson, T. Kistenmacher and G.D. Stucky, J. Chem. Phys. 57, 3780 (1972).

11. E.R. Krausz, S.M. Viney and P. Day, to be published.

12. D.W. Hone and P.W. Richards, Ann. Rev. Mat. Sci. 4, 337 (1974).

13. F.A. Cotton and G. Wilkinson, Advanced Inorganic Chemistry, New York, John Wiley & Sons.

14.   B.G. Anex and K. Krist, J. Amer. Chem. Soc. 89, 6114 (1967).

15.   T.A. Dessert, R.A. Palmer and S.M. Horner, ch. in Ref. 5.

16.   M. Textor and H.R. Oswald, Z. anorg. Chem. 407, 244 (1974).

17.   P. Day, A.F. Orchard, A.J. Thomson and R.J.P. Williams,
      J. Chem. Phys. 42, 1973; 43, 3763 (1965).

18.   B.G. Anex and N. Takeuchi, J. Amer. Chem. Soc. 96, 4411 (1974).

19.   B.G. Anex, M.E. Ross and M.W. Hedgcock, J. Chem. Phys. 46,
      1090 (1967).

20.   L. Atkinson, P. Day and R.J.P. Williams, Nature 218, 668
      (1968).

21.   L.V. Interrante, Chem. Commun. 302 (1972).

22.   P.S. Gomm, T.W. Thomas and A.E. Underhill, J. Chem. Soc. A
      2154 (1971).

23.   F. Mehran and B.A. Scott, Phys. Rev. Lett. 31, 99 (1973).

24.   B.A. Scott, F. Mehran, B.D. Silverman and M.A. Ratner,
      ch. 22 in Ref. 5.

25.   E. Fishman and L.V. Interrante, Inorg. Chem. 11, 1722 (1972).

26.   K. Monteith, L.F. Ballard, C.G. Pitt, B.K. Klein, L.M.
      Slifkin and J.P. Collman, Solid State Commun. 6, 301 (1968).

27.   C. Moncuit and H. Poulet, J. Phys. Radium 23, 353 (1962).

28.   J.N. Paine, Chemistry Part II thesis, Oxford, 1976
      (unpublished).

29.   D.P. Craig and S.H. Walmsley, Excitons in Molecular Crystals,
      New York, Benjamin, 1968.

30.   e.g. D.P. Craig and P.C. Hobbins, J. Chem. Soc. 539 (1955).

31.   Y. Hara, I. Shirotani, Y. Ohashi, K. Asaumi and S. Minomura,
      Bull. Chem. Soc. Japan 48, 403 (1975).

32.   S.B. Piepho, P.N. Schatz and A.J. McCaffery, J. Amer. Chem.
      Soc. 91, 5994 (1969).

33.   B.G. Anex, ch. 19 in Ref. 5.

34.   L.E. Lyons, J. Chem. Soc. 5001 (1957).

35.   R.E. Merrifield, J. Chem. Phys. 34, 1835 (1961).

36.   R.F. Kroening, L.D. Hunter, R.M. Rush, J.C. Claroty and
       D.S. Martin, J. Chem. Phys. 77, 3077 (1973).

37.   M.B. Robin and P. Day, Adv. Inorg. Chem. and Radiochem. 10,
       247 (1967).

38.   S.T. Wilson, J. Bauman, R.F. Bondurant, T.J. Meyer and D.J.
       Salmon, J. Amer. Chem. Soc. 97, 2285 (1975).

39.   D.O. Cowan, J. Park, C.V. Pittman, Y. Sasaki, T.K. Mukherjee
       and N.A. Diamond, J. Amer. Chem. Soc. 95, 7873 (1973).

40.   J.S. Miller and A.J. Epstein, Prog. Inorg. Chem. 20, 2
       (1976).

41.   P. Day, ch. in 'Low Dimensional Cooperative Phenomena', ed.
       H.J. Keller, NATO-ASI Series B7, New York, Plenum, 1975.

42.   R.J.H. Clark, M.L. Franks and W.R. Trumble, Chem. Phys.
       Letters (1976).

43.   B. Mayoh and P. Day, J. Chem. Soc. Dalton Trans. 846 (1973).

44.   H.J. Buser, A. Ludi, P. Fischer, T. Studach and B.W. Dale,
       Z. phys. Chem. 92, 354 (1974); B. Mayoh and P. Day, J.
       Chem. Soc. Dalton Trans. 1483 (1976).

45.   F. Herren, A. Ludi, H.U. Gudel, K. Ziebeck and P. Day, to
       be published.

46.   S.P. Ionov, G.V. Ionova, E.F. Makarov and A.Y. Aleksandrov,
       Phys. Stat. Sol. (b) 64, 79 (1974); S.P. Ionov, G.V. Ionova,
       V.S. Lubinov and E.F. Makarov, Phys. Stat. Sol. (b) 71, 11
       (1975).

47.   e.g. H.R. Zeller, Festkorperprobleme 13, 31 (1973).

48.   K.D. Keefer, D.M. Washecheck, N.P. Enright and J.M. Williams,
       J. Amer. Chem. Soc. 98, 233 (1976); A.H. Reis, S.W. Peterson,
       D.M. Washecheck and J.S. Miller, ibid, 98, 236 (1976).

49.   K. Krogmann, Angew. Chemie, Int. Ed., 8, 35 (1969).

50.   A.P. Ginsberg, R.L. Cohen, F. DiSalvo and K.W. West, J.
       Chem. Phys. 60, 2657 (1974).

51.   B.D. Cutforth, W.R. Datars, R.J. Gillespie and A. van
       Schyndel, ch. in 'Unusual Properties of Inorganic Complexes',
       ed. R.B. King, Amer. Chem. Soc. Adv. in Chem. No. 150 (1975).

52.   D. Cohen, J.A. Ibers and J.B. Wagner, Inorg. Chem. 13, 1377
       (1974).

53.   D. Cohen, J.A. Ibers and R.D. Shannon, Inorg. Chem. 11, 2311
       (1972).

54.   J.E. Mee and J.D. Corbett, Inorg. Chem. 4, 88 (1965); for
       an excellent review on metal rich phases see A. Simon,
       Chem. Unserer Zeit, 10, 1 (1976).

55.   U.T. Mueller-Westerhoff and P. Eilbracht, J. Amer. Chem. Soc.
       94, 9272 (1972); C. LeVanda, K. Bechgaard, D.O. Cowan, U.T.
       Mueller-Westerhoff, P. Eilbracht, G.A. Caudela and R.L.
       Collins, J. Amer. Chem. Soc. 98, 3181 (1976).

56.   Chapters by N.R. Zeller, R. Comes, H. Launois and K. Krogmann
       in 'Low Dimensional Cooperative Phenomena', ed. H.J. Keller,
       NATO-ASI Series B7, New York, Plenum, 1975.

THE STRUCTURE OF LINEAR CHAIN TRANSITION METAL

COMPOUNDS WITH 1-D METALLIC PROPERTIES

K.KROGMANN

Institut für Anorganische Chemie der Universität

D 7500 Karlsruhe 1, GFR

Our knowledge of "really" 1-D metallic compounds on the basis of planar complexes of transition metals has been increased during the last two years by structural work on systems already known, at least in principle. There are many interesting columnar compounds listed and discussed in a comprehensive recent review by MILLER and EPSTEIN (1) but those for which 1-D metallic behavior is proved or to be expected are formed by tetracyanoplatinates, dioxalatoplatinates or halocarbonyliridates. Thus, on the following pages I shall try to discuss new and advanced structural knowledge in this field. Readers who need more introductory information are refered to reviews (1,2,3).

I. Tetracyanoplatinates

1. $K_2Pt(CN)_4X_{0.3} \cdot 3 H_2O$ (KCP)   (X = Cl, Br)

During the last two years several teams have refined the old picture of the KCP structure (4) by modern x-ray and neutron techniques (5,6,7,8,9,10,11,12,13). This was, by the way, not the first time of parallel work on KCP. After our preliminary report of the structure and nature of this compound given at the VIII. ICCC in Vienna (14), structure determinations of KCP were carried out in Liége (15) and in Stuttgart (4). The former reported about a wrong compound $K_2 \sqrt{}Pt(CN)_5 \sqrt{} \cdot 3 H_2O$ in the correct space group, while the latter described the correct analytical composition in a more symmetrical but wrong space group.

All the authors of the above mentioned more recent papers finally agreed on some important conclusions :

a) K^+ ions are <u>not</u> disordered in two planes of the unit cell between Pt(CN)$_4$ layers, as was supposed before (4), but fixed in one plane, and the structure is acentric;

b) therefore, Pt-Pt distances must not necessarily be equal, yet they almost are so;

c) water molecules connect the Pt chains by hydrogen bonds to N or halogen atoms;

d) the defect occupancy of the halide site is confirmed, and no halide ordering is observed, even at low temperatures;

e) water molecules may fill up the unoccupied halide sites.

2. $\underline{/}C(NH_2)_3 \underline{\,}J_2 \underline{/}Pt(CN)_4 \underline{\,}J Br_{0.25} \cdot H_2O$ (Guanidinium salt)

It was prepared and studied by neutron diffraction by J.M.WILLIAMS et al. (16). A preliminary report gives the Pt-Pt-distance of 2,92 Å, and disorder for bromide.

3. Cation deficient   tetracyanoplatinates.

WILLIAMS et al. (17) determined the crystal structure of the compound $K_{1.75} \underline{/}Pt(CN)_4 \underline{\,}J \cdot$ 1.5 H_2O. Previous reports on salts of different composition or other triclinic cell constants (18,19) were shown to be the same. The most interesting feature in this structure is the deviation of the platinum atom chains from strict linearity by forming zig-zag chains. These are in accordance with lattice dimensions, leading to an identity period of four Pt complexes, and differ in this respect from the "aperiodic" superstructures reported earlier (20). There is still some disorder in the structure, but no halide is present.

Similar cation deficient salts are formed with other cations than K^+ (21,22).

4. "Alloy" systems (mixed crystals).

Contrary to 3-D metal systems, it is impossible in 1-D to prepare alloys by substitution of other central atoms in the complexes. Usually, the individual structures show not the least tendency to accept, e.g., Pd(CN)$_4$ units instead of Pt(CN)$_4$. Until now, it has also been unsuccessfully tried to change the degree of partial oxidation in any of these compounds. Yet physicists are interested to study the properties as a function of the FERMI level in one and the same structure.

The KCP structure is so well balanced in the co-ordination requirements of K^+, H_2O and halide, that little effect can be drawn from the exchange of Br^- by Cl^- : the partial oxidation increases from 0.30 to 0.32. But if one starts to substitute K^+ by Rb^+, several consequences occur. Firstly, the greater radius of Rb^+ favors strongly the presence of Cl^- instead of Br^- in the cell center. Secondly, the Pt-Pt distance increases while the degree

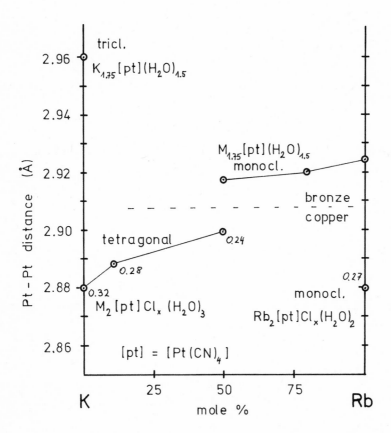

Fig. 1   Metallic distance and composition of partially
oxidized tetracyanoplatinates of K[+], Rb[+] and
mixed crystals. (Numbers at points are DPO.)

of partial oxidation (DPO) goes down. The tetragonal KCP type structure remains intact until about 50 % substitution, when the Pt-Pt distance is 2.90 Å and DPO is 0.24, which means a considerable change in comparison to pure KCP. Fig. 1 shows that distance as a function of mole % Rb in cation sites.

More phases exist in this "alloy" system: The above mentioned K^+ deficient salt, and a similar, but different monoclinic phase with Rb^+, which may also incorporate up to 50 % K^+. This time, the change in DPO is less than observable, and the change in distance is small. Furthermore, there is a monoclinic Rb phase similar to KCP in its Cl^- content, but lower DPO. We were unable, until now, to isolate mixed crystals of this phase, but this may depend on finding the proper crystallizing conditions. Lattice constants and compositions of several compounds are given in Table I.

Table I. Composition and cell constants of some MCP structures

Compound	cell constants (Å)		Pt-Pt (Å)	ref
$K_2Pt(CN)_4Cl_{0.32} \cdot 2 H_2O$	a =	9.88		(4)
	=	b		
	c =	5.76	2.88	
$Rb_2Pt(CN)_4Cl_{0.27} \cdot 3 H_2O$	a =	27.04		(22)
	b =	5.76	2.88	
(from powder data)	c =	17.93		
	β =	107.4°		
$K_{1.75}Pt(CN)_4 \cdot 2 H_2O$	a =	10.36		(17)
	b =	11.83	2.96	
	c =	9.30		
	α =	102.4°		
	β =	106.4°		
	γ =	114.7°		
$Rb_{1.75}Pt(CN)_4 \cdot 1.5 H_2O$	a =	10.52		(22)
	b =	33.17		
	c =	11.70	2.93	
	β =	114.2°		
$Rb_{0.8}K_{0.95}Pt(CN)_4 \cdot 1.5 H_2O$	a =	10.42		(22)
	b =	32.95	2.92	
	c =	11.67		

The similarity of the a/a, b/c, and γ/β values between the K^+(def) and Rb^+ (def) structures hints at similar arrangements in the lattice, though the Rb phase is monoclinic.

It should be noted that the tetragonal phase favors K^+ : a solution containing Rb^+ : K^+ in a molar ratio of 2:1 gives crystals with Rb^+ : K^+ as 2:3. The reverse is true, but not as striking, for the monoclinic (def) phase. All compounds having Pt-Pt distances greater than 2.9 Å exhibit a change in colour from copper to bronze, which must be connected with the frequency of the plasma edge of the 1-D electrons.

II. Dioxalatoplatinates (DOP)

Crystal structure determinations of oxidized dioxalato-platinates are rare. These compounds crystallize rather badly if compared to tetracyanoplatinates. Most crystals are extremely thin, twinned, exhibit defects and distortions, are therefore unusable for quantitative x-ray work. The structure of a Mg^{2+} deficient DOP, $Mg_{0.82}$ Pt $(C_2O_4)_2 \cdot 5.3 H_2O$ was determined (23), in which the cations are supposed to occupy statistically 41 % of suitable lattice sites. This might be doubted after the refinements of the KCP structure, where K^+ got well defined places. However, a re-examination of the old data did not favour any more ordered arrangement of Mg^{2+} ions.

Although (or because) there are so many phases and modifications of these compounds, we are still lacking exact information about possible stacking modes of complexes as well as distribution of cations and water molecules. More material of that kind is needed, before one can try to understand the reported "aperiodic" superstructures (20).

III. Halocarbonyliridates

A few years ago, we reported our results on 1-D metallic phases in which the complex units are $[\,Ir(CO)_2Cl_2\,]$, which are also partially oxidized and are formed with a broad variety of cations (20). Those analyzed are of the cation deficient type, and while the DPO varies considerably from one cation to another, the Ir-Ir distance remains almost constant at 2.86 Å. These values were derived by indexing the GUINIER powder pattern, yielding the Ir subcell with one constant being the Ir-Ir distance (24). In the meantime, single crystal work confirmed some of these unit cell data, which are summarized in Table II. All these compounds are badly crystallizing, and some are hydrolyzed by moist air. By using special crystal growth techniques, we were recently able to prepare small, but undisturbed crystals of $(H_3O)^+$ and K^+ phases, which are now under investigation.

Table II

Unit cell constants of 1-D metallic halocarbonyliridates

Compound	cell constants ($\overset{o}{A}$)		Ir-Ir distance ($\overset{o}{A}$)
$(H_3O)_{0.38}Ir(CO)_2Cl_2 \cdot 2\ H_2O$ (single crystal)	a =	16.83	
	b =	5.72	2.86
	c =	12.24	
	β =	91.7^o	
$K_{0.58}\ Ir(CO)_2Cl_2$ (single crystal)	a =	9.32	
	b =	5.73	2.86
	c =	12.56	
	β =	97.6^o	
$Cs_{0.48}\ Ir(CO)_2Cl_2$ (powder)	a =	9.73	
	b =	5.72	2.86
	c =	12.60	
	β =	97.6^o	
$(Nme_4)_{0.55}\ Ir(CO)_2Cl_2$ (powder)	a =	12.13	
	= b		
	c =	5.72	2.86

REFERENCES

1. J.S.MILLER, and A.J.EPSTEIN, Progr.Inorg.Chem. $\underline{20}$, 1 (1976).
2. H.J.KELLER, ed. "low Dimensional Cooperative Phenomena",
   (Plenum, New York, 1975) NATO-ASI Series B, Vol. 7.
3. K.KROGMANN, Angew.Chem.Int.Ed. (engl.) $\underline{8}$, 35 (1969).
4. K.KROGMANN, and H.D.HAUSEN, Z.Anorg.Allg.Chem. $\underline{358}$, 67 (1968).
5. H.J.DEISEROTH, and H.SCHULZ, Phys.Rev.Lett. $\underline{33}$, 963 (1974).
6. J.M.WILLIAMS, J.L.PETERSEN, H.M.GERDES, and S.W.PETERSON,
   Phys.Rev.Lett. $\underline{33}$, 1079 (1974).
7. G.HEGER, B.RENKER, H.J.DEISEROTH, and H.SCHULZ,
   Mat.Res.Bull. $\underline{10}$, 217 (1975).
8. H.J.DEISEROTH, and H.SCHULZ, Mat.Res.Bull. $\underline{10}$, 225 (1975).
9. C.PETERS, and C.F. EAGEN, Phys.Rev.Lett. $\underline{34}$, 1132 (1975).
10. J.M.WILLIAMS, M.IWATA, F.K.ROSS, J.L.PETERSEN, and
    S.W.PETERSON, Mat.Res.Bull. $\underline{10}$, 411 (1975).
11. J.M.WILLIAMS, F.K.ROSS, M.IWATA, J.L. PETERSEN,S.W.PETERSON,
    S.C.LIN, and K.KEEFER, Solid State Commun. $\underline{17}$, 45 (1975).
12. G.HEGER, B.RENKER, H.J.DEISEROTH, and H.SCHULZ,
    Solid State Commun. $\underline{18}$, 518 (1976).
13. J.M.WILLIAMS, M.IWATA, S.W.PETERSON, K.A.LESLIE, and
    H.J.GUGGENHEIM, Phys.Rev.Lett. $\underline{34}$, 26 (1975).
14. K.KROGMANN, P.DODEL, and H.D.HAUSEN,
    Proceed. 8.Intern.Conf.Coord.Chem. 157 (1964).
15. A.PICCININ, and J.TOUSSAINT, Bull.Soc.Roy. Sci. Liége,
    $\underline{36}$, 122 (1967).
16. J.M.WILLIAMS, T.F.CORNISH, D.M.WASHECHECK, and P.L.JOHNSON,
    Abstracts Am.Cryst.Ass.Summer Meeting (1976).
17. K.D.KEEFER, D.M.WASHECHECK, N.P. ENRIGHT, and J.M.WILLIAMS,
    J.Amer.Chem.Soc. $\underline{98}$, 233 (1976).
18 M.J.MINOT,J.H.PERLSTEIN, and T.J.KISTENMACHER,
    Solid State Commun. $\underline{13}$, 1319 (1973).
19. K.KROGMANN, and H.D.HAUSEN, Z.Naturforsch. B, $\underline{23}$, 1111 (1968).
20. K.KROGMANN in H.J.KELLER, ref. (2), p.284.
21. J.M.WILLIAMS, private communication.
22. K.KROGMANN,H.HECK, and H.WICKENHÄUSER (to be published).
23. K.KROGMANN, Z.Anorg.Allg.Chem. $\underline{358}$, 97 (1968).
24. H.J. ZIELKE, Dissertation Stuttgart 1973.

THE CHEMISTRY OF ANISOTROPIC ORGANIC MATERIALS

F. Wudl

Bell Laboratories

Murray Hill, New Jersey 07974

INTRODUCTION

Our research over the past two years has been guided by two approaches: 1) systematic testing of various parameters suggested to be important by current theories and 2) synthesis of new compounds (ignoring the theorists) with the hope of obtaining a solid having new, exotic properties. Naturally, this presentation will be divided into two major parts along the above guidelines.

Rather than present a detailed analysis of one particular molecule or compound, I will show a spectrum of materials and just mention some of their properties which are relevant to the theme of this NATO school.

While in the oral presentation it was easy to connect the various subsections of each talk, in the written form, because of lack of space, the style is more staccato. I am aware of it and apologies to the reader.

A. Salts of Tetrathiafulvalene

1. <u>Simple Anions</u> - Since this section is limited to compounds of only one molecule (TTF^+), it will deal with attempts to find a systematic change in a physical property; e.g. conductivity, or magnetic susceptibility, as a function of the counterion.

As was stated before (1,2), there are several constraints imposed by nature on the chemist who proposes to systematically

233

study transport and magnetic properties of series of anisotropic solids; briefly, the members of a series to be studied should be isomorphous, isostoichiometric and devoid of the practically uncurable problem of crystal defects. We have also mentioned in the past (1,2) the types of anions which could, in principle, be incorporated into a TTF salt lattice. In that classification, we omitted other inorganic anions (such as SO_4^{2-}, HSO_4^-, HPO_4^{2-}, $H_2PO_4^-$, NO_3^-, etc.) which might produce highly conducting TTF salts (e.g. HSO_4^-, $SO_4^=$)(3a). In this section we will briefly mention one of the major conclusions we were able to draw from a systematic study of counterion variation.

The salts $TTF(SCN)_x$, $TTF(SeCN)_x$, and $TTFI_y$ (x = 0.56-0.58; y = 0.71) are very nearly isomorphous (2) ($TTFSeCN_x$, $TTFSCN_x$, tetragonal) and consist of eclipsed TTF stacks where the solid state magnetic and transport properties are due to the partially filled TTF stacks only. In that sense they are simpler than TTF TCNQ where there are two types of stacks, each of which exhibits a solid state transition(s) at its own characteristic temperature.

In collaboration with F. J. DiSalvo and G. A. Thomas we were able to demonstrate a direct relation between the transition temperature (T_0) of the magnetic susceptibility (cf. Fig. 1) and the interplanar distance between TTF molecules within a stack. The latter varies as a function of counterion, but not monotonically with size (SCN > SeCN > I). This is the first case in which a correlation of solid state molecular parameters with a particular property has been demonstrated. Other systematic studies of this nature should prove useful in determining quantitatively the effect of molecular parameters of individual molecules on the properties of their assembly in a solid.

In these tetragonal pseudohalide salts, the long axes of the anions are parallel to the stacking axis of the TTF's. The anions are disordered; their dipoles randomly point either up or down. In order to test the effect of this dipole disorder on T_0 and conductivity, we attempted to prepare a TTF salt containing a linear, symmetric anion. Attempts to prepare the azide ($\bar{N}=\overset{+}{N}=\bar{N}$) and silver dicyanide (NC-$\bar{Ag}$-CN) failed. We then tried dicyanamide (NC-$\bar{N}$-CN). Fully ionized dicyanamide should have the central nitrogen sp hybridized (linear) rather than sp^2 hybridized (bent) as is the case in its covalent compounds. Metathesis of $TTF_3(BF_4)_2$ with $\phi_4As[(NC)_2N]$ afforded black, prismatic, highly reflecting needles of ca 0.5-2 mm length whose single crystal room temperature conductivity was the highest recorded to date for a TTF salt (σ > 1,000 $(\Omega\ cm)^{-1}$). Elemental analyses for C, H and N showed no N! In addition, replacement of $\phi_4P[(NC)_2N]$ for $\phi_4As[(NC)_2N]$ in the metathesis, afforded no product under identical conditions! Qualitative X-ray fluorescence analysis of "pure" $\phi_4As[(CN)_2N]$

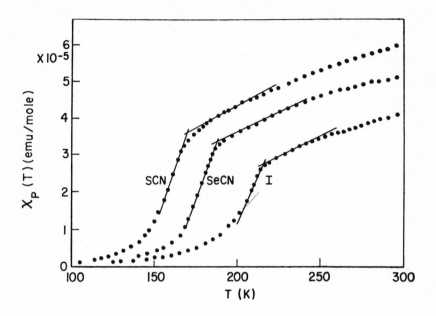

Figure 1:   Temperature dependence of the magnetic susceptibility
of $TTF(SCN)_x$, $TTF(SeCN)_x$ and $TTF I_y$ (x = 0.56 - 0.58;
y = 0.71)

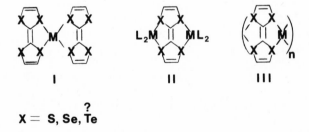

Figure 2:   Possible structures of metal complexes with TTF as a
dithiolene-type ligand

(correct elemental analysis for C, H, and N to within 0.1%!) showed the presence of chloride and bromide. The complete elemental analysis of product obtained from a metathesis with commercial $Na[(NC)_2N]$ (the original starting material also used to prepare the tetraphenylarsonium and tetraphenylphosphonium salts), corresponded to $TTF(BF_4)_{.07}Br_{.43}Cl_{.46}$. Preliminary X-ray crystallographic analysis indicates that this salt belongs to the same family (tetragonal) as TTF bromide, having eclipsed, ordered TTF stacks with interpenetrating anion sublattices (3). Single crystal conductivity was temperature independent down to ca 250 K where a broad metal to insulator transition was observed. We are currently testing the range of possible stoichiometries in these anionic "alloys" of TTF.

2. <u>Transition Metal Complex Anions</u> - On the basis of geometry and redox potentials, TTF could form compounds with transition metals in which it could act as (a) ligand, (b) cation, and (c) cationic ligand. We could, therefore expect different families of materials based on the above considerations; these are represented in Figs. 2-5. Compounds depicted in Fig. 2 implicate TTF as a dithiolene-type ligand. To date we have failed to demonstrate the existence of this type of TTF compound with, for example, zero-valent transition metal complexes.

Figure 3 depicts Krogmann type compounds in which TTF acts as a counterion. Recently A. P. Ginsberg prepared an iridium carbonyl chloride chain with TTF as a counterion. It appears, from spectroscopic evidence (4), that the TTF cations are not stacked uniformly but are dimerized. On the other hand, the metal carbonyl chloride forms a mixed valence chain which is responsible for the anisotropic properties of this TTF salt. This was not an unexpected result since the interplanar distance between atoms in Krogmann compounds is about 2.89-3.0 Å, whereas the interplanar distance in TTF salts varies from 3.47 to 3.62 Å. Consequently, because of this mismatch, the TTF chain, if it forms at all, would be expected to be distorted.

The use of dithiolene complexes as counterions (Fig.4) for TTF-based conductors was first demonstrated (5) for $Ni(Mnt)_2^=$ (Mnt = $(NC)_2C_2S_2^=$). Of the many dithiolene-TTF salts prepared to date, the 1:1 TTF $[(CF_3)_2C_2S_2]_2Cu^I$ is the most interesting because it exhibits a spin-Peierls transition at 12°K as observed by magnetic susceptibility measurements (6).

The third type of transition metal anionic system, depicted in Fig. 5, is based on the well documented (7) one-dimensional ligand-bonded magnetic chains. The known materials contain water molecules at the additional coordination sites; beyond the waters there is a row of large organic cations (e.g. $(CH_3)_4N^+$). If the coordinating

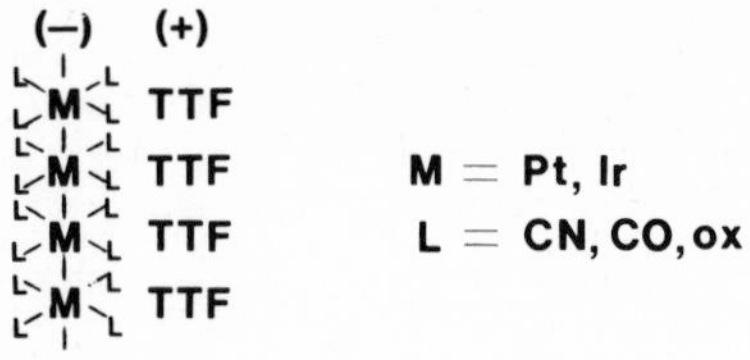
(–)
(+)
M
TTF
TTF
TTF
TTF
M = Pt, Ir
L = CN, CO, ox

Figure 3: TTF^+ as a counter-ion in Krogmann type compounds

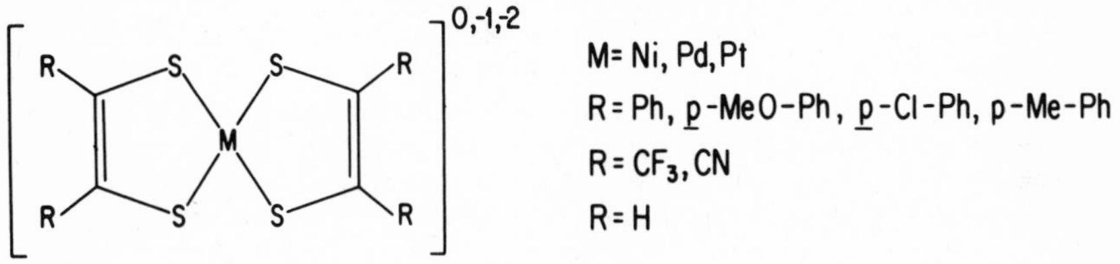
0,-1,-2
R
S
M
M= Ni, Pd, Pt
R= Ph, p-MeO-Ph, p-Cl-Ph, p-Me-Ph
R= CF3, CN
R= H

Figure 4: Dithiolene complexes which could act as counterions for TTF-based conductors

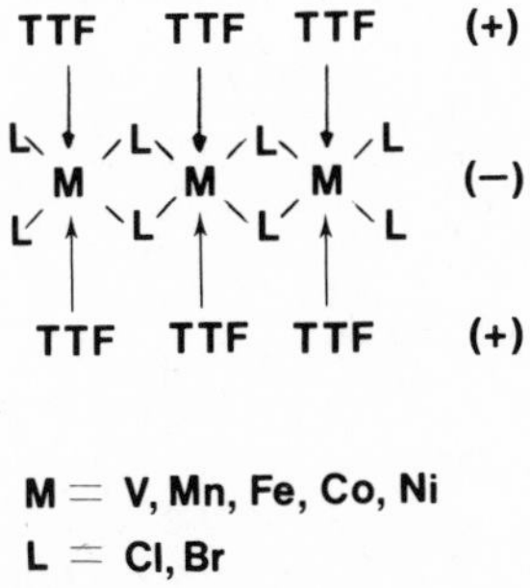
TTF TTF TTF (+)
M M M (–)
TTF TTF TTF (+)
M = V, Mn, Fe, Co, Ni
L = Cl, Br

Figure 5: Schematic structure of compounds with TTF coordinated to one-dimensional ligand-bridged metal chains

water and the large cation could be replaced by TTF^{+}, the TTF's would be able to form a uniform stack (the distances between anionic and cationic sites are now more evenly matched) and generate two types of long range interactions, an organic conducting stack and an inorganic magnetic "stack".

We envisioned the synthesis of these materials according to reaction (1) where M could be Co^{II}, Fe^{II}, Mn^{II}, etc; X could be Cl,

$$TTF + MX_n \xrightarrow{[0]} TTF_x MX_m \qquad (1)$$

Br, I; and [0] any mild oxidizing agent. Another approach depicted in Eq. (2) would involve direct reaction between an oxidizing transition metal and TTF:

$$TTF + MX_n \rightarrow TTF_x MX_m \qquad (2)$$

Here M could be Fe^{III}, Pt^{IV}, etc. and X the same as above.

When a solution of TTF and $CoCl_2$ in acetonitrile was exposed to oxygen, a change in color followed by deposition of black, shiny, prismatic needles was observed. The product exhibited a wide range in stoichiometry. In Fig. 6, the conductivity and magnetic susceptibility of a salt of one particular stoichiometry are shown. The magnetic behavior is dominated completely by high spin d_7 cobalt. While the room temperature conductivity is not as high as TTF TCNQ or KCP, it is still substantial (~ 10 (Ω cm)$^{-1}$) and is probably due to TTF stacks. Infrared and far infrared spectroscopy indicate the existence of Co-Cl, Co-O, and O-O bonding. Unfortunately the complete solid state structure of this material is unknown.

An example of reaction (2) is depicted below:

$$TTF + Pt^{IV}Cl_4 \rightarrow TTF_5 Pt_2 Cl_{10.6} \qquad (3)$$

Again, in this case, a wide range of stoichiometries exist. Elemental analyses vary from batch to batch but the above stoichiometry corresponds to the compound whose properties are depicted in Fig. 7. In contrast to the cobalt chloride-TTF case, there is long range magnetic interaction in this compound. The temperature dependence of the magnetic susceptibility and also the conductivity are identical to Ginsberg's TTF $Ir(CO)_2Cl_2$ mentioned above. So far we have been unsuccessful in our efforts to grow large, single crystals of this compound.

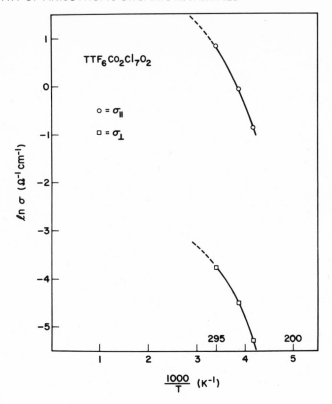

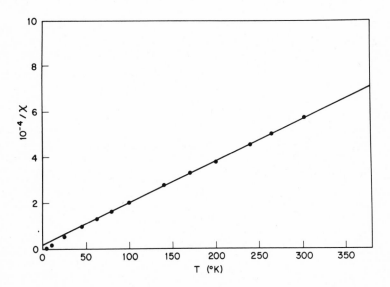

Figure 6:   Temperature dependence of the dc-conductivity (a) and magnetic susceptibility (b) of $TTF_6 Co_2 Cl_7 O_2$

As can be seen from the preliminary nature of this discussion, the exploration of inorganic compounds of TTF is still in its infancy. We hope that more chemists (8) and physicists will join in the research of this fertile area of mixed organic-inorganic solids.

B. Methylated Tetrathiafulvalenes

The purpose for the preparation of methylated (particularly unsymmetric) TTF's rests on the hypothesis that the introduction of disorder into a one-dimensional metal may prevent a uniform periodic distortion of its lattice and thus may stabilize the metallic state. Eventually we intend to dope TTF X_n (where X = Cl, Br, I, SCN, SeCN, etc.) with one or more of these methylated molecules in order to generate "stereoalloys".

Of all the methylated materials depicted in Fig. 8, we prepared UDMTTF, SDMTTF, and TMTTF. Of course, the most interesting would be MTTF because this is the simplest unsymmetric modification of TTF; it has not been isolated to date. The UDMTTF and tSDMTTF (t = trans) were prepared by the usual approach (9). They were isolated by repeated fractional crystallization and their purity was monitored by nmr and mass spectroscopy.

Preliminary results with these materials containing counter-ions other than TCNQ are shown in Table I. Clearly, simple methylation has dramatic effects on stoichiometry and hence on solid state properties. Of these, we studied $TMTTF_{17}(BF_4)_8$ in detail. The temperature dependence of the magnetic susceptibility and single crystal, four probe, conductivity are shown in Fig. 9. The observed hysteresis in the conductivity may be suspect since it appears as though the crystals develop fractures on cycling. Also, closer examination of crystal morphology revealed that they were hollow. In one specimen, the central cavity appeared to extend the entire length of the crystal.

Initial results from X-ray structure analysis, currently in progress in Professor P. Coppens' laboratory indicate a triclinic structure with two molecules per unit cell of dimensions 7.1×12.9×14.3 Å and 92.7, 99.15, 93.84° angles. Probably the most striking result is the difference in transition temperatures between the magnetic susceptibility ($\sim$40 K) and the conductivity ($\sim$200 K)!

We are currently attempting to grow better $TMTTF_2BF_4$ crystals as well as TTF TCNQ and TTF pseudohalides doped with these methylated TTF's.

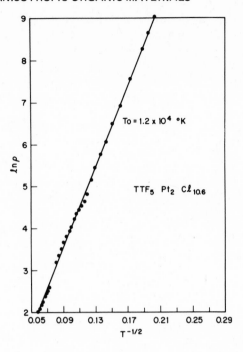

Figure 7:   dc-conductivity of $TTF_5 Pt_2Cl_{10.6}$

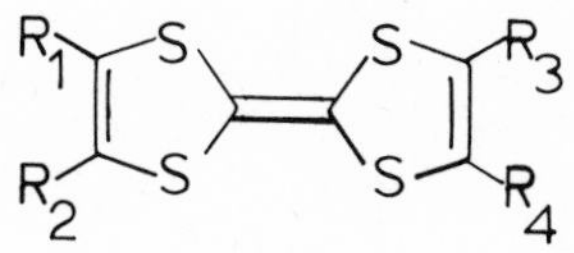

MTTF; $R_1 = CH_3$, $R_2 = R_3 = R_4 = H$

<u>t</u> SMDMTTF; $R_1 = R_4 = CH_3$, $R_2 = R_3 = H$

USDMTTF; $R_1 = R_2 = CH_3$, $R_3 = R_4 = H$

TMTTF; $R_1 = R_2 = R_3 = R_4 = CH_3$

Figure 8:   Methylated TTF derivatives and their abbreviations
             used in this text

Table 1:   Physical properties of different methylated TTF-
            tetrafluoroborates

COMPOUND	i. r. (KBr) cm^{-1}	U.V-vis CH$_3$CN $\lambda_{max}(\varepsilon)$	Compaction Resistance $\sim$ Ω	e. p. r. (CH$_3$CN)
(TTF)$_3$(BF$_4$)$_2$ Black Needles	2950(w) 2370(w) 1320(vs, broad) 1240(m) 1110(m) 1075(s) 1020(vs, broad) 828(s) 740(s) 800(w) 695(m)	250(sh) 290(sh) 305(sh) 316(19,800) 335(sh), 432(34,100) 375(sh), 576(9600) 397(sh)	insulator	quintent g=2.00838 a$_H$=1.26G
(TMTTF)$_{17}$(BF$_4$)$_8$ Black Needles	2320(w) 940(w) 1650(w) 915(s) 1575(w) 1550(m) 1470(w) 1450(w) 1430(m) 1250(vs, broad) 1080(s) 1050(vs)	297(sh) 310(14,600) 325(14,800) 390(sh) 420(sh) 459(12,300) 525(1820) 650(4660)	1	Thirteen lines g=2.0076 a$_H \simeq$ 0.77G
(UDMTTF)$_{11}$(BF$_4$)$_5$ Black Needles	1250(s, broad) 1080(w) 1030(m, broad) 900(m, broad) 815(m, broad) 725(m, broad)	298(sh) 310(sh) 322(15,300) 385(sh) 410(sh) 450(15,400) 505(sh) 615(5470)	20	Nonet g=2.0078 a $\simeq$ 0.956G
(SDMTTF)$_{11}$(BF$_4$)$_5$ Black Powder	1300(s, broad) 1110(w) 1080(m) 1055(m) 985(w) 815(m, broad)	295(sh) 310(sh) 319(17,100) 350(sh) 408(sh) 447(15,700) 500(sh) 610(5490)	3	Thirteen lines(?) g=2.0078 a $\simeq$ 0.68G

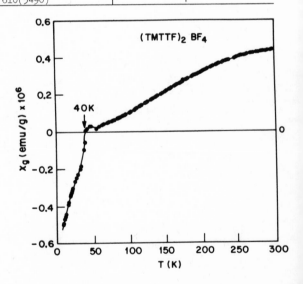

Fig. 9a

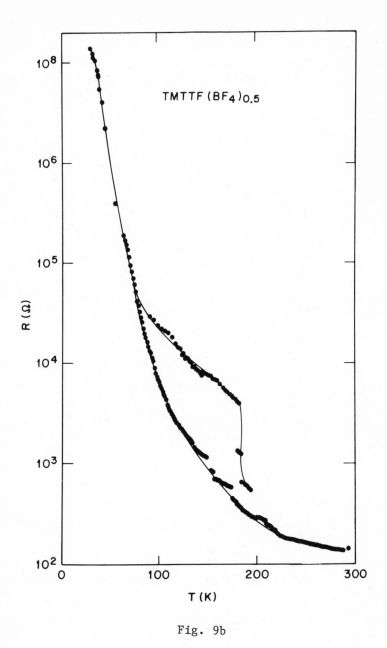

TMTTF (BF$_4$)$_{0.5}$

Fig. 9b

Figure 9:   Temperature dependence of the magnetic susceptibility(a)
and four probe dc-conductivity (b) of (TMTTF)$_2$BF$_4$

NONFULVENOID DONORS, NEW MOLECULES, AND POLYMERIC COMPOUNDS

A) Oxidative Additions to TTN and TTT

The fulvenoid structure of TTF is not the only one which can
lead to highly conducting organic solids. It has been known for
some time that TTT and, more recently, TTN can also produce highly
conducting solids (10).

TTN   TTT

Because of the disulfide functionality in these molecules, we
expected to be able to prepare interesting one and two dimensional
solids based on the oxidative addition of transition metals across
the S-S (or Se-Se in TSeT) link.

B) New Acceptors

Since the advent of TCNQ and TNAP, no stable, larger acceptors
have been prepared.

While TCNDQ is quite unstable, a pyrene analog (B) would be
expected to be more stable, because biphenyl inter-ring hydrogen
repulsions are eliminated.

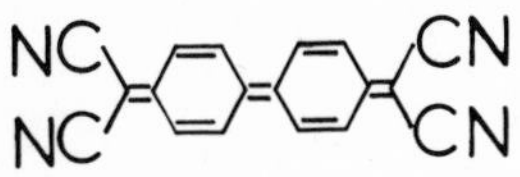

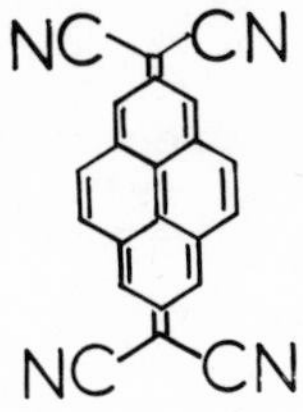

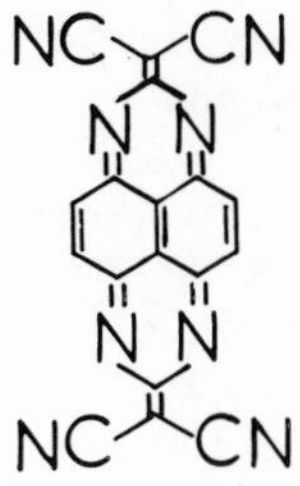

A

TCNDQ

B

1

Rather than attempt to prepare **B**, we decided to prepare **1** for the following reasons: (a) a projected synthesis of **1** appeared more straightforward than that of **B**; (b) substitution of N for C was expected to enhance the electron affinity of the acceptor and also to increase the number of available oxidation states (cf **1-5**).

$$1 \rightleftharpoons \qquad \rightleftharpoons \qquad \rightleftharpoons \qquad \rightleftharpoons$$

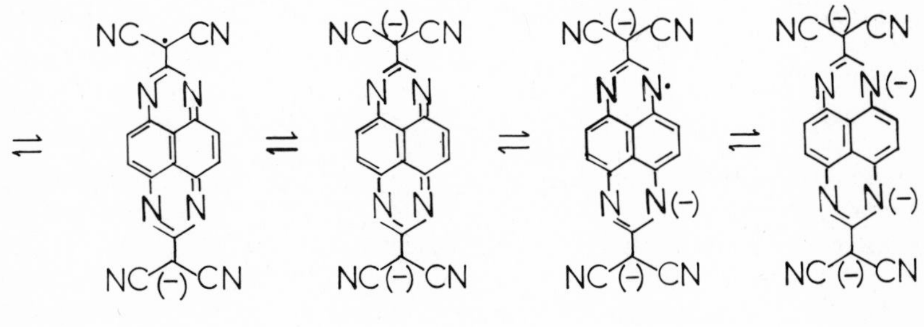

M^+ $2M^+$ $3M^+$ $4M^+$

2   3   4   5

We expected to enhance the probability of partially filled band formation.

Here we report on the preparation of 3a,b and some of its properties. The dianion was prepared by the sequence of reactions depicted below:

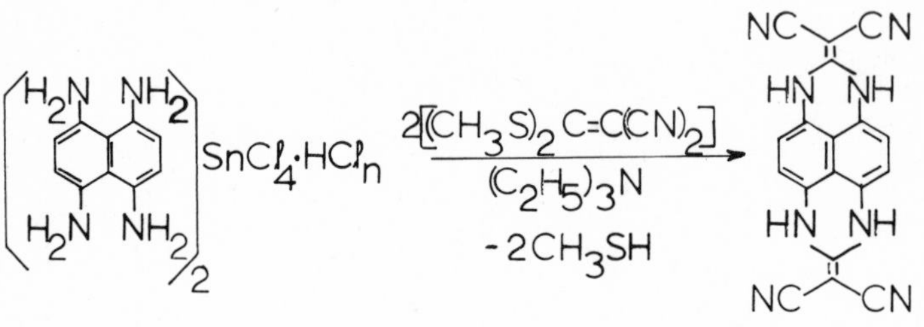

$$(4)$$

$$\underline{C}$$

$$C + 2NaH + \left\{ \begin{array}{c} \phi_4 AsCl \\ or \\ Bu_4 NCl \end{array} \right\} \rightarrow \underline{3a} \text{ or } \underline{3b} \qquad (5)$$

Since attempts to purify the crude reaction mixture obtained from reaction (4) failed, it was treated with base in acetonitrile in the presence of either tetrabutylammonium or tetraphenylarsonium chloride. Even when both reactions (4 and 5) were carried out under strictly anaerobic conditions, 3a(b) was the only characterizable product isolated.

After several abortive trials to obtain characterizable products from attempted oxidations of 3a to 2 or 1, we decided to examine the solution electrochemistry of 3 (cf. Fig. 10). Electrolysis at voltages more negative than -1.6 v vs. SCE produced a species which exhibited two irreversible oxidation waves at +0.16 and +0.56 v. The reversible reduction at -1.55 v could be assigned to the couple 3=4 on the basis of esr experiments.

Electrolysis of 3a in CH_3CN at -1.35 v in an esr cavity generated a relatively stable species (g = 2.0033, $t_{1/2}$ ca 2 min) whose spectrum is depicted in Fig. 11. From the analysis of the normalized intensities we deduced that the radical had four equivalent hydrogens and four equivalent nitrogens (calcd:

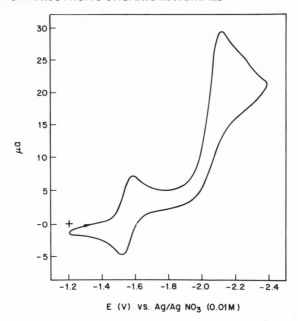

Figure 10:   Cyclic voltammogram of a degassed 10^{-3} M solution of $(Ph_4As^+)_2[(NC)_2CCN_2C_{10}H_4N_2CC(CN)_2]^{2-}$ in CH_3CN using 0.1 M $(n-C_4H_9)_4N^+ClO_4^-$ as supporting electrolyte, Ag/0.01 M $AgNO_3$ as reference electrode and platinum bead as working electrode. Scan rate, 200 mv/sec. The irreversible peak at -2.12 volt is due to $(Ph_4As)^+$

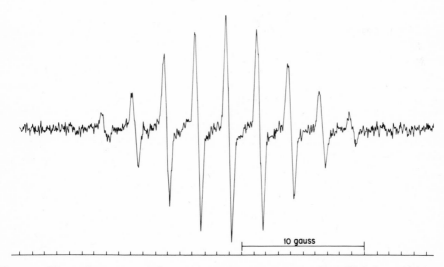

Figure 11:   Electron spin resonance spectrum of radical trianion generated during electrolysis of <u>3a</u> in acetonitrile (-1.35 v <u>vs</u>. SCE)

262:232:160:84:32:8:1; found: 262:235:164:82:33:8:1). There is practically no coupling to the nitrile nitrogens.

It is interesting to note that the most stable species among 1-5 are 3 and 4 and not 2 or 1. This, of course, does not mean that the radical anion derived from B will also be unstable.

Current studies on the solid state structure of 3a and the metathesis products of it with radical cations are in progress.

C) Stable π-Radicals

Over the years, we have restricted ourselves to the preparation of charge-transfer solids in which we have problems of coulomb interactions between donor and acceptor and within donor and/or acceptor stacks. In addition, the unpredictable formation of relatively uninteresting D-A mixed stacks could be very discouraging, particularly if several years of research were spent in the synthesis of a new donor or acceptor.

In order to alleviate these problems, we decided to attempt to prepare stable, planar, neutral π-delocalized radicals. These could be "intercalated" into neutral TCNQ or neutral TTF stacks. For example, in order to generate a "true" 1/4 filled organic metal band, one may proceed as follows:

In this scheme Q could be TCNQ and R a neutral radical similar in shape and size to TCNQ. One way to form a neutral molecule is to have positive and negative charges in the same species; for example, at their isoelectric point, neutral amino acids have a zwitterionic structure:

$$
\begin{array}{c}
H \\
| \\
R-C-CO_2^{-} \\
| \\
NH_3^{+}
\end{array}
$$

With these points in mind, we settled on aza-TCNQ (ATCNQ):

ATCNQ·   TCNQ⁻

Since nitrogen and carbon have very similar covalent radii,
we expected to produce only a minor perturbation on the size and
shape of TCNQ. However, because nitrogen is trivalent and is more
electronegative than carbon, we expected ATCNQ to be electronically
different from TCNQ. Scheme I shows the synthesis of a potassium
salt of ATCNQ. The tetrabutylammonium and tetraphenylarsonium
salts of ATCNQ can be prepared by metathesis. Scheme II contains
some very recent results on the preparation of new materials based
on ATCNQ. Attempts to reversibly oxidize ATCNQ$^{(-)}$ in solution
have, so far, failed. This indicates that (contrary to our
expectations), ATCNQ is a rather unstable radical probably because
b is a major contributor to the overall structure of ATCNQ. In b

b   a

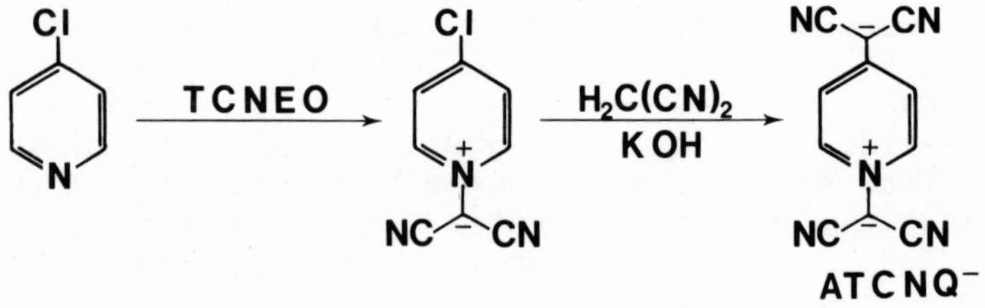

SCHEME I

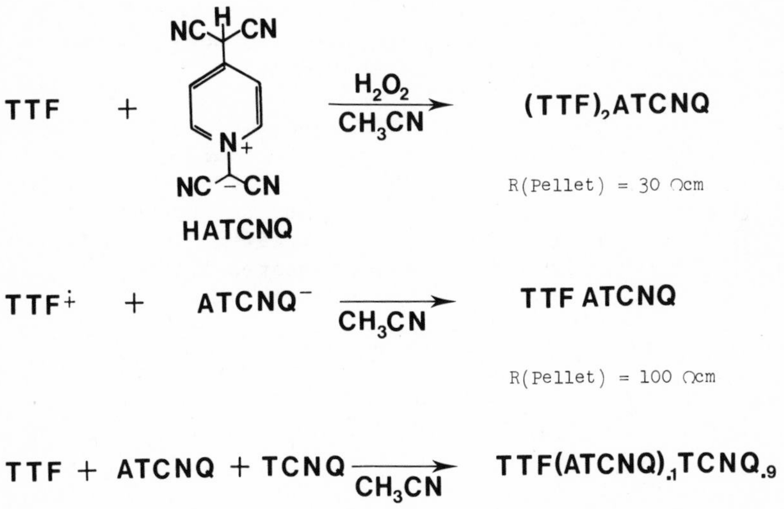

TTF
+
NC H CN
N+
NC − CN
HATCNQ
H_2O_2
CH_3CN
$(TTF)_2ATCNQ$
R(Pellet) = 30 Ωcm
TTF^{+}
+
ATCNQ^{-}
CH_3CN
TTF ATCNQ
R(Pellet) = 100 Ωcm
TTF + ATCNQ + TCNQ
CH_3CN
$TTF(ATCNQ)_{.1}TCNQ_{.9}$

SCHEME II

the spin is <u>localized</u> to one dicyanomethylene moiety and consequently dimerization and disproportionation are not unexpected
stabilization pathways. However, ATCNQ can be incorporated into
TTF TCNQ (cf Scheme II). Whether it exists as ATCNQ$^-$ or ATCNQ$^{\cdot}$
within this novel solid is not known.

    This is a relatively new area of research and therefore other
structures which would tend to form stable π-radicals (such as the
carboxamidopyridyls) (14) should be designed and tested.

D) Polymers

    In this section we will describe our very recent attempts to
prepare two-dimensional organic materials (other than graphite
itself) and also heteroatom chain materials.

    1. <u>Thiographite</u> - We based our research on the assumption
that the electronic properties of graphite could be modified by
introduction of heteroatoms into the layers themselves, rather in
between the layers (intercalation), making the heteroatom an
integral, covalently bonded, part of graphite. Indeed, Ciusa (11)
claimed earlier this century that he prepared tiographite (C_4S) by
the following reaction:

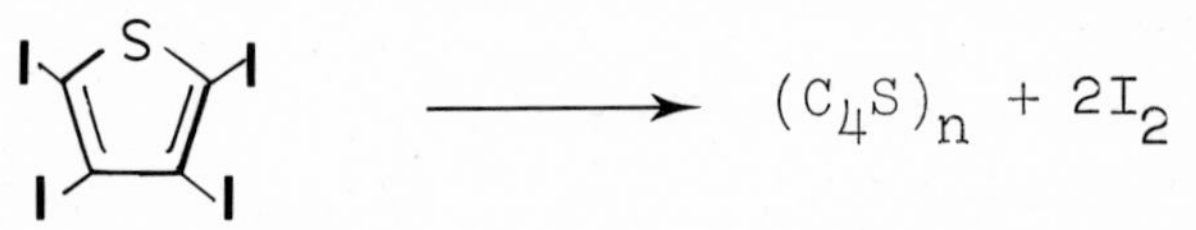

$$(C_4S)_n + 2I_2$$

    Another chemist (12) attempted to repeat Ciusa's experiments
but failed to obtain iodine-free material. We also found that,
depending on the experimental conditions, we could obtain material
with varying amounts of iodine but never completely iodine-free.
The more iodine remaining in the sample, the lower the room
temperature conductivity. One sample had a room temperature compressed pellet resistivity of .01 Ωcm and the latter was invariant

to ca. $77°K$; below this temperature, it increased by a factor of
ca 2 (near $4°K$). However, X-ray powder patterns revealed that
this material was apparently amorphous.

Attempts to elucidate the mechanism of polymerization and to
catalyze this reaction are in progress.

2. Polymethyleneditelluride. Since the discovery of the
superconducting state of $(SN)_x$, so elegantly shown by R. Greene, it
has obviously become important to determine if indeed $(SN)_x$ is a
substance that is part of a family of materials or whether it is
unique. It would appear that $(SeN)_x$ should be the next member in
a series; however, the weakness of the Se-N bond and the explosive
nature of Se_4N_4 [an "obvious" precursor to $(SeN)_x$] discouraged us
from attempting syntheses of this polymer. On the other hand,
$(SeCH)_n$ or $(TeCH)_n$ may very well be preferable and may indeed have
properties similar to $(SN)_x$.

A literature search revealed that Morgan and Drew studied the
following reactions (13):

$$(CH_3\overset{\underset{\|}{O}}{C})_2O + TeCl_4 \rightarrow Cl_3TeCH_2CO_2H + (Cl_3TeCH_2\overset{\underset{\|}{O}}{C})_2O + CH_2(TeCl_3)_2 + $$

$$HCl + \text{other products} + Te$$

$$CH_2(TeCl_3)_2 \rightarrow (CH_2Te_2)_n$$

The product of the last reaction was described as follows:
"...produced as a dark red, amorphous powder which...changed
slowly at the ordinary temperature, but rapidly at, or near $30°$ to
a black modification which melted gradually from 50 to $90°$ without
decomposition, forming viscous, pitch-like drops. On cooling this
pitch solidified to a dense, brittle solid having a silvery lustre
and resembling a heavy fusible metal."

We decided to repeat Morgan and Drew's work, to characterize
the product a little better, and to measure its transport properties.

At first we had some difficulty preparing the bis trichloro-
telluromethane but soon learned to improve on the original recipe.
Reduction of this species afforded the original red material which
transformed on standing at temperatures $>25°C$ to the silver compound. One may speculate that the red material consisted mostly of
rings which slowly polymerized by a ring-opening polymerization.
The results of elemental analysis of the final product were in

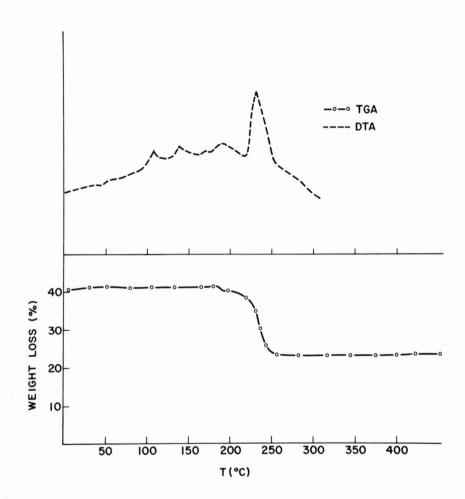

Figure 12: TGA and DTA data of the polymer $(CH_2 Te_2)n$

excellent agreement with the calculated values for $(CH_2Te_2)_n$. In Figure 12 we depict results of thermal and gravimetric analyses of this polymer. The small endothermic transition at ∿25 C corresponds to a glass transition temperature and the exothermic transition at 100°C corresponds to an insulator-to semiconductor transition (room temperature resistivity of 2,000 Ωcm). The conductivity <u>decreases</u> reversibly with an <u>increase</u> in pressure!

At <u>ca</u>. 50°C, $(TeCH_2Te)_n$ can be extruded into long, flexible fibers. These have the appearance of lead or silver wire. Attempts to obtain X-ray diffraction patterns from these fibers failed. The material was apparently still amorphous; also, its appearance changed upon x-irradiation. Finally, since the material is completely insoluble in all common solvents, we could not establish its molecular weight.

Conclusions and Future Directions

We have demonstrated (a) that higher room temperature conductivities than those of TCNQ salts are achievable with organic solids based on TTF and (b) that TTF, TTN, and TTT are versatile molecules for the generation of hybrid organic-inorganic solids with interesting physical properties.

We have already shown in several of the above sections the paths in which we are directing our research. In general, we expect to see chemists concentrate their efforts in the generation of new molecules which will yield solids where the interchain coupling is even larger than in HMTSeF TCNQ (<u>cf</u> A. Bloch and D. O. Cowan's chapters) in order to obtain a truly "ground state" organic metal.

It is a pleasure to thank G. A. Thomas, W. M. Walsh, Jr., F. J. DiSalvo, M. L. Kaplan, and A. Kruger for their stimulation, patience, and collaboration.

REFERENCES

1. F. Wudl, D. E. Schafer, W. M. Walsh, Jr., <u>Mol. Cryst. Liq. Cryst.</u>, <u>32</u>, 147 (1976).
2. F. Wudl, D. E. Schafer, W. M. Walsh, Jr., L. W. Rupp, F. J. DiSalvo, J. V. Waszczak, M. L. Kaplan, and G. A. Thomas, <u>J. Chem. Phys.</u>, in press.
3. S. LaPlaca, Private communication.
3a. R. Shumacher, Private communication, F. Wudl, unpublished results.
4. A. P. Ginsberg, J. W. Koepke, J. J. Hauser, K. W. West, F. J. DiSalvo, C. R. Sprinkle, R. L. Cohen, <u>Inorg. Chem.</u>, <u>15</u>, 514 (1976).

5. F. Wudl, C. H. Ho, A. Nagel, <u>Chem. Commun</u>., 923 (1973).
6. S. W. Bray, H. R. Hart, Jr., L. V. Interrante, I. S. Jacobs, J. S. Kasper, G. D. Watkins, S. H. Wee, and J. C. Bonner, <u>Phys. Rev. Letters</u>, <u>35</u>, 744 (1975).
7. See several examples in "Extended Interactions Between Metal Ions in Transition Metal Complexes", <u>Amer. Chem. Soc. Symposium Series</u> #5, L. V. Interrante, Ed., 1974.
8. J. Wilson, et al. (Monsanto) prepared a $CuCl_2$ salt with tetragonal crystal structure. J. Wilson, private communication.
9. F. Wudl, M. L. Kaplan, E. J. Hufnagel, E. W. Southwick, <u>J. Org. Chem</u>., <u>39</u>, 3608 (1974); A. Kruger, M. L. Kaplan, unpublished results.
10. Y. Matsunaga, <u>J. Chem. Phys</u>., <u>42</u>, 2248 (1965); E. A. Perez-Alberne, H. Johnson, Jr., and J. Trevoy, <u>Ibid</u>., <u>55</u>, 1547 (1971), I. F. Schegolev, "Conference on Organic Conductors and Semiconductors", Siófok, Hungary (1976). F. Wudl, D. E. Schafer, B. Miller, <u>J. Amer. Chem. Soc</u>., <u>98</u>, 252 (1976).
11. R. Ciusa, <u>Gazz. Chim. Italiana</u>, <u>55</u>, 385 (1925).
12. J. P. Wibaut, <u>Z. Angew. Chem</u>., <u>40</u>, 1136 (1927).
13. G. T. Morgan and H. D. K. Drew, <u>J. Chem. Soc</u>., <u>127</u>, 531 (1925).
14. E. M. Kosower, A. Teuerstein, <u>J. Amer. Chem. Soc</u>., <u>98</u>, 1586 (1976) and references within.

SUPERCONDUCTIVITY AND SUPERCONDUCTING FLUCTUATIONS

W. A. Little

Physics Department

Stanford University, Stanford   CA   94305

The first part of this paper will be devoted to an outline
of the theory of superconductivity in order to develop sufficient
background to appreciate what effects fluctuations have upon superconductivity. The second part by Dr. H. Gutfreund will discuss in
more detail superconductivity and the more general problem of instabilities of one-dimension systems. In the third part we will
discuss jointly the excitonic model of superconductivity and the
results of a recent analysis of a particular model which we have
proposed which appears to offer the possibility of achieving superconductivity at comparatively high temperatures.

In view of the vast amount of data presented at this meeting
on the one-dimensional metals compared to that on superconductors,
one might ask why so much time is then given to theoretical considerations of superconductivity. The justification for this is
that undoubtedly the driving force behind much of the work on the
Krogmann salts, the organic metals and similar exotic materials is
the search for superconductivity. The quest for a truly organic
superconductor remains a major goal of many research efforts. In
a very real sense this is what keeps much of our research afloat.
(See Fig. 1). A fortunate by-product has been the large amount
of exciting physics and chemistry which has resulted from these
endeavors. This has more than justified the effort to date. But,
perhaps the most valuable consequence and one which will have the
greatest long term impact has been the emergence of research groups
where true interdisciplinary work is done, where physicists and
chemists have learned to talk and work with oneanother in close
effective collaboration. The impact of this on material science
can be expected to be great.

Fig. 1.   Relationship of research on organic metals to that on
          superconductivity.

        First, I would like to comment on the prospects of making
organic materials with conductivities of the order of that of
copper <u>without</u> invoking superconductivity. I think the chance of
doing this is small. The reason for this is that carbon, upon which
such materials are based, requires a comparatively large energy to
obtain charge separation which is a pre-requisite for conduction.
For $C + C \rightarrow C^+ + C^-$ about 11eV is required whereas in the heavier
metals the analogous reaction requires about 5eV. The difference
arises from the greater screening of the electron-electron interaction by the higher inner electron density of the heavier elements.
        By building larger molecular structures such as TTF or TCNQ
one can reduce the energy for the charge separation between <u>molecules</u>.
But by so doing many of the available electrons from the carbon atoms
are tied up in covalent bonds. These electrons are not available
for conduction. A compromise must thus be made between these two
factors to get the maximum conductivity.
        In the structures prepared to date the molecular units, while
tightly bound in themselves, are only weakly bound to one another
by the van der Waals' force. The Debye frequency is thus rather
low. A resistivity vs T curve of a typical metal (Fig. 2) shows
that as one approaches the Debye temperature from below the

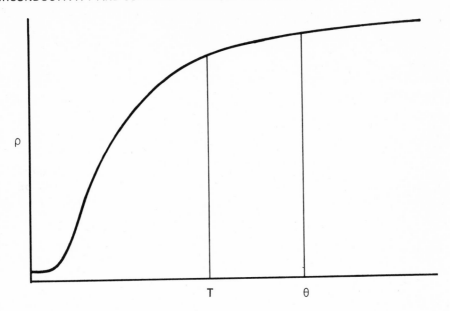

Fig. 2.   Resistivity vs T for a metal.   The Debye temperature
          is θ.

resistivity rises abruptly.   A low Debye temperature, θ leads to
a high resistivity at room temperature.
     As I see it the one hope of obtaining a very highly conducting
organic metal lies in the control of θ.   If organic structures
could be built which are covalently bound throughout as in a
"organic diamond-like" structure, but with some conduction electrons
present then the small mass of the carbon plus the rigid bonding
arrangement of the framework would give a high Debye temperature
and the possibility of very high conductivity at room temperature.
Fred Wudl's mention of a "thio-graphite" compound is, perhaps, a
step in this direction.[1]   Needless to say the chemistry required
for the preparation of such materials is likely to be an order
of magnitude more difficult than that for the charge transfer salts
known today.
     The other possibility is to produce a high temperature super-
conductor and with this in mind we turn to the theory of superconductivity.

THEORY OF SUPERCONDUCTIVITY

     The Hamiltonian of the system can be described in terms of
the creation and annihilation operators $c_k^{\dagger}$ and c_k, respective:

$$H = \sum_{k} \varepsilon_k c_k^\dagger c_k + \sum_{k,\bar{k},q} V(q) c_{k+q}^\dagger c_{\bar{k}-q}^\dagger c_{\bar{k}} c_k \tag{1}$$

where ε_k is the energy of the state, k measured relative to the Fermi energy and $V(q)$ is the electron-electron interaction. Using the wave function

$$\psi = \sum_{k} (u_k + v_k c_k^\dagger c_{-k}^\dagger)\phi_0 \tag{2}$$

ϕ_0 is the "vacuum" (absence of electrons) and v_k and u_k are constants which satisfy

$$u_k^2 + v_k^2 = 1 \tag{3}$$

one can calculate the expectation value of the total energy from

$$H\psi = E\psi \quad . \tag{4}$$

By now minimizing E with respect to u_k and v_k subject to the constraint (3) one obtains the familiar energy gap equation of Bardeen, Cooper & Schrieffer (BCS). The details of this were given at the previous NATO-ASI at Starnberg and will not be repeated.[2] Here we wish to focus on the nature of the BCS wave function (2). If we expand the product we obtain a series of terms of the form:

$$\psi = \left[u_k u_{k'} \cdots + \sum_{k} u_k \cdots v_{k'} c_{k'}^\dagger c_{-k'}^\dagger + \sum_{k',k} u_k \cdots v_{k'} c_k'' c_{-k}'' c_{k'}^\dagger c_{-k'}^\dagger \right.$$
$$\left. + \text{etc.} \cdots \right]\phi_0 \quad . \tag{5}$$

The first term gives no electrons, the second creates a linear combination of Slater determinants containing pairs of electrons in opposite momentum states, the third creates two such pairs etc. This wave function, containing a variable number of electrons, corresponds to a system in the Grand Canonical ensemble. States of a fixed number of electrons can be projected out from this and gives

$$\psi_N = \prod_N \left(\sum_{k} (v_k/u_k) c_k^\dagger c_{-k}^\dagger \right)\phi_0 \tag{6}$$

We see that all the terms in the product are identical and thus all the pairs are in exactly the same state. This shows that the BCS-state represents a Bose Einstein-like condensate of pairs of electrons. This feature endows the superconductor with transport properties which differ in an essential way from those of a normal conductor.

First a simple generalization can be made of the wave function (6), to a state in which each pair has a center of mass momentum, Q

$$\psi_N = \prod_N \left[\sum_k \left(\frac{v_k}{u_k} \right) c^+_{k+Q/2} c^+_{-k+Q/2} \right] \phi_0 \quad . \tag{7}$$

We note that in both (6) or (7) the pairs are all in the same state. It is thus impossible to form a wave packet from particles in this state because a wave packet corresponds to a combination of states of different momentum. Because of this one cannot localise the pairs and classical concepts based on such localization such as the Hall effect or simple pictures of transport have no meaning. How then does a superconductor cary current? This occurs as follows.

Once condensation has occured the pairs remain in a state of fixed momentum, Q. However, this momentum for a charged particle is the canonical momentum,

$$\vec{Q} = (m\vec{v} - e\vec{A}^*/c) \tag{8}$$

where $\vec{A}$ is the vector potential of any magnetic field which may be present, and e^* is the charge of the pair. Because Q is fixed, the current, which is proportional to the velocity, $\vec{v}$ is determined by $\vec{A}$. This is the foundation for the London equation[3] which tells us that the current density j(r) is proportional to this vector potential. This relationship is entirely different from the expression for j(r) in a normal metal and accounts for the remarkable electromagnetic behavior of the superconductor.

FLUCTUATIONS

The number of pairs, N which is contained in the condensate described by the wave function (6) is not necessarily a constant but can fluctuate. The local value of the phase of the wave function also can fluctuate. To discuss these effects it is best to describe the superconductor by a simplified form of the wave function, (8) namely the order parameter[3]

$$\psi(r) = |\Delta(r)| e^{i\theta(r)} \quad . \tag{9}$$

Where the square of the amplitude, $\Delta(r)$ gives the local density of pairs and $\theta(r)$ is the local value of the phase.

It can be shown from the BCS theory that the free energy of the superconducting state is given by:

$$F(\psi) = \int \left(a|\psi(r)|^2 + b|\psi(r)|^4 + c|\nabla\psi(r)|^2 \right) d^3r \tag{10}$$

where

$$a \approx (T - T_c)$$

If $T > T_c$ then, for a homogeneous solution, $\psi(r)$ = constant, the plot of $F|\psi|$ vs $|\psi|^2$ is as in Fig. 3a. The state of lowest free energy corresponds to $|\psi|^2 = 0$. For $T < T_c$, 'a' becomes negative and the state of lowest free energy has a finite amplitude for $|\psi|^2$. (Fig. 3b). Configuration in which $\psi(r)$ varies with r have a higher free energy because the term, c is positive.

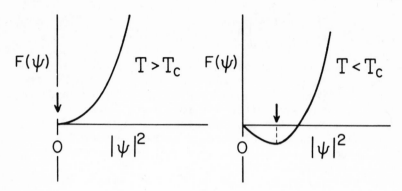

Fig. 3.   (a) Free energy vs $|\psi|^2$ for supercondutor for $T > T_c$.
          (b) Free energy vs $|\psi|^2$ for $T < T_c$.

The system does not necessarily reside in the state of minimum
free energy, but can fluctuate about this minimum. Each configuration, ψ can occur and will have a weight in the average of
$e^{-F(\psi)}/kT$. It is these fluctuations which destroy the long range
order in a superconductor.

LONG RANGE ORDER
Fluctuations can occur in the amplitude and in the phase of
the order parameter. One can readily show that fluctuations in
the phase destroy the long range order of the superconductor.[4]
This can be seen as follows. A fluctuation of the amplitude of
$\psi(r)$ in a given region has associated with it a fluctuating current
of particles to and from this region. One can show that this
current is proportional to the gradient of the phase, $\theta(r)$. So
the fluctuating currents give rise to fluctuations in $\theta(r)$. This
is illustrated in Fig. 4 where the current $j(r)$ causes the phase
$\theta(r)$ to advance from r to r'. As $j(r)$ fluctuates, so does the
gradient of the phase and as the distance between r and r' is
increased so the amplitude of this phase fluctuation will grow.
At a sufficiently large separation the order parameter at $\psi(r')$
will be as often out of phase as in phase with that at $\psi(r)$. Thus

$$\underset{|r-r'| \to \infty}{\text{Lt}} \; < \psi^*(r)\psi(r')> = 0 \tag{11}$$

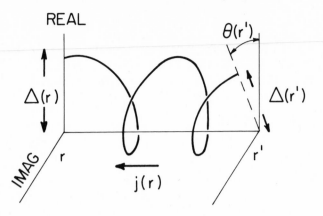

Fig. 4.   Phase variation of order parameter as result of current j(r).

This is the condition for the absence of long range order in a superconductor and explains why in a long thin (one-dimensional) system this order is destroyed. Because of the absence of long range order one does not have a true phase transition in such a one-dimensional system and the mean field transition temperature must be viewed with caution. On the otherhand certain properties of the one-dimensional "superconductor" do undergo dramatic changes near the mean field temperature. Experimentally this may be in-distinguishable from a true phase transition. Thus the statement that no phase transition can occur in a one-dimensional system, while rigorously true in the strict mathematical sense, should not be taken too seriously, for some physical properties of the system can reveal changes which are essentially the same as those of a system which does have a true phase transition. We illustrate this with a discussion of the decay of a persistent current.

Consider a long "one-dimensional" system bent back on itself to form a closed loop.[3,4] If now a current is circulating in this loop then the phase of the order parameter must advance as one goes round the loop. Any wave function must be single valued and in particular, this is true of the order parameter $\psi(r)$, so the total phase change round the loop must be an integral multiple of 2π. If one examines the phase of some point r' relative to, say, the starting point r then, as in our argument above, then as r - r' becomes large, fluctuations cause the loss of the long

range order. We see that these fluctuations cause the phase
helix to fluctuate so as to wind up tighter in some regions and
looser in others. See Fig. 5. However, these fluctuations occur
such that the total number of turns of the helix remain fixed
because of the constraint of the single value requirement on the
order parameter. Because of this, the total circulating current
(which is determined by the total phase change round the loop)
remains unchanged. So we see the phase fluctuations which destroy
the long range order do <u>not</u> cause the decay of a persistent current.

     The only way the circulating current can decay is by an
<u>amplitude</u> fluctuation which drives the amplitude of the order
parameter to zero. This allows one to lose one loop of the helix
by pulling it through the origin - as illustrated in Fig. 6. It
requires a finite amount of free energy, ΔF for this to occur so
the probability of it occurring is proportional to
$e^{-\Delta F/kT}$ too. The resistance of the loop should therefore
decrease exponentially with temperature in a like manner. Results
which yield a power law rather than an exponential dependence
result from calculations which ignore the boundary conditions on
the phase. Thus though there would be no long range order and no

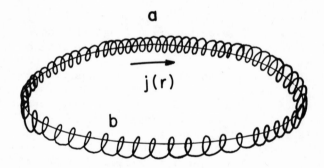

Fig. 5. Helix describing the phase variations of the order para-
         meter in a superconducting loop. Fluctuations can occur
         such that the phase variations may be rapid in some regions
         (a) and slow in others (b). The total number of turns of
         the helix round the loop is constrained to be an integer
         by the requirement of the order parameter to be single
         valued.

phase transition in a one-dimensional system nevertheless the
resistance of the system should fall extremely rapidly as one goes
below the mean field transition temperature.

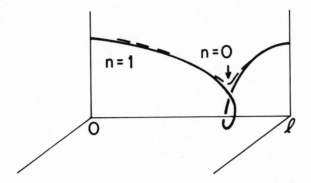

Fig. 6.   Amplitude fluctuation which causes a change in the number
of turns of the phase helix from n = 1 to n = 0.   The
perimeter of the loop is ℓ.

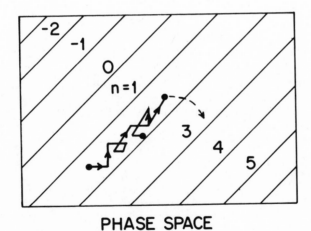

Fig. 7.   Phase space which represents the order parameter for the
superconducting loop may be subdivided into regions corresponding to the number, n of turns of the phase helix.
Fluctuation within a region of given n can occur readily
(–) but between regions (----) are comparatively rare.

It is useful to note that this has certain consequences in regard to the use of statistical mechanics in such systems.[4] We see that fluctuations can readily occur <u>within</u> each region of phase space in which the total phase change round the ring remains constant, but that fluctuations between these regions becomes highly improbable as $T \to 0$. See Fig. 7. One can show that the time between such transitions from one region to the other can become so long that it can exceed the age of the Universe! In that case the usual assumption of statistical mechanics of equating the ensemble average to the time average breaks down and the results of calculations using an ensemble average give results which are irrelevant to the interpretation of any experiment done in a finite time span. This point must be kept in mind in considering the results of exact or rigorous calculations of one-dimensional models. This underlines once again the need for examining with care the assumptions one makes in applying conventional theoretical techniques to systems of limited dimensionality.

REFERENCES

1.   F. Wudl,   (these proceedings)

2.   W. A. Little, "Low Dimensional Cooperative Phenomena" Ed. H. J. Keller, Plenum Press, New York (1975) p.35.

3.   M. Tinkham, "Intro to Superconductivity" McGraw Hill, New York (1975).

4.   W. A. Little, Phys. Rev. <u>156</u>, 396 (1967).

This work was supported in part by the Binational Fund, National Aeronautical and Space Agency, Contract JPL 953752 and National Science Foundation, Grant DMR 74-00427-A03.

INSTABILITIES IN ONE-DIMENSIONAL METALS

H. Gutfreund

The Racah Institute of Physics

The Hebrew University, Jerusalem, Israel

I. INTRODUCTION

There exist several lines of approach in the study of instabilities in one-dimensional metals. One line of approach is based on the Frohlich Hamiltonian

$$H = \Sigma \varepsilon_k a_k^+ a_k + \Sigma \omega_q b_q^+ b_q + \Sigma g_q a_{k+q}^+ a_k (b_{-q} + b_q^+), \qquad (1)$$

where a_k, b_q are the electron and phonon destruction operators, ε_k and ω_q are their respective energies, and g_q is the electron-phonon coupling constant. This Hamiltonian is a convenient starting point for the discussion of the phonon softening and the Peierls instability (1-5), the sliding conductivity (6,7) and the possibility of paraconductivity above the Peierls transition (8,9), the competition between superconductivity and the Peierls transition (10,11), and of various properties of the Peierls state. All the discussions based on this Hamiltonian are within the scope of mean field theory or, at best, mean field theory with corrections due to thermodynamical fluctuations.

Another line of approach is based on the electron gas Hamiltonian

$$H = \Sigma \varepsilon_k a_{k,\sigma}^+ a_{k,\sigma} + \Sigma V(q) a_{k+q,\sigma}^+ a_{k'-q,\sigma'}^+ a_{k',\sigma'} a_{k,\sigma}, \qquad (2)$$

where we have added to the electron creation and destruction operators the spin index σ. This Hamiltonian contains only the electron-electron interaction and it is assumed that the electron-phonon interaction is effectively included in the latter. The ground state properties of this Hamiltonian in one-dimension can be derived exactly (12-15) and it is, therefore, a convenient starting point for a discussion of the single chain.

A third approach is based on the Hubbard model and it is particularly appropriate in the case U/4t>1 (U - the Coulomb interaction between two electrons on the same site, t - the hopping integral). This approach was discussed in greater detail by Dr. Emery (16) and I shall not mention it in my lecture.

The first part of this lecture will be devoted to the discussion of the single metallic chain based on the electron-gas Hamiltonian of eq.(2). This will supplement certain aspects of the problem mentioned only in passing by Dr. Emery. It will also prepare the grounds for the discussion of the possibility of superconductivity in a quasi-one-dimensional system (17). Strictly speaking the single chain cannot undergo a phase transition at any finite temperature since the fluctuations drive the transition temperature to zero. Real systems, however, are composed of many parallel chain with some kind of interaction between them and these may exhibit phase transitions just because of this interaction, which constitutes a deviation from one-dimensionality. Thus, the discussion of phase transitions involves the problem of coupled chains. Some aspects of this problem will be discussed in the second part of this lecture.

II. THE ONE-DIMENSIONAL ELECTRON GAS

A one-dimensional Fermi system is characterized by a Fermi "surface" consisting of two points at $\pm k_F$ (k_F being the Fermi momentum). This results in a striking difference between the three-dimensional and the one-dimensional Fermi systems. While in the first case there exist low-energy electron-hole excitations with momenta between 0 and $2k_F$, in the second case one finds such excitations only in the neighborhood of q=0 and $q=2k_F$. This means that the important electron-electron interaction processes are those which involve momentum transfers in the neighborhood of these two values. There are four relevant scattering processes which are represented in fig. 1. The q≈0 processes may involve either electrons on one side of the Fermi "surface" (g_4) or on both sides (g_2). In both cases the electrons remain on the same side after their interaction, namely, continue to move in the same direction. Therefore, these processes are referred to as forward scattering processes. In the q≈$2k_F$ processes two electrons on both sides of the Fermi "surface" interact and change sides (directions of motion). This is the backscattering process. In the case of one electron per atom (half filled band) there is also an Umklapp process in which two electrons on the same side of the Fermi "surface" scatter together to the other side (g_3). Various treatments of one-

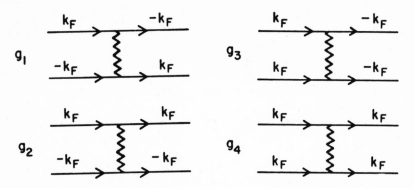

Fig. 1. Diagrammatic representation of the four interaction
processes in a one-dimensional Fermi system. The
forward scattering processes are denoted by g_2 and
g_4. The backscattering – by g_1 and g_3. The latter
exists only for a half filled band.

dimensional systems in the literature may be characterized by the
assumptions made on these coupling parameters and this whole approach
to instabilities is generally referred to as one-dimensional
"g-ology".

The basic difficulty in the theoretical treatment of order in
one-dimensional Fermi systems is the fact that such systems possess
two inherent instabilities. These show up as divergences in higher
order electron-electron interaction processes. One divergence
occurs in the particle-particle channel (Cooper channel) and it
indicates the onset of superconductivity. The simplest example of
a process in this channel is shown in fig. 2a. This process
diverges when $k_2 = -k_1$ and this result is valid in any number of
dimensions. The other divergence occurs in the particle-hole
channel ("zero-sound" channel) and it indicates the onset of the
Peierls or Overhauser instabilities which result in charge or
spin-density waves. The simplest example of a process in this
channel is shown in fig. 2b. In the one-dimensional case it
diverges when $k_1 - k_3 = 2k_F$. There is a competition between the
divergences in the two channels and one has to treat them
simultaneously. This was first done by Bychkov et al (18), who
considered a model in which $g_1 = g_2 = g$, neglecting g_3 and g_4.
They summed the so-called parquet diagrams (which amounts to
making the mean field approximation) and found that in the
logarithmic approximation the transition temperatures to the

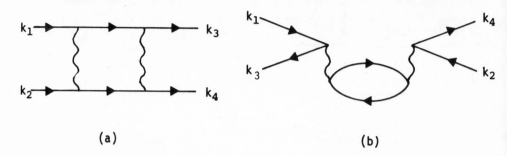

Fig. 2. Diagrammatic representation of the simplest
       interaction processes in the particle-particle
       channel (a), and in the particle-hole channel (b).

Peierls state and to the superconducting state are equal. They
predicted a single phase transition to an ordered state of condensed
quartets of two-electrons and two-holes (in contrast to the superconducting state characterized by the condensation of Cooper pairs).
This is a pioneering work in this field, however, it suffers from
two significant disadvantages. One is the restriction of the
interaction parameters to $g_1 = g_2$, and the other is the mean field
approximation. The results of recent work which is free from
these disadvantages are described in the next section.

III. TYPES OF ORDER IN THE SINGLE CHAIN

Although there are no phase transitions in a strictly one-
dimensional system, the question of order in a single chain is
meaningful in the sense of a tendency towards long range order as
$T \rightarrow 0$. Any type of order in the ground state would show up in the
response of the system to a corresponding generalized external
field. If the coupling constants are not too strong (19), there
are four possible types of ordering as $T \rightarrow 0$: charge density
wave (CDW) and antiferromagnetic or spin density wave (SDW), and
singlet and triplet pairing (SS and TS). To each of these types
of order there corresponds a generalized susceptibility. For
example in the case of CDW

$$\chi_{CDW}(x,t) = -i\Theta(t)<[\Psi_{1+}(x,t)\Psi_{2+}^+(x,t), \ \Psi_{2+}(0)\Psi_{1+}^+(0)]> , \qquad (3)$$

where $\Psi_{1+}(x,t)$ is the field of electrons on one side of the Fermi
surface (denoted by the index 1) and spin projection + . Eq.(3)
describes the propagation of electron-hole pairs. A similar
expression can be written for the case of SDW, but in this case

the electron and the hole have opposite spins. The susceptibilities
corresponding to the superconducting types of order describe the
propagation of Cooper pairs in the singlet or triplet states. The
imaginary part of the Fourier transform of these susceptibilities
behaves as $\text{Im}\chi(\omega) \propto \omega^{\alpha}$, and the system will show a tendency to
develop those types of order, as $T \to 0$, for which the exponent α is
negative.

We shall now summarize the basic results of the extensive
study of these susceptibilities carried out in recent years. The
first discussion of order in a one-dimensional system that went
beyond the mean field approximation is due to Solyom (12). He
applied the renormalization group method to calculate the
susceptibilities corresponding to the CDW, SDW and SS types of
order, and found that for $g_1 \geq 0$, the line $g_1 = 2g_2$ separates
between regions of superconducting and charge (spin)-density wave
behaviour (only g_1 and g_2 were taken into account). This approach
was extended by Fukuyama et al (12) who also calculated the triplet
superconductivity response function. The renormalization group
method apparently fails for $g_1 < 0$, except in the neighbourhood
of the origin.

The results for the negative values of the backscattering
coupling constant g_1 are based on the work of Luther and Emery (13),
who found a remarkable solution for the one-dimensional electron
gas model, for a particular negative value of g_1. They showed that
the one-dimensional Hamiltonian is described in terms of two kinds
of degrees of freedom: charge and spin density oscillations. The
first give rise to gapless phonon-like excitations, while the latter
result in an excitation branch with a gap. Luther and Emery were
able to diagonalize this Hamiltonian for the case $g_1 = -6/5$ (in
units of πv_F , v_F - Fermi velocity) and to calculate explicitly the
low frequency behaviour of the response functions. Their results
were then extended to all values of $g_1 < 0$ by P. Lee (14).

Our present understanding of the types of order in the single
metallic chain is summarized in fig. 3, taken from ref.(15). It
shows the regions in the (g_1,g_2)-plane of the different types of
behaviour of the response functions. The basic feature of this
diagram is the line $g_1 = 2g_2$ which essentially separates between
pairing and charge(spin)-density waves. In the lower half plane
there are regions in which the low frequency behaviour of the
CDW- and SS-susceptibilities both diverge, however the CDW-
-divergence is stronger on the right-hand side of this line and
the SS-divergence is stronger on its left-hand side. Note that
the SDW and the TS types of order can exist only in the upper half
plane. The picture in the lower half plane differs in some details
from other treatments of this region (for example (14)) because
it contains the effect of the g_4-process (in a real system $g_4 = g_2$;

they are denoted differently because they play a different role in
the theory). This process does not affect the $g_1 = 2g_2$ line, but
it changes the lines separating the region in which both CDW and
SS response functions diverge from those regions in which only one
of them diverges, and it also modifies the boundaries beyond which
the whole approach breaks down (broken lines in fig. 3). The
general conclusion from this picture is that the sign of $2g_2-g_1$
distinguishes between superconducting (negative) and
charge(spin)-density wave behaviour (positive). One can understand
why this combination of coupling constants is in one-dimension the
effective interaction which determines the low frequency behaviour
of the response functions. First, one can show that the g_4 process
only serves to modify the "sound"-velocity of the charge density
oscillations. The g_2- process appears with a factor 2 for the two
possible relative spin arrangements of the two electrons. The
backscattering may also occur either with parallel or antiparallel
spins. The first case, for a one-dimensional system of identical
particles, is indistinguishable from a forward scattering process
apart from being an exchange, rather than a direct process. It
therefore appears in the effective interaction with a minus sign.
The reason that the backscattering process with antiparallel spins
does not contribute to the low-frequency behaviour of the response
function is due to the fact that this process may be described by
the spin-density degrees of freedom and these have a gap in their
excitation spectrum.

It is interesting to ignore for a while all the difficulties
connected with one-dimensional systems mentioned before, and
perform a simple mean field calculation of the transition
temperature to each of the four types of order. One gets a BCS-like
expression

$$T_c = E_o \exp(1/\lambda_i) \qquad , \qquad (4)$$

where E_o is some electronic cutoff and the λ_i are the effective
coupling constants corresponding to the various types of order:

$$\lambda_{SS} = \tilde{g}_2 + \tilde{g}_1 \qquad ,$$

$$\lambda_{TS} = \tilde{g}_2 - \tilde{g}_1 \qquad ,$$

$$\lambda_{CDW} = 2\tilde{g}_1 - \tilde{g}_2 \qquad ,$$

$$\lambda_{SDW} = -\tilde{g}_2 \qquad , \qquad (5)$$

where $\tilde{g}_i$ is g_i in units of πv_F. If one now asks in what regions
of the (g_1, g_2)-plane is one of these transition temperatures higher
than all the others, one gets the picture in fig. 4. Although the
mean field treatment is incorrect in several respects, the results

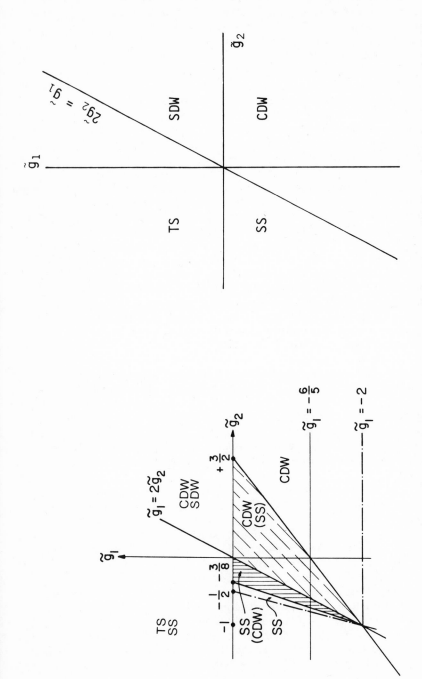

Fig. 3. Regions in the (g_1, g_2)-plane in which the response functions for the indicated types of order diverge. The dotted and dashed lines represent the limits for which the model can be solved.

Fig. 4. Regions in the (g_1, g_2)-plane with the highest mean field transition temperature to the indicated type of order.

indicate the regions in the space of the coupling parameters in
which the system tends to develop large fluctuations corresponding
predominantly to one of the possible types of order, and these
regions are remarkably similar to the qualitative picture
obtained from the more rigorous treatment.

IV. THE PROBLEM OF COUPLED CHAINS

Many of the previous treatments which attempted to describe
the experimental results in real materials like KCP or TTF-TCNQ
applied the mean field theory in one dimension. In such an approach
one assumes implicitly that some interchain coupling of an
unspecified nature exists, which is sufficiently strong to suppress
the one-dimensional thermodynamic fluctuations, and that the actual
results depend only weakly on the strength of this coupling. This
approach is in general insufficient because it ignores the specific
effects of different interchain coupling mechanisms, and it is
also expected to fail completely in the case of weak interchain
coupling, when the transition temperature should depend strongly
on the coupling strength. It is therefore desirable to include
the interchain interactions explicitly.

There are two different approaches to this problem. One
approach starts from the solutions for the single chain and
introduces interchain coupling as a perturbation. Another approach
is to consider the system of coupled chains as a three-dimensional
but very anisotropic system. The advantage of this approach is that
for a sufficiently strong interchain coupling, namely, when the
anisotropy is not too large, the mean field theory is a good
approximation. One can therefore compute the transition temperatures
in the mean field approximation and then check a posteriori, by
calculating the fluctuations, when this approach breaks down.

The first approach, that starting from the single chain
properties, was pursued in ref. (20). The system considered is a
lattice of parallel chains characterised by the coupling constants
g_2, g_4 and $g_1 < 0$. The order parameter which develops on each chain
as $T \to 0$ is determined by these coupling constants, as described
in fig. 3. The interchain coupling serves to stabilize the relative
phases of the order parameter on different chains and thereby to
produce a three-dimensional long range order at finite temperatures.
One can obtain such a phase transition only to a state which is
compatible with the type of order allowed by the single chain
parameters. It can be shown that if the electrons on different chains
interact through the Coulomb potential and are not allowed to move
from one chain to another, then the only possible phase transition
is to the CDW-state. The charge density waves on adjacent chains
are then aligned either in phase (when the interchain interaction

is negative) or out of phase (when it is positive). When hopping between chains is allowed one can also get a superconducting phase transition and the final result depends on the single chain parameters, the interchain Coulomb interaction and the hopping integrals. Another study of the coupled chain problem is based on an extension of the renormalization group method (21). In this case it is possible to include only the interchain Coulomb interaction and again one finds only phase transitions to the CDW-state.

When the interchain coupling is sufficiently strong it may be more appropriate to adopt the other approach, mentioned above, and start from a three-dimensional anisotropic system. Let me illustrate this approach by the discussion of the Peierls transition in an electron-phonon system described by the Hamiltonian in eq. (1). Let us assume that the intechain coupling is represented by an anisotropic electron dispersion of the form

$$\varepsilon(k) = \varepsilon(k_z) + \eta\varepsilon_F(\cos a p_x + \cos a p_y), \qquad (6)$$

where a is the distance between adjacent chains and η is a measure of the interchain coupling. The details of this model are described in ref. (3) and here we quote only the pertinent results. The mean field transition temperature T_p is almost independent of η up to a certain critical η_c, depending on the nature of the electron dispersion $\varepsilon(k_z)$, at which it drops rapidly to zero. There exists a characteristic temperature $T_o \sim \eta T_F/4$, which plays an important role in the model. At this temperature the inverse transversal correlation length crosses the Brillouin zone boundary. If the mean field T_p happens to be smaller than T_o, the thermodynamic fluctuations will have a small effect, suppressing the actual transition temperature by at most 20%. In this case the mean field theory makes a reasonable approximation. If, however, the mean field T_p exceeds T_o, we expect large fluctuations and mean field theory is then invalid. Fig. 5 shows the Peierls transition temperature as a function of η for several values of the dimensionless electron-phonon coupling constant s, defined as $s = 2N(0)g^2/\omega$, where $N(0)$ is the density of states at the Fermi energy, ω is the phonon frequency and g is the electron-phonon coupling constant in the Frohlich Hamiltonian. The full line is the mean field T_p and the dashed line is the actual T_p, depressed from its mean field value by thermodynamic fluctuations. The latter is shown only as long as this shift does not exceed 20%. Wether a given system may be described reasonably well by mean field theory depends, in this model, on the values of s and η. I shall not speculate here on these values in the various materials discussed at this Summer School. I would , however, like to point out that this whole picture was found useful in analyzing the recent pressure experiments in KCP (22). Pressure is expected to increase η and these experiments indicate that KCP is "climbing" uphill on one of the dashed lines in fig. 5.

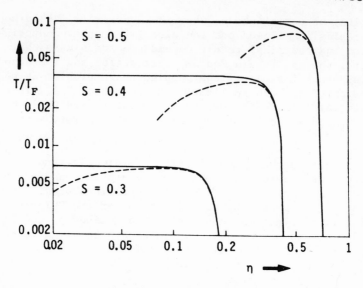

Fig. 5. The Peierls transition temperature as function of
the interchain coupling η for several values of the
electron-phonon coupling s. Full lines- mean field
T_p, dashed lines – actual T_p shifted by fluctuations.
The latter is shown only when the shift is less than
20%.

V. DISCUSSION

In this lecture I have summarized our theoretical understanding
of the question of order in a single chain described by the
Hamiltonian in eq. (2) and discussed briefly the even harder
problem of coupled chains. Although I believe that the picture
described here is relevant to the understanding of the qualitative
aspects of order in real quasi-one-dimensional materials, it is
very difficult to relate it directly to specific compounds. First,
because real materials are always more complicated than systems
of parallel electron-gas chains. This, however, is a trivial remark.
The question is wether the electron gas model is sufficient to
describe the basic features of a single stack in a real compound.
One thing which this model neglects, for example, are the
retardation effects which are ignored because the electron-phonon
interaction is replaced by an effective static electron-electron
interaction. Another feature missing in this model is the effect
of impurities. It is not clear how these two effects would modify
the picture described above. It is my belief that they do have
an important effect on the actual value of the transition

temperatures, but the basic features described in fig. 3 do
represent "real physics". One reason why it is hard to relate this
figure to real materials is because there is no way to measure
directly the interaction constants g_i. Nevertheless, it is tempting
to speculate on the position of compounds like KCP or TTF-TCNQ
in the (g_1, g_2)-plane. If I had to, I would roughly put KCP
somewhere in the second quadrant and TTF-TCNQ in the first quadrant
of this plane. Without being more specific, I think that one can
safely assert that these two, and all the presently discussed
organic and metallo-organic compounds, are on the right-hand side
of the line $2g_2 = g_1$. The question is whether there exist materials
on the other side of this line and how does one proceed to find them.
The compounds proposed for the realization of the excitonic
superconductor in a one-dimensional system (23,17) are characterized
by a small and positive g_1 and a considerably larger negative g_2.
This places such compounds in the region of triplet-pairing
fluctuations and a coupled system of such chains is expected to
undergo a phase transition to the triplet superconducting state.
However, it is not clear that such systems can be synthesized
and recent calculations impose severe conditions on their structure.

REFERENCES

1. H. Frohlich, Proc. R. Soc. A223, 296 (1954); C. G. Kuper, Proc.
   R. Soc. A227, 214 (1955).
2. B. Horovitz, M. Weger and H. Gutfreund, Phys. Rev. B9, 1246 (1974).
3. B. Horovitz, H. Gutfreund and M. Weger, Phys. Rev. B12, 3174 (1975).
4. M. J. Rice and S. Strassler, Solid State Comm. 13, 125 (1973).
5. P. A. Lee, T. M. Rice and P. W. Anderson, Phys. Rev. Lett. 31,
   462 (1973).
6. P. A. Lee, T. M. Rice and P. W. Anderson, Solid State Comm. 14,
   703 (1974).
7. M. J. Rice, S. Strassler and W. R. Schneider, p. 282 in Lecture
   Notes in Physics Vol. 34 (ed. H. G. Schuster, Springer Verlag,
   (1975).
8. D. Allender, J. W. Bray and J. Bardeen, Phys. Rev. B9, 119 (1974);
   M. Weger, B. Horovitz and H. Gutfreund, Phys. Rev. B12, 1086
   (1975).
9. B. R. Patton and L. J. Sham, Phys. Rev. Lett. 33, 638 (1974).
10. H. Gutfreund, B. Horovitz and M. Weger, J. Phys. C7, 383 (1974);
    A. Birnboim and H. Gutfreund, J. Phys. Lett. 35, L147 (1974);
    B. Horovitz and A. Birnboim, Solid State Comm. 19, 91 (1976).
11. M. J. Rice and S. Strassler, Solid State Comm. 13, 697 (1973).
12. J. Solyom, J. Low Temp. Phys. 12, 547 (1973); H. Fukuyama,
    T. M. Rice, C. M. Varma and B. J. Halperin, Phys. Rev. B10,
    3775 (1974).
13. A. Luther and V. J. Emery, Phys. Rev. Lett. 33, 589 (1974).
14. P. A. Lee, Phys. Rev. Lett. 34, 1247 (1975).

15. H. Gutfreund and R. Klemm, Phys. Rev. B14, 1073 (1976).
16. V. J. Emery, present volume.
17. H. Gutfreund and W. A. Little, present volume.
18. A. Bychkov, L. P. Gorkov and I. E. Dzyaloshinski, Soviet Physics - JETP 34, 422 (1972).
19. Other phases are possible for sufficiently large values of the g_i-s; see, B. Horovitz, Solid State Comm. 18, 445 (1976).
20. R. Klemm and H. Gutfreund, B14, 1086 (1976).
21. L. Mihaly and J. Solyom, J. Low Temp. Phys. (in press); N. Menyhard, Proceedings of the Conference on Organic Conductors and Semiconductors, Siofok, Hungary, 1976.
22. D. Jerome and M. Weger, present volume.
23. D. Davis, H. Gutfreund and W. A. Little, Phys. Rev. B13, 4766 (1976).

PHYSICAL CONSIDERATIONS AND MODEL CALCULATIONS FOR

ONE-DIMENSIONAL SUPERCONDUCTIVITY

H. Gutfreund and W. A. Little[*]

The Racah Institute of Physics, The Hebrew U., Israel

Physics Department, Stanford U., Stanford CA [*]

I. INTRODUCTION

The present era of intensive research in the field of one-dimensional conductors which started with the work of the Penn group[1] on TTF-TCNQ was preceded by the discussion of the possibility of superconductivity in one-dimensional organic materials which was suggested by one of us[2] in 1964. In that paper a new mechanism of superconductivity was proposed. In this mechanism the effective attraction between electrons at the Fermi surface is induced by the exchange of an electronic excitation (broadly referred to as an exciton), rather than by phonons as is believed to be the case in all presently known superconductors. The main attraction of this, so called, exciton mechanism is that it apparently implies high temperature superconductivity. The one-dimensionality came in only because the system of a linear conducting spine with polarizable side chains seemed to be a suitable structure for the realization of the exciton mechanism, which requires a spatial separation between the conduction electrons and the electrons of the exciton system. We are sure that everybody would be just as happy to make the exciton mechanism work in a different geometry. Indeed, Ginzburg[3] has discussed extensively the exciton mechanism in two-dimensional structures. Allender et al[4] applied Ginzburg's ideas to a specific system consisting of a thin metallic layer coated by a semiconductor with a high dielectric constant. Their work was followed by an unsuccessful attempt to find superconductivity with an enhanced transition temperature in such a system[5]. We believe that the one-dimensional structure still has the best prospects for a successful realization of the exciton mechanism. The reasons for this opinion will be specified in the course of this lecture.

A completely unrelated reference to superconductivity in quasi-one-dimensional systems was made by Weger[6], who suggested that the relatively high transition temperatures in the A-15 compounds are due to their one-dimensional structure. One of the basic questions in the excitonic model is whether the one-dimensionality does not destroy the high transition temperatures warranted by the exciton mechanism. On the otherhand, in Weger's theory it is precisely the one-dimensional structure which gives rise to enhanced transition temperatures, mainly because of the high density of states at the Fermi energy E_F obtained in such systems when E_F lies near a band edge.

The idea of the exciton mechanism in a one-dimensional system deserves a careful and critical examination as it is based on an attempt to apply the BCS theory of superconductivity in a predictive manner in a different context. Such an examination involves three types of problems:

a) Problems associated with the one-dimensionality: the absence of phase transitions in one-dimension, the competition with other instabilities, localization;

b) Problems associated with the mechanism itself: the effect of the exchange interaction between the electrons on the spine and on the dyes, the question of vertex corrections;

c) Quantitative problems: the balance between the exciton exchange interaction and the Coulomb repulsion, the sensitivity to structural features such as the number of dyes per unit cell, their distance from the spine etc.

In addition to these questions there remain the questions of chemistry. To discuss these one must focus on a specific model. The one we have proposed[7] is illustrated in Fig. 1. It involves a chain of partially oxidized Pt-atoms tightly coupled to a polarizable ligand system. The chemical questions which may be asked of this model include the following: Will such structures be stable? Can the Pt-spine be oxidized without destroying the ligands? Will the monomers stack? Will there be sufficient metal-metal interaction at the 3.4Å separation required by the π-system of the bulky ligands?

The answers to some of the qualitative, and certainly the quantitative, questions raised above depend upon certain specific features of the model. We shall therefore begin with a description of the model, continue with the calculation of the interactions between electrons on the spine and derive the transition temperature from a simple BCS-like equation. When that is done we shall examine this entire scheme along the lines drawn above.

II.   THE MODEL

The model which we propose and the work leading up to this proposal are described in detail in the paper[7] by D. Davis, H.

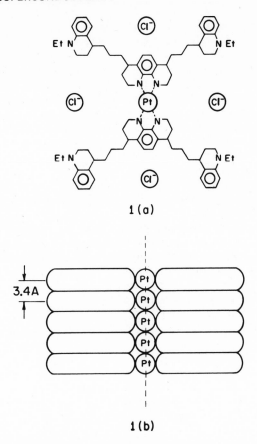

1 (a)

1 (b)

Fig. 1.   Proposed model of the structure of an excitonic superconductor. (a) Top view of square planar phenanthroline-
dye ligands complexed to Pt. Double bonds in the chromophore are omitted for simplicity. Et stands for ethyl.
(b) Side view of chain.

Gutfreund and W. A. Little (DGL). As shown in Fig. 1 the proposed
model consists of a chain of Pt atoms surrounded by a sheath of
highly polarizable dye-like molecules. Each unit of the conductive
chain of Pt-atoms consists of a bis-phenanthroline ligand system
complexed to the metal atom. To each of the 1-10 phenanthroline
ligands are attached two cyanine-dye chromophoric units at the 4 and
7 positions.
    In the proposed structure the polarization of the chromophore
results in the movement of a positive hole from one nitrogen atom

remote from the Pt-site to the nitrogen adjacent to the Pt.
Because of the large movement of charge a strong electron—exciton
interaction can be expected.

The d_{z^2} orbitals of the Pt-atoms of the chain overlap with
one another to give a linear conductive pathway. Because of the
repulsion between the π-electrons of the bulky ligands the Pt
atoms along the chain can only come to within about 3.4Å of
one another. We assume that sufficient overlap occurs at this
separation to yield a conductive chain.

One requires four counterions (Cl$^-$) for each square-planar
ligand system plus two additional negatively charged ions for
Pt in the Pt(II) oxidation state. Then the Pt-chain needs to
be partially oxidized to give a partially filled d_{z^2} conduction
band as in $K_2 Pt(CN)_4 Br_{0.3} \cdot 3H_2O$.

In this paper we present details of a calculation on a
simplified version of the system illustrated in Fig. 1. Instead
of working with the large phenanthroline groups we use a skeleton
structure of the chromophore units alone as illustrated in Fig. 2.
This gives the essence of the results of calculations of the more
complex system done earlier[8] and illustrates more clearly the
physics of the problem.

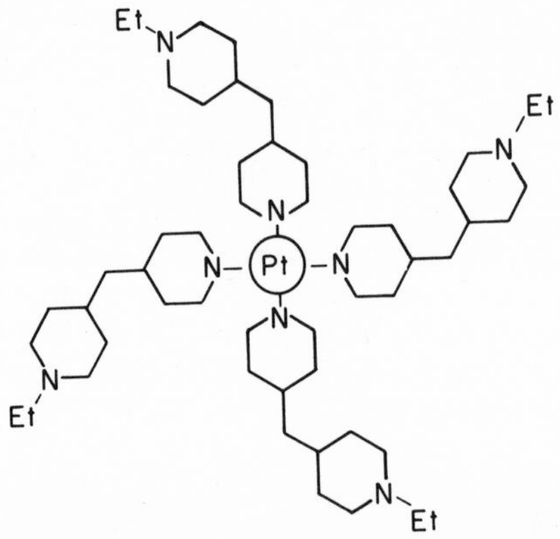

Fig. 2.   Simplified version of structure of Fig. 1 for which
           detailed calculations are presented in this paper.

III.   EQUATION FOR T_c

To calculate the transition temperature T_c we have adopted
the method of Kirzhnits, Maximov and Khomski.[9] This method applies

to a weak coupling superconductor and results in the following
BCS-like equation for the gap function $\phi(\underline{k})$

$$\phi(\underline{p}) = -\int \frac{d^3k}{(2\pi)^3} \frac{U(\underline{p},\underline{k}) \tanh (\xi(\underline{k})/2T_c)}{2\xi(\underline{k})} \phi(\underline{k}) \quad , \quad (1)$$

$\xi(\underline{k})$ is the electron energy measured from the Fermi-surface and
all the interactions are represented by

$$U(p,k) = V_0(\underline{p-k}) \left[1-2\int_0^\infty \frac{\rho(p-k,\omega) d\omega}{\omega + |\xi(\underline{p})| + |\xi(\underline{k})|} \right] \quad , \quad (2)$$

with $V_0(\underline{p-k})$ being the bare Coulomb interaction and $(V_0\rho)$ being
essentially the imaginary part of the total interaction. It is
worth pointing out that this kernel is a smooth function of its
variables unlike the interaction itself which has a complicated
resonant structure. In addition it should be noted that the
kernel decreases when either one of the energies $\xi(p)$, $\xi(k)$ departs
from the Fermi surface. The merit of (1) is that it brings out
explicitly the relationship between the kernel and the microscopic
properties of the system such as the electron band energies, the
exciton band energies, the Coulomb interaction and the electron-
exciton matrix elements. Our estimates of the transition temperature for the model described in section (V) will be derived from
the one-dimensional form of eq. (1) with k integrated between
$-\pi/a$ and π/a . This requires further justification in view of
the questions raised in the last paragraph of the introduction.
This justification will be provided in section (VI).

Let us now be more specific about the kernel $U(p,k)$. We
separate it into the contribution of the Coulomb interaction

$$U_c(p,k) = V_0(p-k) \left[1-2 \int_0^\infty \frac{\rho_c(p-k,\omega) d\omega}{\omega + |\xi(p)| + |\xi(k)|} \right] \quad , \quad (3)$$

and the exciton-exchange interaction

$$U_{ex}(p,k) = -2 \sum_\alpha \frac{|Q_\alpha(p-k)|^2}{E_\alpha(p-k) + |\xi(p)| + |\xi(k)|} \quad , \quad (4)$$

where $Q_\alpha(q)$ is the electron exciton coupling constant and $E_\alpha(k)$
the exciton band energy. This separation is discussed in greater
detail in DGL. It should be noted that in contrast to the case
of the phonon-mechanism the exciton-exchange interaction has the
same cutoff as the Coulomb interaction.

IV.   THE INTERACTIONS

The calculation of the kernel $U(p,k)$ requires a knowledge of the electron-band energies $\xi(p)$, the exciton band energies $E_\alpha(q)$, the coupling constants $Q_\alpha(q)$ and the Coulomb energy between two electrons on the spine. The actual values and shapes of the first two quantities do not have a drastic effect on the results within a wide range of reasonable parameters. We shall therefore not discuss them in further detail and refer the reader to DGL. On the otherhand, the interaction energies are crucial and it is therefore extremely important to estimate them in a reliable way.

IVa.   THE ELECTRON-EXCITON INTERACTION

The electron-exciton coupling parameter has the form

$$Q_\alpha(q) = \langle 1_\alpha q, \; k-q \; |V| \; 0,k \rangle \quad , \tag{5}$$

which corresponds to the scattering of an electron from momentum k to $k-q$, accompanied by the creation of an exciton of momentum q and band index α. When the electron states on the spine are described in the tight-binding approximation and the exciton-band states by a linear combination of terms in which always a single molecule is in an excited state, one obtains (DGL)

$$Q_\alpha(q) = \sqrt{\frac{2}{N}} \sum_{m,n,\nu} \int |\phi(r_1)|^2 \; V(r_1,r_2) \; c_{\alpha n}^\nu(q) e^{iqR_m} \rho_\nu(r_2,R_{nm}) d^3r_1 d^3r_2 \tag{6}$$

where $|\phi(r_1)|^2$ is the density of the electron on the spine, $\rho_\nu(r_2,R_{nm})$ is the transition density between the ground state and the excited state ν of the dye at site n in the m-th unit cell, and $c_{\alpha n}^\nu$ are the coefficients of the exciton band states. The calculation of ρ is described in ref. 10 and the result for the lowest excitation band of the pyridine cyanine is represented in Fig. 3.   It shows a clear oscillating dipole pattern. The typical dependence of $|Q_\alpha(q)|^2$ upon q for the mode α which couples most strongly to the spine is illustrated in Fig. 4.   Its distinctive feature is the sharp fall off in $|Q_\alpha(q)|^2$ with increasing q. This occurs at values of $q \sim 1/b$, where b is of the order of the distance of the nearest terminal group of the dye from the spine. In conventional superconductors the electron-ion interaction has a very short range, which results in a weak momentum dependence of the electron-phonon interaction. We believe that the strong momentum dependence of the electron-exciton coupling is the most significant difference between the exciton and the phonon mechanisms of superconductivity, and as we shall see it has important consequences.

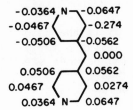

PYRIDINE CYANINE

Fig. 3. Linear–combination–of–atomic–orbitals calculated values of the transition density for the principal low-lying absorption band for the pyridine cyanine.

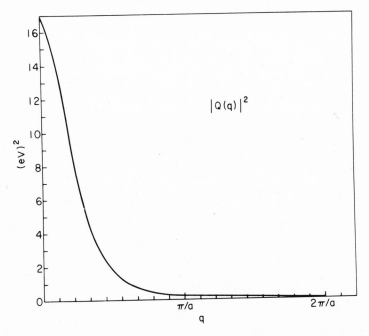

Fig. 4. Calculated electron-exciton interaction $|Q(q)|^2$ as a function of momentum transfer q.

IVb. THE COULOMB INTERACTION

The Coulomb interaction is the hardest quantity to estimate in a reliable way. The starting point are the parameters γ_n which measure the bare interaction between electrons on two atoms on the spine separated by a distance $r_n = na$, where a is the interatomic spacing. For simplicity we use the Nishimoto-Mataga[11] form for these parameters.

$$\gamma_n = \frac{e^2}{r_n + b} \tag{7}$$

The parameter γ_0, which corresponds to the interaction of two electrons on the same atom, may be derived from experimental values of the ionization energy and the affinity, and the result for platinum is $\gamma_0 = 6.03eV$. This determines the length b = 2.4Å.
The bare interaction is modified by the screening of the electrons in the same filament and in neighboring filaments. In addition, it is also screened by the dielectric constant of the surrounding organic medium. The first of these contributions to screening may be estimated on the basis of the calculations of Davis[12] on filamentary compounds like KCP. He found, in the Thomas-Fermi approximation that the screening was approximately isotropic, but with a screening length about ten times that of platinum metal. We use his approximate expression to obtain the screened parameters $\bar{\gamma}_n$

$$\bar{\gamma}_n = e^2 \exp\left[-\lambda(r_n+b)\right] / (r_n+b) \tag{8}$$

with

$$\lambda = 0.14\text{Å}^{-1} \quad .$$

The Fourier transform of the partially screened interactions (suitably corrected at large q for errors introduced by the discrete atomic representation (see DGL)) is plotted in Fig. 5 (upper curve).
Let us now discuss the screening effects of the organic medium. The interaction of the spine electrons with the low lying highly polarizable states of the dyes was singled out and treated dynamically in the exciton part of the kernel, exactly as the electron-phonon interaction in conventional superconductors. There still remains the interaction with the higher exciton bands which in the cyanine dyes lie at substantially higher energies (5-15eV). In view of this separation it is convenient to include their effect in the Coulomb part of the interaction, because they simply contribute to the overall static dielectric constant. The higher excitation bands are generally relatively narrow and one can therefore neglect the momentum dependence of their energies. One can also neglect the much lower electron band excitation energies $\xi(p)$ and with these approximations we obtain for the Coulomb part of the kernel (eq.3)

$$U_c(p-k) = V_o(p-k) - 2 \sum_{E>E_{ex}} \frac{|Q_E(p-k)|^2}{E} \quad , \tag{9}$$

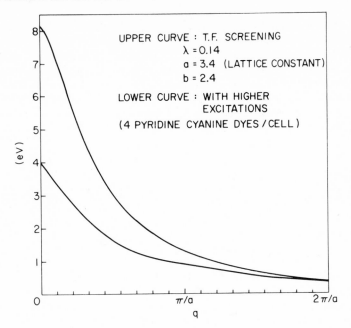

Fig. 5.   Screened Coulomb interaction calculated using Thomas-
          Fermi screening due to electrons in the same and neighbor-
          ing filaments (upper curve), and with the addition of
          dielectric screening from the neighboring organic environ-
          ment (lower curve).

where the sum extends over the exciton-band with energies E larger
than the lowest exciton-band energy E_{ex}. The first term represents
the Coulomb interaction screened by the electrons in the filaments
and the second is the contribution of the screening by the organic
medium. This second term may be expressed in an equivalent form
as

$$V_{hex}(r_1-r_2) = \sum_{n,R_1^n,R_2^n} V_0(r_1,R_1^n) \, \Pi_0(R_1^n,R_2^n) \, V_0(R_2^n,r_2) \qquad (10)$$

where R_1^n, R_2^n are the atomic coordinates of the nth polarizable
ligand and Π_0 is the lowest order contribution of the proper
polarization of the dye in the static ($\omega = 0$) approximation.
The Fourier transform of this expression is subtracted from $V_0(q)$
to give the lower curve of Fig. 5.
     Note that the total Coulomb interaction $V_c(r) = V_0(r) + V_{hex}(r)$
can be written in the form

$$V_c(r) = \frac{V_0(r)}{1 - V_{hex}(r)/V_c(r)} \qquad , \qquad (11)$$

the denominator playing the role of an effective dielectric constant ε. We found that the numerical value for this dielectric constant came out to be about 2. This is reasonable for the electrons in the spine may be considered to be buried in an organic environment whose dielectric constant would be of this order of magnitude.

V. NUMERICAL RESULTS

Using the geometry of the system illustrated in Figs. 1 and 2 we have calculated numerically the transition temperature T_c. It was convenient to use instead of eq. (1), the zero temperature equation for the gap

$$\phi(p) = -\int_{-\pi/a}^{\pi/a} \frac{dk}{4\pi} \frac{U(p-k)\ \phi(k)}{[\xi^2(k) + \phi^2(k)]^{\frac{1}{2}}} \tag{12}$$

where the integration is over the 1D Brillouin zone. Then we obtain T_c from the gap at $T = 0$ from the relation

$$kT_c = 3.5\ \phi(k_F)_{T=0} \tag{13}$$

For singlet superconductivity the spins of the electrons of the pair are antiparallel and $\phi(k) = \phi(-k)$. Eq. (12) can thus be rewritten as

$$\phi^s(p) = \frac{-1}{4\pi} \int_0^{\pi/a} dk \frac{[U(p,k) + U(p,-k)]\ \phi^s(k)}{\left\{\xi^2(k) + [\phi^s(k)]^2\right\}^{\frac{1}{2}}} \tag{14}$$

On the otherhand, for triplet superconductivity the orbital symmetry requires $\phi(k) = -\phi(-k)$ and eq. (12) gives

$$\phi^t(p) = \frac{-1}{4\pi} \int_0^{\pi/a} dk \frac{[U(p,k) - U(p,-k)]\ \phi^t(k)}{\left\{\xi^2(k) + [\phi^t(k)]^2\right\}^{\frac{1}{2}}}. \tag{15}$$

It should be noted that because of the denominator, the integral in both cases is dominated by the contribution of the kernel at $k = k_F$ and thus by the terms $U(p_F,k_F)$ and $U(p_F,-k_F)$. In Fig. 6 we show a plot of $U(p_F,k_F-q)$ vs q calculated for the model of Fig. 1. We see that $U(p_F,k_F)$ is negative, due to the strong exciton contribution of equ. (4) resulting from the large value of $|Q_\alpha(q)|^2$ near $q = 0$. On the otherhand $U(p_F,-k_F)$ is positive because of the dominance of the Coulomb repulsion for large momentum $(2k_F)$ transfers. Because of this the combination $[U(p_F,k_F) - U(p_F,-k_F)]$ which occurs in the triplet case yields a more attractive contribution than does $[U(p_F,k_F) + U(p_F,-k_F)]$ which occurs in the singlet case. Consequently we expect a larger zero temperature gap, and T_c for triplet state superconductivity compared to that for the singlet case. This is borne out by the numerical result for the two gaps illustrated in Fig. 7.

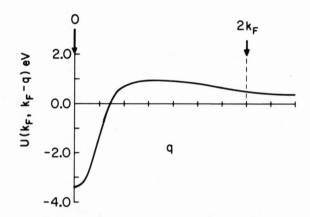

Fig. 6.   Plot of $U(k_F, k_F-q)$ vs q for the model system.   The major
          contributions to superconductivity come from the inter-
          action at $q = 2k_F$ (g_1) and at $q = 0$ (g_2)

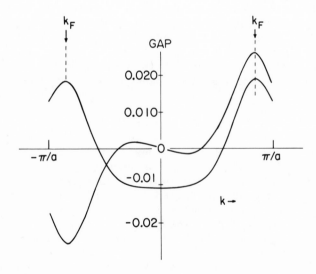

Fig. 7.   Calculated results for the singlet (symmetric) and triplet
          (antisymmetric) gap functions for the model system of
          Fig. 1.

We have calculated T_C for various values of the electron band
width E_O and Fermi momentum, k_F. For band widths of the order of
2-3eV and $k_F \approx (\frac{5}{6}) \pi/a$ we found $T_C \approx 10^3$ ^0K for singlet superconductivity, and T_C for triplet superconductivity a little higher
in each case. This indicates that, indeed, superconductivity at
or above room temperature appears to be possible in such an excitonic
system. We find that T_C is reduced if the density of states at
the Fermi surface is reduced by varying E_O or k_F in an appropriate
way. A similar calculation on a model with only two polarizable
dyes per unit cell gave $T_C = 0$ as did a calculation in which the
Pt-N distance was increased from 2 to 3Å . This indicates how
important is the dense packing and close coupling of the exciton
system to the conductive spine.

VI. DISCUSSION

In the introduction we listed several problems associated
with the one-dimensional excitonic superconductor. To answer the
quantitative questions raised there, we have performed detailed
calculations of the various interactions, as described in sections
(IV) and (V). The conclusion is that in the proposed structure
the exciton mediated electron-electron interaction is strong enough
to overcome the Coulomb repulsion. At the same time, one learns
from these calculations that only those structures with the
excitonic system bonded directly to the conductive spine and
containing three or four dyes per atom of the spine appear to have
any chance of exhibiting superconductivity due to the exciton
mechanism. This emphasizes the advantage of the one-dimensional
over the higher dimensional models of an excitonic superconductor.
In a one-dimensional structure it is much easier to bring a high
density of excitonic medium in close contact with the conduction
region.

The most striking result of the calculation of the electron-
exciton coupling is the strong momentum dependence of this coupling.
This momentum dependence is due to the spatial separation between
the spine electrons and the electrons in the polarizable medium.
This separation serves to reduce the exchange interaction between
the two kinds of electrons and is a characteristic feature of
many of the models proposed for the excitonic superconductor.

The strong momentum dependence of the electron-exciton
interaction effectively restricts the excitons which play any
role in the anticipated superconductivity to those of low momenta,
those below a certain critical momentum q_c. At higher momenta the
electron-electron interaction is determined by the coulomb repulsion.
This specific momentum dependence of the interaction plays an
important role in the discussion of two problems mentioned in the
introduction, namely, the problem of the competition with other
instabilities and the problem of vertex corrections.

The various instabilities in one-dimensional systems were
discussed by one of us (H. G. in the present proceedings).   It
is shown there that if the interactions are not too strong, each
system may be characterized by two coupling parameters, g_1 and g_2,
which measure the interaction with momentum transfer q = $2p_F$ and
q = 0, respectively.   The basic result of the investigation of the
various instabilities is that the line $2g_2$ = g_1 separates the
region of charge and spin density wave instabilities which lie
on the right of this line from the pairing instabilities on the
left.   It is believed that all the one-dimensional organic and
metalo-organic compounds known at present lie to the right of this
line.   Explicit calculations on the system proposed here show
that it is characterized by a substantial and negative g_2 and by
a significantly smaller and positive g_1.   (See Fig. 6)   This
places the system in the region of (g_1,g_2) - plane, where both
rigorous analysis and the simple treatment by means of the BCS
equation (section V) predict triplet superconductivity.   See Fig.
8). Having established that in our case superconductivity dominates,
we feel that without claiming accuracy of the calculated T_c, it
is not necessary to calculate similtanously the pairing and

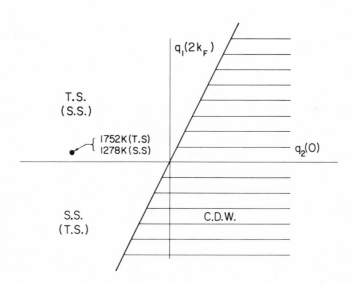

Fig. 8. g_1 - g_2 plane showing region of charge density wave
          instabilities (CDW) and singlet (triplet) superconcutivity
          S.S. (T.S.)   The calculated values of T_c for the model
          are indicated.

insulating instabilities as in the coupled channel equations of
Bychkov, Gorkov and Dzyaloshinskii.[13]

We note that if the spine electrons are also coupled to
phonons then this will affect mainly the value of g_1, and if this
coupling is sufficiently strong, it will result in a negative
value for g_1. The system will then lie in the region of singlet-
rather than triplet-superconductivity. (See Fig. 8)

Let us now discuss the problem of vertex corrections. The
formulation of the theory of superconductivity depends on the
validity of Migdal's theorem which asserts that vertex corrections
are small and may be neglected.[14] The lowest order correction to
the electron-phonon vertex is shown in Fig. 9. For an incoming
phonon of phase velocity ω/q, much smaller than the Fermi velocity
v_F, this correction is of order $\omega_F/E_F \approx 10^{-2}$. Most of the phonons
involved in conventional superconductivity have momentum $q \simeq p_F$
and, therefore, a very small phase velocity. In the present case
only excitons with small momentum and hence, phase velocities
much greater than v_F are involved in the conjectured superconducting transition. It was shown[15] that for phonons with a high phase
velocity the vertex correction in the figure is of the order of
$g^2 N(o)/\omega_D$, where g is the electron-phonon coupling constant. This
is a crude estimate of McMillan's parameter λ and is in our case
(for excitons) about 0.2. However, this result for the vertex
correction was obtained under the assumption of a momentum
independent g. In our case the coupling constant $Q(q)$ is

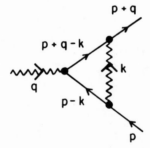

Fig. 9.   First vertex correction to the Migdal approximation
          which we show is small for the particular model proposed
          here.

strongly peaked around $q = 0$, which results in a further reduction
of the vertex correction by the factor $\Delta E/E$, where E is a typical
exciton energy and $\Delta E = \varepsilon(p_F + q_c) - \varepsilon_F$. The net lowest order

correction for the system proposed is $\simeq 0.05$. One should point
out that in a strictly one-dimensional system the lowest vertex
correction diverges[16] for phonons (excitons) of momentum $2p_F$.
Again, this does not pose a problem in our case because such
excitons are essentially decoupled from the spine electrons.

We have used the characteristic momentum dependence of the
interaction to discuss two of the basic problems raised by the
present model with conclusions which justify the use of the simple
BCS equation to obtain the order of magnitude of T_c. There, is
however one additional aspect to the use of the BCS equation. It
is a mean field equation and neglects the effects of fluctuations.
We believe that a small degree of interchain coupling is sufficient
to bring the actual transition temperature close to its mean field
value; and by writing the BCS equation we implicitly assume such
coupling. However, the quantitative properties of such systems
will depend strongly on the nature and strength of the interchain
coupling which affects the fluctuation amplitudes. We feel that
it would be premature to try to estimate these effects at present.

Perhaps the most important lesson which can be learned from
this calculation is the need for a large number of polarizable
substituents (four in this case) for each atom of the spine;
and, extremely close bonding of the substituents to the spine.
Calculations on a system with only two substituents per atom gave
$T_c = 0$ as did a calculation with polarizable ligands on every
other metal atom. Likewise when one increased the Pt-N distances
from 2Å to 3Å, again T_c fell to zero. This shows how critical is
the balance between the exciton and Coulomb interactions.

An important consequence of this last result is that in a
system with a relatively diffuse electron distribution such as
the π-system in TTF or TCNQ it appears impossible to bring a set
of polarizable substituents close enough to these electrons to
yield a significant exciton attraction. This we believe is a
compelling reason for focussing on the transition-metal type
complexes.

Before discussing the chemical problems which are unique to
this model it is appropriate to ask whether the model is the best
possible candidate. What about 2D layer systems for example? A
calculation somewhat similar in nature to ours on a layer structure[17]
shows the exciton interaction in that case to be substantially
weaker than the Coulomb term. This result follows because of the
lower packing density that one can achieve in such a 2D geometry
as compared to 1D. For this reason and the above arguments against
the TTF-TCNQ systems we feel that the model proposed is among the
best possible as far as satisfying the theoretical criteria for
excitonic superconductivity. Chemical modifications, some of which
we will discuss, certainly should be considered, provided, of
course, they do not change substantially the requirements of close
packing and polarizability of the ligands.

The first chemical problem relates to the assumed properties
of the cyanine dye ligands. In our calculation we have treated
the dye as a symmetric cyanine in which each ligand system would
require four counter ions for the dyes, two for the platinum atoms
plus any required for oxidation of the spine. Such a highly
charged ligand system would tend to oppose a stacked configuration.
For this reason we believe a neutral merocyanine type of substituent
would make a better choice of ligand. Moreover, in these the polar
nature of the environment and the nature of the end groups on the
chromophore can be used to make the electronic structure of the
merocyanine essentially symmetric[18] and thus compensate for the
chemical asymmetry of the dye. The transition density is relatively sensitive to this asymmetry. The question is how small
can the asymmetry be made in practice?

A second question relates to the oxidation of the chain. Can
the Pt-chain be oxidized without destroying the double bonds of
the chromophore system? Evidence from cyanine - iodide complexes
with TCNQ indicate that simple cyanine systems are likely to be
stable while carbo- or dicarbo cyanines are not. In any case it
would be necessary to keep in mind the relative oxidation potentials
of the ligands and of the chain.

A third uncertainty lies in the strength of the metal-metal
interaction at the 3.4Å separation required of the π-orbitals in
the bulky ligands. Extrapolation of the results of an X-α calculation[19] on a Pt-complex suggest that the band width at this
separation would be of the order of 1eV. This would be adequate
but is probably too weak to be a significant directive force in
crystal packing. It would be valuable though to obtain direct
experimental evidence of this band width.

A major problem is whether these complexes will stack. The
structure was chosen so that at least some factors would favor
such a stacked array. The van der Waal's interaction between the
dye complexes would be strong and directive and the planar space-
filling structure should favor a compact stacked configuration.
However, if such a system does not stack what could be done about
it? Fortunately something is known about packing forces.[20] It
is known too, that dichloro-substituted aromatic systems favor
face-to-face packing.[21] The favored structure must not contain
voids. If it does the chance of obtaining it are increased by
allowing the voids to be filled with solvent molecules.[22] Hydrogen
bonds can play an important role in determining the overall crystal
structure. Could they be used to favor packing of the merocyanine
units?

This is a small subset of the type of problems that remain to
be answered. Much more experimental work is needed in molecular
engineering before these types of problem can be answered with
anything like engineering precision.

We have emphasized in the introduction the caution one has to take in trying to use the BCS theory in a predictive manner in a completely new physical regime. On the otherhand, there has been one successful use of the BCS theory in this way – the prediction of the superfluidity in He^3. If He^3 were a one-dimensional system we would place it in the (g_1,g_2)-plane in the neighborhood of the systems discussed above. The interaction in He^3 has a short range repulsion (hard core) and a long range attraction, which in momentum space corresponds to negative g_2 and positive g_1, and places it in the region of triplet super-conductivity, and indeed, He^3 is a triplet "superconductor".

REFERENCES

1. L. B. Coleman, M. J. Cohen, D. J. Sandman, F. G. Yamagishi, A. F. Garito and A. J. Heeger, Solid State Comm. 12, 1125 (1973).

2. W. A. Little, Phys. Rev. 134, A1416 (1964).

3. V. L. Ginzburg, Sov. Phys.-JETP 20, 1549 (1965); Contemp. Phys. 9, 355 (1968); Ann. Rev. Mater Sci. 2, 663 (1972); V. L. Ginzburg and D. A. Kirzhnits, Phys. Ref. 4, 345 (1972).

4. D. Allender, J. Bray and J. Bardeen, Phys. Rev. B7, 1020 (1973).

5. M. Strongin, Sol. St. Comm. 14, 88 (1974).

6. M. Weger, Rev. Mod. Phys. 36, 175 (1964).

7. D. Davis, H. Gutfreund and W. A. Little, Phys. Rev. B13, 4766 (1976).

8. D. Davis; Ph.D. dissertation (Stanford University, 1974) unpublished.

9. D. A. Kirzhnits, E. G. Maximov and D. I. Khomskii, J. Low Temp. Phys. 10, 79 (1973).

10. H. Gutfreund and W. A. Little, J. Chem. Phys. 50, 4468 (1969).

11. K. Nishimoto and N. Mataga, Z. Phys. Chem. (Frankf) 13, 140 (1957).

12. D. Davis, Phys. Rev. B7, 129 (1973).

13. Yu Bychkov, L. P. Gor'kov, and I. E. Dzyaloshinskii, Zh. Eksp. Teor. Fiz 50, 738 (1966) Sov. Phys. - JETP 23, 489 (1966) .

14. A. B. Migdal, Sov. Phys. - JETP 7, 996 (1958). See also
     "Quantum Theory of Many Particle Systems," A. L. Fetter &
     J. D. Walecka, McGraw Hill Book Co., N.Y. (1971) p. 406.

15. S. Engelsberg and J. R. Schrieffer, Phys. Rev. 131, 993
     (1963).

16. A. M. Alfanas'ev and Yu. Kagan, Zh. Eksp. Teor. Fiz. 43,
     1456 (1962) Sov. Phys. - JETP 16, 1030 (1963) .

17. W. A. Little, J. Low Temp. Phys. 13, 365 (1973).

18. K. Mees, "Theory of the Photographic Process" (Mac Millan,
     New York, (1966)).

19. A. Abarbanel, Ann. Phys. 91, 356 (1975); D. Whitmore, Phys.
     Lett. 50A, 55 (1974).

20. A. I. Kitaigorodsky, "Molecular Crystals and Molecules,"
     (Academic Press, New York, 1973).

21. M. Cohen and B. S. Green, Chem. Brit. 9, 490 (1973).

22. H. Kuroda, "Energy and Charge Transfer in Organic Semi-
     conductors," Ed. K. Masuda and M. Silver, (Plenum Press,
     New York, 1974, p. 177).

     We wish to acknowledge support for this work from the
Binational Fund, National Aeronautical and Space Agency, Contract
JPL 953752 and National Science Foundation, Grant DMR 74-00427-
A03.

ORGANIC LINEAR POLYMERS WITH CONJUGATED DOUBLE BONDS

G. Wegner

Institut für Makromolekulare Chemie der Universität
Stefan-Meier-Straße 31
D-7800 Freiburg (West Germany)

DESIGN OF A SYNTHESIS FOR POLYMER SINGLE CRYSTALS

Solid state physicists like to work with single crystals as
large and as defect free as possible. Polymer chemists and physicists have learned to live with the fact, that most polymers cannot
be obtained in form of true single crystals but rather as semicry-
stalline materials in which part of the macromolecules remains
amorphous where as other parts of the same molecules are part of
microscopic crystallites embedded in the amorphous matrix. For some
time it was even believed that perfect polymer single crystals do not
exist at all, mainly because the phenomenon of chain-folding observed in polymer crystallization is so general and structure independent that formation of extended chain crystals of polymers
seemed to be impossible. It was therefore felt to be a particular
challenge to device methods how to synthesize large, nearly defect
free polymer single crystals in order to study the behaviour of such
crystals which - by principle - have to be highly anisotropic materials, the anisotropy, of course, arising from the fact that all
atoms of the polymer backbone are covalently linked together in
chain direction and laterally the crystal is held together by van-
der-Waals interactions only.

With the arrival of Little's ideas on the possible chemical
structure of a high-temperature superconducting material (1,2) an
additional challenge was felt by the synthetic chemists, namely
to synthesize a long chain molecule with a backbone of conjugated
carbon double bonds, highly polarizable (dyestuff) side groups and
the ability to crystallize in form of acceptable single crystals.
So far the conventional methods of organic and metal organic chemistry were not successful (3) to prepare molecules of the kind

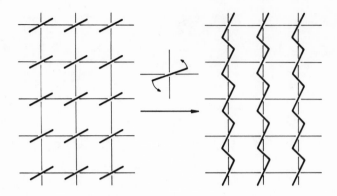

Fig. 1. Scheme of a topochemical polymerization: transformation of
a single crystal of monomers into a single crystal of polymers.

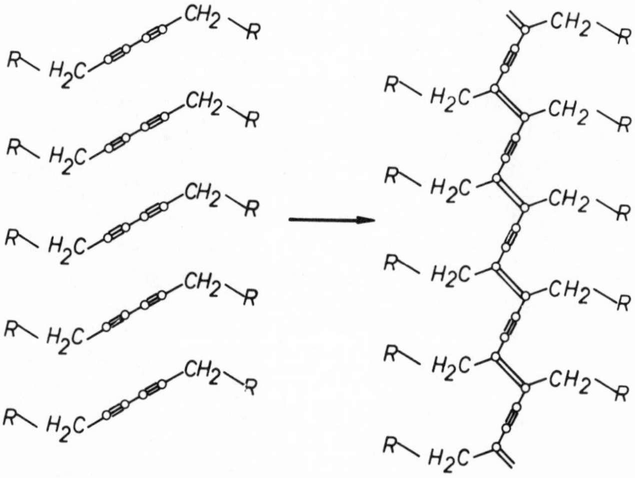

Fig. 2. Topochemical polymerization of diacetylenes; the structure
of R is explained in Tab. 1.

asked for, if one wants to test Little's ideas experimentally.

In the course of our studies on reaction mechanisms, structure
and morphology in solid state polymerizations (4) we came more or
less accidentely across a reaction which gives rise to nearly defect free macroscopic single crystals of macromolecules with a
backbone of conjugated C-C-double and -triple bonds. This was first
reckognized (5) in 1969 and has since been shown to be a reaction
with many interesting facets both with regard to solid state

chemistry and physics.

The basic principle of the reaction which is generally named "topo-chemical polymerization of monomers with conjugated triple bonds" ("diacetylenes") is described by the chemical formulation

$$R-C\equiv C-C\equiv C-R \longrightarrow \left\langle\!\!{}^{R}_{}C-C\equiv C-C{}^{}_{}\!\!\right\rangle_{\!n}^{\!\!R} \longleftrightarrow \left\langle\!\!{}^{R}_{}C=C=C=C{}^{}_{}\!\!\right\rangle_{\!n}^{\!\!R}$$

where R is any of the substituents indicated in Tab. 1.

The true nature of the reaction is better understood by in-spection of Fig. 1 and 2. Fig. 1 shows schematically what is meant by the term topochemical polymerization: a diffusionless solid-state transformation of a single crystal of a suitable monomer in-to the corresponding single crystal of a polymer such that the centers of gravity of the monomers have the same crystallographic position and symmetry as the base units of the polymer. All reac-tivity comes about by very specific rotations of the monomer on its lattice site determined by the packing properties of the molecules. Thus inevitably a crystal of extended polymer chains is formed.

The idea that such reactions should occur goes back to G.M.J. Schmidt (6) who developed a number of general rules relating to organic solid state chemistry but the reaction of the diacetylenes described more closely in Fig. 2 is the only example where really polymer single crystals can be obtained.
In the monomer crystal the molecules are arranged in a ladder like fashion such that the ends of one triple bond system approach the beginning of the triple bond system of the next unit to a dis-tance $\leqslant$ 0.4 nm. Polymerization occurs by successive tilting of each molecule along the ladder without moving the center of gravity. Thus the mode of packing of the side groups R, the specific volume and the lattice symmetry can be retained throughout the reaction.

Polymerization is simply brought about by annealing the co-lorless monomer crystals below their melting point or by high ener-gy or uv-irradiation. Due to the formation of the conjugated back-bone polymerizing crystals turn deeply colored as polymerization starts and finally turn deep red with typical metallic luster. In many cases quantitative conversion is reached within a few hours. The details of preparation and polymerization of a number of mono-mers have been published previously (7-10). No exact molecular weight determinations have been performed so far but based on viscosity measurements of the dissolved polymers, the very high mechanical strength of the polymer crystals in chain direction (11) and the perfection of the polymer crystals regarding number and types of dislocations (12,13) it can be safely assumed that very long chains are formed extending essentially from one defect site to the next over distances of up to 1 µm or more.

Table 1: Some examples of diacetylenes which undergo topochemical polymerization

a) symmetrical (R = R') R	b) unsymmetrical R	R'
$-(CH_2)_n-CH_3$	$-CH_2OH$	$-C{\equiv}C-CH_2OH$
$-(CH_2)_n-OH$	$-CH_2-O-CO-NH-Ph$	$-C{\equiv}C-CH_2-O-CO-NH-Ph$
$-(CH_2)_n-CO_2H$ and salts	$-(CH_2)_n-OH$	$-(CH_2)_m-CH_3$
$-(CH_2)_n-O-CO-NH-Ph$	$-(CH_2)_n-CO_2H$	$-(CH_2)_m-CH_3$
$-(CH_2)_n-O-SO_2-$⬡$-CH_3$		
-⬡ (o-, m-, <u>not</u> p-) NH-Ac		

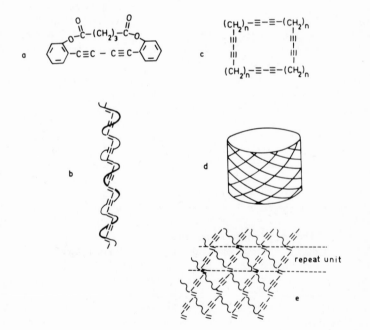

Fig. 3. Formation of helical (b) or cylindrical (d) structures by polymerization of cyclic monomers such as a or c. The architecture of molecules like d is best understood, if the cylinder is thought to be cut and rolled into a plane (e).

As already mentioned the polymerization of diacetylenes which
is restricted to the solid state and does not occur in the melt,
gas phase or solution is a general method for synthesis of polymers
with a sequence of unsaturated carbon bonds in the backbone as indicated in Tab. 1 for linear diacetylenes. Polymerization of cyclic
diacetylenes gives rise to helical (10) or cylindrical (14) structures as explained by Fig. 3.
The packing conditions necessary to bring about reactivity in the
lattice are summarized in Fig. 4 and can be used as a guiding
principle by those who want to develop new systems by putting
special substituents into the side groups. This scheme which is
based on chemical experience as well as on a number of crystal
structure analyses of non-polymerizable monomers and of solid-state
polymerized crystals was more quantitatively developed by R.H.
Baughman,

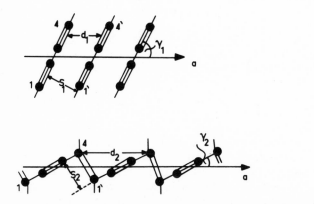

$$S_1 = d_1 \sin \gamma_1$$
$$0.34 < S_1 < 0.40 \, nm$$
$$\gamma_1 \approx 45°$$
$$d_2 = 0.49 \pm 0.01 \, nm$$
$$\gamma_2 = 13.5°$$
$$S_2 = 0.12 \, nm$$

Fig. 4. Packing requirements in the monomer characterized by parameters S_1, d_1 and γ_1 necessary to bring about polymerization to
nearly perfect crystals whose chains are characterized by S_2, d_2
and γ_2.

who considered the effect of deviations from the best set of parameters given in Fig. 4 in terms of the Free Energy of reaction.

Besides the important relation between packing of the molecules
and chemical reactivity there is the question to solve by what specific chemical mechanism chain growth does occur inside the lattice.
Based on observations obtained, if one looks to the spectral changes occuring in a polymerizing single crystal such as the ones depicted in Fig. 5 it was proposed that polymerization proceeds by
carbenes as active intermediates (8,16). Possible steps of the
reaction sequence are: thermal excitation of the conjugated triple
bonds to dicarbene-like intermediates, coupling of carbenes with

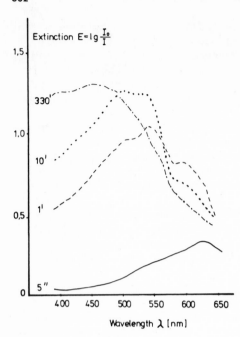

Fig. 5. Spectra of a polymerizing single crystal of
$H_3C-(CH_2)_{11}-C\equiv C-C\equiv C-(CH_2)_{11}-CH_3$
(low temperature modification)
as polymerized by uv-irradiation at $\lambda=250$ nm for various
times.

formation of double-bonds and growth of the macromolecule by further
addition of excited monomer to the carbene-like chain end. The very
defined absorption maxima of transients seen in polymerizing single
crystals around 600 nm and 560 nm were atributed to already long
chains having either one or two carbene-chain-ends. On addition of
a solvent to the polymerizing crystal which is able to dissolve
the monomer without distracting the polymer crystalline order the
active chain ends are destroyed and the only remaining absorption
is generally centered around 500 nm. This is the absorption due to
the final "dead" polymer. Sometimes, expecially at very high con-
versions the absorptions of the transients persist even on pro-
longed treatment with solvent indicating that some chain ends are

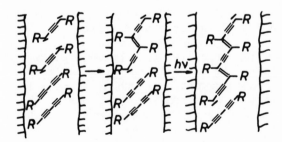

Fig. 6. Growth of the polymer chain by carbenes as active
intermediates.

inaccessibly trapped in the otherwise perfect polymer matrix.
This effect has to be kept in mind in further discussions of the
intrinsic optical and electrical properties of such materials. Recent ESR – Spectroscopic investigation of thermal polymerization of
hexadiine-1,6-diol-bis(p-toluene sulfonate) has clearly demonstrated that paramagnetic triplet species of the kind described in
Fig. 6 do exist indeed in partially converted single crystals and
the following zero field splitting parameters were derived:
$D/hc = + 0.2731 \text{ cm}^{-1}$ and $E/hc = - 0.0048 \text{ cm}^{-1}$ (17,18).

CRYSTAL PERFECTION AND MECHANISM OF PHASE TRANSITION IN SOLID STATE SYNTHESES

Solid state polymerizations can be regarded as special types
of phase transitions in which a solid monomer is changed into a
solid polymer. It is not surprising that the mechanism of phase
change shows profound impact on the perfection of the polymer
phase thus produced. In the case of diacetylene polymerization
it was demonstrated that the reaction proceeds homogeneously inside
the monomer crystal starting at points distributed at random
throughout the lattice (19-21). Thus, a solid solution of extended
chain macromolecules dispersed in the monomer matrix is formed at
first as shown in Fig. 7 b. Consequently, the coherence between all
parts of the crystal is retained and the single-crystal character
is never destroyed.
This behaviour is quite uncommon. Normally, solid state reactions
proceed by nucleation of a new phase at defects, surfaces or imperfections of the mother-phase as indicated in Fig. 7 a. It is
quite obvious that in these cases large polymer crystals cannot be
obtained because coherence between the various nuclei is soon going
to be lost because inherent differences in specific volume between
mother- and daughter-phase lead to a breakdown of the single crystal into polycrystalline material. Various aspects of such considerations and impacts onto the question under what circumstances
oriented, fiber like product phases can be obtained have been reviewed recently (4). These considerations are important in order
to understand the morphologies obtained for example in the solid
state synthesis of $(SN)_x$ or $(CH_2-O)_n$ (22,4).
It should be noted that topotaxy quite often observed in solid-state
syntheses, that is coincidence of certain crystallographic directions
of mother and daughter phase is not necessarily a consequence of
topochemical effects. The fibre like growth of some polymers is
rather a consequence of simultaneous polymerization and crystallization at internal surfaces (23,4). If nucleation occurs at the
surface of a crystalline matrix orientational effects are expected
because of the impact of surface free energy onto the shape of the
nucleus. There is a separate equilibrium form for every possible
relative orientation of the two sets of axes. Most relative

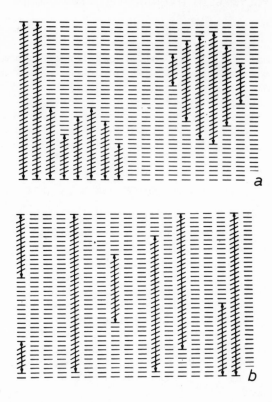

Fig. 7. Two different mechanisms of phase transformation in solid-state polymerization. a) heterogeneous growth by nucleation; b) homogeneous reaction in form of solid solution. Mechanism b is found in diacetylene polymerization.

orientations will give high-angle boundaries, hence relatively sphe-rical equilibrium forms. A few orientations will always involve some very low energy boundaries and will give highly non-spherical equilibrium forms such as fibrils or elongated ribbons as they are observed in $(SN)_x$ or $(CH_2-O)_n$.

Furtheron, it should be stressed, that there is very little hope to device methods of crystallization of long chain molecules from solution into macroscopic crystals. Due to the kinetics of crystallization a macromolecule must form a folded chain crystal, if it crystallizes from dilute solution or melt (24,25). All attempts to prepare a polymer backbone first by one or the other methods of polymer chemistry e. g. in order to simulate polymer charge-transfer-complexes of the type TTF-TCNQ or polyconjugated molecules or more ambigous molecules such as the ones proposed by Little in the hope that these products can be crystallized in a second step are therefore highly unrealistic. In addition to solid-

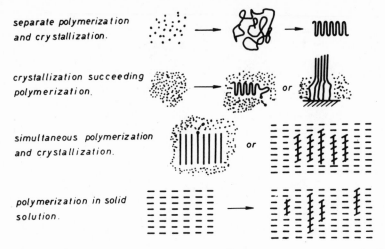

separate polymerization
and crystallization.

crystallization succeeding
polymerization.   or

simultaneous polymerization
and crystallization.   or

polymerization in solid
solution.

Fig. 8. Schematic representation of the four different methods how
to produce crystals of polymers.

state reactions such as the topochemical polymerization of diacetylenes true single crystals of macromolecules can only be formed
by simultaneous polymerization and crystallization either from solution or from the solid-state if one succeeds to sufficiently control the nucleation step (4,23).
Although such reactions have considerable importance in the technical production of some polymers the nucleation processes are not
well understood so that polycrystalline materials are obtained
only.
For sake of clarity, the four different methods of polymer crystallization are again depicted and summarized in Fig. 8 (4).

CRYSTAL AND MOLECULAR STRUCTURE OF POLYDIACETYLENES

Single crystals of polydiacetylenes can be analyzed by the
conventional direct methods of x-ray structure analysis. The fol-
lowing compounds have been analysed:
Poly (hexadiine-1,6-diol-bis-p-toluene sulfonate) 1 (26)
Poly (hexadiine-1,6-diol-bis-phenylurethane)   2 (27)
Poly (5,7-dodecadiine-1,12-diol-bis-phenylurethane 3 (28)

R: $-CH_2-O-SO_2-$ ⬡ $-CH_3$ 1

$-CH_2-O-\underset{O}{\overset{\|}{C}}-NH-$ ⟨⟩   2

$-(CH_2)_4-O-\underset{O}{\overset{\|}{C}}-NH-$ ⟨⟩   3

The molecular dimensions of polymer 1, the most investigated polydiacetylene because it is so readily synthesized in form of large
crystals, can be seen in Fig. 9. The polymer backbone consists of
a regular sequence of triple-single-double-bonds all arranged
in one plane. The substituents R are arranged in trans-position
with regard to the double bonds. Packing is determined by the aromatic rings of the side groups with their ring planes almost perpendicular to the plane of the backbone. Packing of the polymer
chains is best clarified by looking to a projection along the chain
axis (Fig. 10 a).
There are two symmetry related molecules in the unit cell of space
group, P 2_1/c, a = 1.4493, b = 0.4910, c = 1.4936 nm, β = 118.14°,
D = 1.483 g cm^{-3}. The chain axes extend along b. The crystals show
pronounced cleavage behaviour with (100) and (102) as cleavage
planes (compare also Fig. 12). Crystals of 1 undergo a very interesting reversible phase transition of higher order if cooled to
-150°C (29). At this temperature a doubling of the unit cell occurs
by slight rotation of the side groups of every second layer of molecules placed along the (102) plane as shown in Fig. 10 b. There
are now two kinds of molecules per unit cell giving rise to a splitting of the electronic absorption bands (30,31).

Polymers 2 and 3 have the possibility to form hydrogen bonds.
In polymer 2 the hydrogen bonds extend perpendicular to the plane
of the backbone (27).The bond lengths and angles of the backbone
are very similar to polymer 1 as indicated in Table 2.
In polymer 3, however, the bonds adjacent to the central triple -
bond have considerable double-bond character and the bonds linking
two consecutive base units are of the type of single bonds between

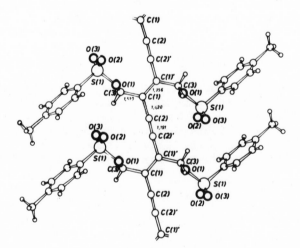

Fig. 9. Structure of polymer 1 as projected on the plane of the
backbone (26).

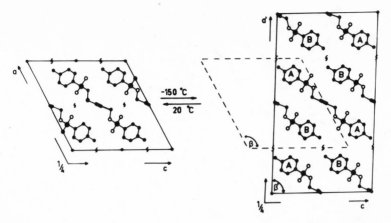

Fig 10 a)   (left).Structure of polymer 1 (high temp. modification)
projected along [010] (26). b)   (right).Low temp. modification;
differently inclined rings are labeled A or B (29).

two double bonds (compare Table 2). The crystal structure of 3 is
shown in Fig. 11. Hydrogen bonds extend in the same plane as the
polymer backbone linking consecutive side groups. (010) is a pronounced cleavage plane. The crystals are monoclinic P 2_1/a,
a = 0.6229, b = 3.9027, c = 0.4909 (chain axis) nm, β = 106.85°,
D = 1.257 g cm^{-3}. There are 4 chains per unit cell.
The bond lengths occuring in the polymer backbone are compared
in Table 2. The backbone of polymers 1 and 2 is thus best described by formula I (Table 2), but the backbone of polymer 3 has
a strong contribution of the electronic structure described by
formula II (Table 2). At present, it is completely unknown what
factors cause this considerable difference in bond lengths. Taking into account various evidences from Raman-resonance spectra
of more than 20 different polydiacetylenes (32,33) it is felt that
the best description of the polymer backbone is given by III
(Table 2). This describes admixing of limiting structures I and II
to the final resonance structure III with extended electron delocalization along the backbone but a potential minimum at the central bond which is always much shorter than all other bonds.

    It seems that there is a fair influence of the side groups on-
to the electronic structure of the backbone arising from two factors, namely packing and effects due to polarization of the backbone electrons by dipolar side groups. Since the side groups must
be packed such that the crystallographic repeat of 0.49 nm of the
backbone is met, the backbone will be either compressed or extended depending on the actual packing dimensions of the side groups.
Furtheron, it is quite often observed that substances with similar
packing properties but different polarizability of the side groups

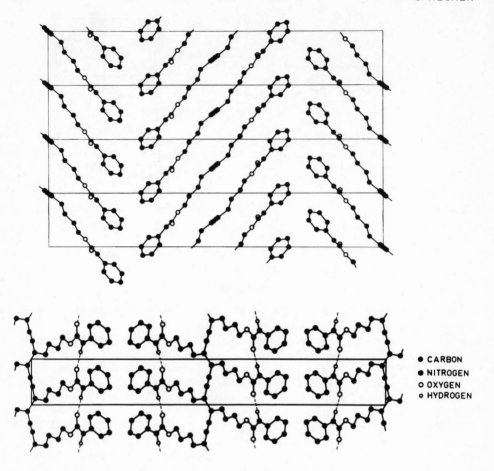

Fig. 11. Crystal structure of poly (5,7-dodecadiine-1,12-diol-bis-phenylurethane)
above: projection along [001]; below: projection along [100]

show different positions of the uv-absorption maxima. As already
mentioned, Raman-spectroscopy is an excellent tool in order to study the structure of polydiacetylenes (32,33,34). Generally the spectra are resonance enhanced, when the exciting radiation is close in
frequency to a real electronic excitation of the polymer. This Ra-
man-spectrum is dominated by those molecular motions which couple
strongly with the electronic states of the backbone. The majority
of the normal vibrational modes of the polymer are associated with
side-group motions which do not interact strongly with the back-

$$\begin{array}{ccc}
\underset{\substack{\bullet\bullet-C}}{\overset{\substack{R}}{\diagdown}}\!\!\!\!\overset{\text{b c}}{\underset{\substack{a}}{C-C\equiv C-C}}\!\!\!\!\overset{\substack{C-\bullet\bullet}}{\diagup}\quad R\;\;I &
\overset{\substack{R}}{\diagdown}\underset{\substack{\bullet\bullet=C}}{C=C\equiv C=C}\overset{\substack{C=\bullet\bullet}}{\diagup}\;R\;\;III &
\overset{\substack{R}}{\diagdown}\underset{\substack{\bullet\bullet=C}}{\overset{\text{b c}}{\underset{\substack{a}}{C=C=C=C}}}\overset{\substack{C=\bullet\bullet}}{\diagup}\;R\;\;II
\end{array}$$

R	Bond Distances (nm)		
	a	b	c
$-CH_2-O-SO_2-\bigcirc-CH_3$ Kobelt and Paulus 1974	0,136	0,143	0,119
$-CH_2-O-\underset{O}{\overset{\|}{C}}-NH-\bigcirc$ Hädicke et al. (1971)	0,136	0,141	0,121
$-(CH_2)_4-O-\underset{O}{\overset{\|}{C}}-NH-\bigcirc$ Enkelmann and Lando (1976)	0,146	0,138	0,117
expected for I	0,134	0,143	0,121
expected for II	0,146	0,132	0,128

Table 2. Comparison of bond lengths occuring in the backbone of polymers 1, 2 and 3.

bone electrons. Thus, the resonant Raman spectrum is much simpler than the non-resonant spectrum and it was consequently reckognized (33,35) as an important method to study the molecular structure as well as interaction of electronic excitations and phonons in the one-dimensional polydiacetylene crystals.

SOME PROPERTIES OF POLYDIACETYLENES

Polydiacetylenes are a new class of materials with strongly anisotropic behaviour. Since the polymer 1, poly (hexadiine-1,6-diol-bis-p-toluene sulfonate), is most readily prepared (36) and is easily obtained in form of very large, nearly defect free single crystals (12,21), a large number of experiments and measurements have been carried out using this material. Care must be taken in generalization of these results to the whole class of polydiacetylenes and considerable work has yet to be invested in order to separate the intrinsic properties of the conjugated main chains arranged in a crystal from the contribution of the side groups to the overall properties of the crystal under investigation.

Crystals of polymer 1 have the typical shape shown in Fig. 12; sometimes they are truncated exhibiting the intrinsic (102) cleavage plane as a growth surface. (100) is the most developed growth surface as well as the second cleavage plane. Most data of reflectivity or electrical conductivity refer to this surface.

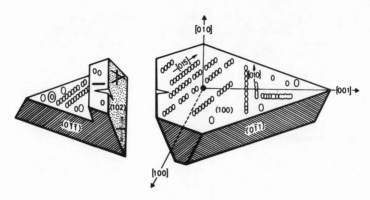

Fig. 12. Relation between habit, crystallographic axes and other
morphological features in typical crystal of poly(hexadiine-1,6-
diol-bis-p-toluene sulfonate) (1). Typical etch-pits appearing on
(100) after short treatment of the monomer or partially polymerized crystal with a poor solvent are also shown (compare Ref. 12).

     The defect structure of these crystals has been characterized
(12) using the etch-pit technique in order to label emergent dislocations. Furtheron, freshly cleaved surfaces were investigated
for sliptraces and other defects. The pertinent observations are
summarized in Fig. 12. The etch-pits reveal that there are very
few families of dislocations left in the crystals and further perfection is reached during thermal polymerization due to simultaneous annealing as demonstrated by X-ray topography (37).

     Crystals of polymer 1 exhibit a sharp absorption edge and a
strong peak in reflectivity at 2 e V for light polarized parallel
to the polymer chain. For light polarized perpendicular to the
chain the reflectivity is low and featureless. At low temperatures
the reflection peaks shift to lower energy and split. Fig. 13 shows
the very high anisotropy in absorption (dichroism) as seen by transmission spectroscopy (38) and Fig. 14 shows the real and imaginary
part of the dielectric constant at room temperature and at 9 K as
obtained by Kramers-Kronig analysis from reflectivity (31). At the
maximum near 2 e V reflectivity amounts up to 75 %. The side bands
of the main electronic peak arise from coupling of the stretching
modes of the carbon single, double and triple-bonds to the electronic transition and are thus identified as phonon side bands in agreement with the data obtained from the resonant enhanced Raman spectra.
The splitting of the bands at low temperature is a consequence of
the phase transition occuring below 200 K (compare Fig. 10). The
band profile is asymmetric, all peaks display sharp edges at the
low-energy side and a weaker decay at the high-energy tail. These
band shapes were analyzed in agreement with recent Hückel π-electron
molecular orbital calculations of Wilson (39) in terms of a van Hove

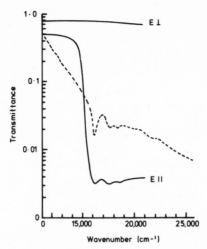

Fig. 13. Transmission spectra of thin film of polymer $\underline{1}$ at 300°K.
Full curves: single crystal with light polarized parallel and perpendicular to the polymer chains; dash curve: partially crystalline
film (according to Bloor et al. (38).

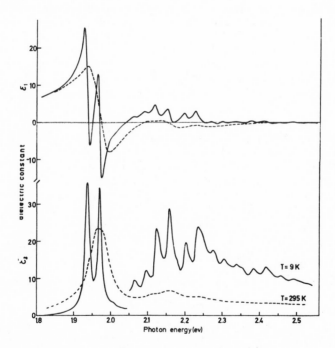

Fig. 14. Real and imaginary part of the dielectric constant
$\varepsilon_1 + i\varepsilon_2$ at 9 K (full curve) and at room temperature (dahsed curve)
(according to Ref. 31). The high energy part of ε_2 is 4 x en-
larged. E_{11} to chains.

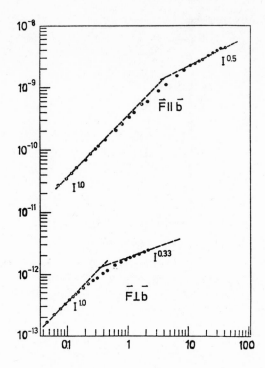

Fig. 15. Photocurrent (amps) VS. light intensity for (100)-surface conductivity of a single crystal of polymer $\underline{1}$ according to (40) $I_{\parallel}/I_{\perp} \approx 800$.

singularity. It was thus assumed that the optical absorption results form transitions between delocalized one-dimensional states of the polymer chain forming broad bands.

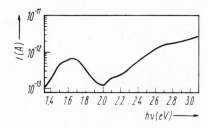

Fig. 16. Action spectrum of the photocurrent i measured in a surface conduction cell according to Ref. 31.

Measurements of dark and photoconductivity are yet ambiguous. The first measurements were performed by Schermann and Wegner (41,42) on a number of then available polydiacetylenes including polymer 1. They showed that polydiacetylenes are poorly conducting in the dark, that conductivity is anisotropic and that photoconductivity is observed. Recent work of Bässler and coworkers has much clarified these observations (31,40,43), but it is still unclear wether conductivity is electronic or excitonic in nature. Fig. 15 and Fig. 16 are taken from the work of Bässler et al. and show quantitative numbers of the electrical properties observed with polymer 1.

The mechanical properties (Youngs modulus, ultimate tensile strength, deformation processes) of polymer 2, poly (hexadiine-1,6-diol-bis-phenylurethane) were investigated by Baughman, Gleiter and Sendfeld (11). Crystals of this polymer are extremely suited for such measurements because they grow as ribbon-like specimens elongated along the chain direction up to several cm in length. The properties observed are those of metal and ceramic whiskers. The per chain modulus obtained is nearly as high as that of diamond.

Acknowledgement

As far as results from the authors own laboratory are reported, he wants to thank all his coworkers, expecially Dr. V. Enkelmann, Dr. M. Steinbach, Dr. G. Lieser, M. Knoch, M. Leyrer, B. Tieke and H.J. Graf. He also acknowledges intensive discussions with Dr. D. Bloor and that the work was supported by the Deutsche Forschungsgemein-schaft and the Fonds der Chemischen Industrie.

REFERENCES

1. W. A. Little, Phys. Rev. A, 134, 1416 (1964)
2. W. A. Little, J. Polymer Sci. C. 17, 3 (1967)
3. E. P. Goodings, Chem. Soc. Rev. 5, 95 (1976)
4. G. Wegner, A. Munoz-Escalona and E. W. Fischer, Makromolekulare Chem. Suppl. 1, 521 (1975)
5. G. Wegner, Z. Naturforschg. 24b, 824 (1969)
6. F. L. Hirshfeld and G. M. J. Schmidt, J. Polymer Sci. A, 2, 2181 (1964)
7. G. Wegner, Makromolekulare Chem. 154, 35 (1972)
8. G. Wegner, Chimia 28, 475 (1974)
9. J. Kiji, J. Kaiser, G. Wegner and R. C. Schulz, Polymer 14, 433 (1973)
10. R. H. Baughman and K. C. Yee, J. Polymer Sci., Polymer Chem. Ed. 12, 2467 (1974)
11. R. H. Baughman, H. Gleiter and N. Sendfeld, J. Polymer Sci., Polymer Phys. Ed., 13, 1871 (1975)

12. W. Schermann, J. O. Williams, J. M. Thomas and G. Wegner, J. Polymer Sci., Polymer Phys. Ed. 13, 753 (1975)
13. D. Bloor, L. Koski and G. C. Stevens, J. Materials Sci. 10, 1689 (1975)
14. R. H. Baughman, private communication
15. R. H. Baughman, J. Polymer Sci., Polymer Phys. Ed. 12, 1511 (1974)
16. K. Takeda and G. Wegner, Makromolekulare Chem. 160, 349 (1972)
17. G. C. Stevens and D. Bloor, Chem. Phys. Letters 40, 37 (1976)
18. H. Eichele, M. Schwoerer, R. Huber and D. Bloor, to be published
19. J. Kaiser, G. Wegner and E. W. Fischer, Israel J. Chem. 10, 157 (1972)
20. R. H. Baughman, J. Appl. Phys. 43, 4362 (1972)
21. D. Bloor et al. J. Materials Sci. 10, 1678 (1975)
22. a) K. A. Mauritz and A. J. Hopfinger, J. Polymer Sci., Polymer Phys. Ed. 14, 1813 (1976)
    b) R. H. Baughman, R. R. Chance and M. J. Cohen, J. Chem. Phys. 64, 1869 (1976)
23. B. Wunderlich, Advances in Polymer Sci. 5, 568 (1968)
24. J. I. Lauritzen u. J. D. Hoffmann, J. Res. Nat. Bur. Stand 64A, 79 (1960)
25. H. G. Zachmann, Kolloid-Z.Z. Polym. 216-217, 180 (1967), 231, 504 (1969)
26. D. Kobelt and H. Paulus, Acta Cryst. B30, 232 (1974)
27. E. Hädicke, E. C. Mez, C. H. Krauch, G. Wegner and J. Kaiser, Angew. Chem. 83, 253 (1971)
28. V. Enkelmann and J. B. Lando, to be published
29. V. Enkelmann and G. Wegner, Makromolekulare Chem., in press
30. D. Bloor, F. H. Preston and D. J. Ando, Chem. Phys. Letters 38, 33 (1976)
31. D. Reimer. H. Bässler, J. Hesse and G. Weiser, Phys. stat. Sol. (b) 73, 709 (1976)
32. R. H. Baughman, J. D. Witt and K. C. Yee, J. Chem. Phys., 60, 4755 (1974)
33. D. Bloor, F. H. Preston, D. J. Ando and D. N. Batchelder in: Structural Studies of Macromolecules by Spectroscopic Methods, p. 91 f., K. J. Ivin, Ed., Wiley 1975
34. A. J. Melveger and R. H. Baughman, J. Polymer Sci. A2, 11, 603 (1973)
35. D. N. Batchelder and D. Bloor, Chem. Phys. Letters, 38, 37 (1976)
36. G. Wegner, Makromolekulare Chem. 145, 85 (1971)
37. J. M. Schultz, to be published
38. D. Bloor, D. J. Ando, F. H. Preston and G. C. Stevens, Chem. Phys. Letters, 24, 407 (1974)
39. E. G. Wilson, J. Phys., C. Solid State Phys., 8, 727 (1975)
40. K. Lochner, B. Reimer and H. Bässler, Chem. Phys. Letters, in press
41. W. Schermann and G. Wegner, Makromolekulare Chem. 175, 667 (1974)
42. G. Wegner and W. Schermann, Colliid & Polymer Sci. 252, 655 (1974)
43. B. Reimer and H. Bässler, Phys. stat. Sol. (a) 32, 435 (1975)

X-RAY AND NEUTRON SCATTERING INVESTIGATION OF THE CHARGE

DENSITY WAVES IN TTF-TCNQ

R. Comès[*]

Laboratoire de Physique du Solide associé au CNRS
Université Paris-Sud - Bâtiment 510
91405   ORSAY   (France)

I. - INTRODUCTION

The earlier theoretical predictions of the existence of a giant Kohn anomaly in 1-d conductors (1) due to the divergent response of the electron gas at the wave vector $2k_F$, and leading ultimately to a particular metal insulator Peierls transition (2), to occur at finite temperature because of the coupling between chains (3, 4), have now received experimental confirmation on several real physical systems including TTF-TCNQ (4-14).

Almost simultaneously with the first structural evidence of such features in the platinum chain compound $K_2Pt(CN)_4Br_{0.30} \cdot xH_2O$ (KCP) and related substances (15-21), it was proposed independently by Coleman et al (22) and Ferraris et al (23) on the basis of well known electrical transport measurements that the organic salt tetrathiofulvalene tetracyanoquinodimethane (TTF-TCNQ) was undergoing such a metal-insulator Peierls transition around 60°K. During several years anomalies observed on various physical properties accumulated on this compound for a phase transition around 54 K (24) and even for a second phase transition at 38 K (25, 26), but conclusive structural evidence

---

[*] Results reported here are part of an active collaboration between the Orsay X-Ray Scattering group with F. DENOYER, S. KHANNA, J.P. POUGET, the Brookhaven Neutron Scattering group with W.D. ELLENSON, S.M. SHAPIRO, G. SHIRANE and the University of Pennsylvania with A.F. GARITO and A.J. HEEGER.

could not be obtained from several conventional X-Ray investigations (26, 28). This created the very frustrating situation in
which the structural features expected from 1-d systems could be
observed in one compound (KCP) while the most intriguing electrical
features were observed on another system (TTF-TCNQ) (29).

Such particular phase transitions indeed do not reveal
themselves by a simple lowering of the symmetry of the lattice
(as in ferroelectric or anti-ferroelectric phase transitions for
example) which gives rise to a splitting of strong Bragg reflexions generally easy to observe in simple powder diffraction
experiments, but through a sinusoidal modulation of the lattice,
arising from the coupling with the 1-d electrons, which manifests
itself only by the development of superlattice reflexions (satellites) surrounding the main parent Bragg peaks of the undistorted
high temperature crystal. These satellites can be extremely weak
in intensity, and depending on the filling of the conduction band
(the Fermi wave vector k_F which is often not known) they can be
located at wave vectors, which cannot be written as simple frac-
tions of the reciprocal lattice vectors of the unmodulated lattice, meaning that the period of the modulation (superstructure)
in real space can be incommensurate with the main lattice.

In fact, the observation of this type of superstructure
reflexions is at the limit of detection of most of the existing
diffraction equipment. In KCP already, which fulfilled almost
the ideal requirements for X-Ray scattering investigations, with
one good X-Ray scatterer (Pt) in an overall light material
(little absorption), of relatively simple structure and which was
available in large good quality single crystals, the satellites
could hardly be detected with conventional diffraction experiments (30). More precisely, compared to the intensity of the
closest parent Bragg peaks, the satellite intensity in KCP was
of the order of a few 10^{-5}, arising from a charge density wave
amplitude determined to be 0,0047 c (c = 5.692 Å = lattice spacing in chain direction at 7 K) (31). Assuming a similar amplitude for the charge density waves in TTF-TCNQ, the absolute
intensity expected from the satellite reflexion close to the
strongest Bragg peak (0 1 3 reflexion) can be estimated to be
about 20 times smaller for X-Rays (a factor of at least 5
comes from the smaller size of the available single crystals and
a factor of about 4 from the smaller structure factor per unit
volume). In addition there are only very few strong Bragg reflexions in TTF-TCNQ and an overall effective intensity of 10^{-2}
compared to the KCP satellites probably gives a better average
figure. In the case of neutrons this figure must be reduced by
another 10^{-1} arising from the size alone of TTF-TCNQ crystals.

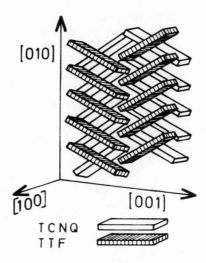

Figure 1 : Schematic representation of the TTF–TCNQ structure.
──────────   Note that the molecular plane is not perpendicular
             to the stacking axis b ([010]) but tilted around
             the a axis ([100]), that there are two identical
             molecules with opposite tilts for one repeat unit
             along the c direction ([001]), and that there are
             alternatively TCNQ and TTF stacks in a direction.

This explains why all the structural results on the metal insulator phase transition in TTF-TCNQ were obtained by scattering techniques (X-Rays or neutrons) which are precisely designed for the detection of much lower signals than conventional diffraction experiments (32). It also explains that even with these techniques the experimental results were most often obtained close to the limit of detection which led to some initial errors or wrong assignments, but also limited the exploitable data to a few satellites or a few Brillouin zones making a detailed analysis of the atomic motions extremely hazardous.

In the present stage, it seems that a good agreement has emerged from the different investigations on the low temperature modulated lattices, the high temperature precursors and the gene-ral trends of their temperature dependence. As will be described below, this already constitutes a very rich ensemble of structural information, and has considerably improoved the understanding of the exceptional properties of TTF-TCNQ. Rather than following the chronological sequence of the successive studies, we shall first consider the high temperature precursors, a second section will deal with the low temperature phases, and a last section will briefly outline the present understanding of these structural features.

For the experimental conditions of X-Ray and neutron scatterings, and the samples, the reader is referred to the already published reports (5-14, 32), mostly untwinned crystal were used, which were 100 % deuterated for the neutron experiments.

II. - THE HIGH TEMPERATURE PRECURSORS

A. The 1-d regime T > 60°K

Let us first recall that the unmodulated room temperature structure is of monoclinic symmetry (space group $P2_1/c$) with a unit cell a = 12.298 b = 3.819 c = 18.468 Å and β = 104°.46 at room temperature (33), which contains 2TTF and 2TCNQ molecules. This structure is characterized by the existence of 4 different molecular stacks of identical molecules each as shown schemati-cally in Figure 1. This phase is stable above the Peierls tran-sition which takes place at 54°K.

Figure 2a shows an X-Ray diffuse scattering pattern from TTF-TCNQ at 60°K, and Figure 2b for comparison a pattern from KCP (taken at room temperature). Besides the usual broad diffuse spots which are due to the small wave vector acoustic phonons and which locate the reciprocal layer lines perpendicular to the stacking direction, diffuse intensity maxima running along these layer lines are clearly visible on both patterns, forming the so

called "satellite sheets". As is well known, such an intensity
distribution is characteristic of 1-d correlations ; the increasing intensity of this diffuse scattering with increasing scattering angle further shows that displacive effects (displacements or
motion of atoms and not impurities) are responsible for it. This
analogy and the existing inelastic neutron scattering data on
TTF-TCNQ (10) clearly demonstrates that giant 1-d Kohn anomalies
in the phonon dispersion spectrum (which are the dynamic manifesta-
tion of charge density waves) are present in TTF-TCNQ as well as
in KCP.

There are however remarkable differences between the
precursor effects of the two compounds :

(i) two types of 1-d scattering are observed in TTF-TCNQ. The
first type assigned to $2k_F$ is found at the wave vector $\pm$ 0.295 b^{*} ;
this scattering only develops below 150°K and corresponds to a
phonon anomaly measured by inelastic neutron scattering which
sharpens in the same temperature range, as shown figure 3. The
second type of scattering is observed at the wave vector 0.59 b^{*}
(or 0.41 b^{*} in the reduced zone), that is to say at twice the
value of the wave vector of the former scattering, it is therefore
assigned to $4k_F$ (12). At first sight, this second scattering can
be thought to be just a higher order diffraction from the $2k_F$
scattering ; this is however ruled out by the comparable intensities of the two scatterings at 60°K (figure 2a), and by the
fact that above 150°K, only the $4k_F$ scattering is visible by
eye on photographic patterns. Figure 4 shows microdensitometer
readings of X-Ray patterns taken between 60°K and room temperature (14), and Figure 5 shows comparable counter measurements
from the independent investigation of Kagoshima et al (13). The
excellent agreement between these two sets of data clearly esta-
blishes the existence of two different precursors in TTF-TCNQ
namely at the wave vectors $2k_F$ and $4k_F$. Noticeable in figures 4
and 5 is also the slight shift in wave vector from 0.41 b^{*} $\pm$ 0.02
to 0.45 b^{*} $\pm$ 0.02 for the $4k_F$ scattering ;

(ii) the diffuse satellite sheets of TTF-TCNQ appear as interrupted lines, in contrast to the continuous lines of KCP. This is
simply due to structural differences : in KCP, most of the intensity is scattered by the Pt atoms, in other words by the modulation of single rows of Pt atoms (1-d regime) and the intensity
does therefore not depend on the wave vector components perpendicular to the Pt strings, producing the continuous intensity observed for the satellite sheets of KCP ; in TTF-TCNQ, the charge
density waves modulate the distances between molecules of light
atoms all contributing appreciably to the scattered intensity,
the intensity therefore strongly depends on the wave vector components perpendicular to the stacking axis, giving to the satellite sheets their aspect of interrupted lines.

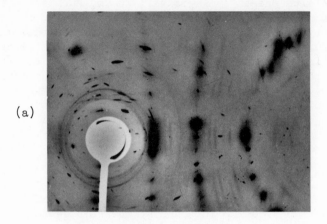

(a)

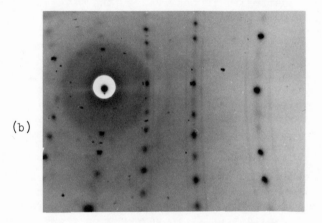

(b)

Figure 2 : X-Ray diffuse scattering patterns.
─────────
b) KCP pattern at room temperature from Comes et al (15)
     the $2k_F$ anomaly gives rise to continuous satellite
     sheets of scattering at $\pm 2k_F$ from the main layer
     lines perpendicular to the chain axis.

a) TTF-TCNQ at 60°K from Pouget et al (12), two series
     of satellite sheets at ± 0.0295 $b^{\times}$ ($2k_F$) and
     ± 0.59 $b^{\times}$ (± 0.41 $b^{\times}$ in the reduced zone) assigned
     to $4k_F$ are clearly visible. The satellite sheets
     appear here as interrumpted lines due to the mole-
     cular form factor.

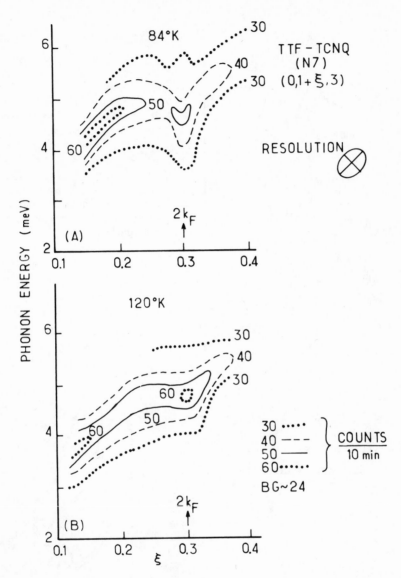

Figure 3 : (from Shirane et al (10)). – Intensity countours of the TA branch mainly polarized along c*, showing the development of a sharp $2k_F$ anomaly at low temperature.

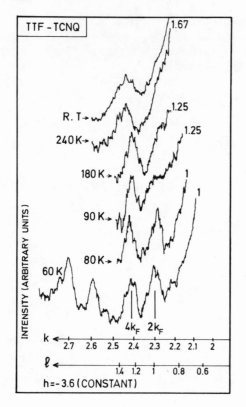

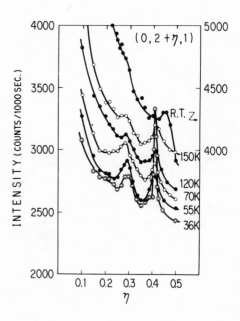

Figure 4 : (from Khanna
et al (14)). Microdensi-
tometer reading of a series
of X-Ray patterns between
60°K and room temperature.
Note the different tempe-
rature dependence of the $2k_F$
and $4k_F$ scattering. Intensity
at 90 and 180 K has to be
multiplied by 1.25 and by
1.67 at 240 K and room temperature in order to scale
with the lower temperature
recordings.

Figure 5 : from Kagoshima et
al (13)). X-Ray counter measurements performed in the same zone.

It was already shown that in first approximation, the general aspect of the intensity distribution in the "satellite sheets" of TTF-TCNQ can be reproduced assuming independent modulations of the different stacks (1-d regime) and translations of rigid molecules (12) ; but small contributions from other types of motions as observed for example in TTF_7I_5 (34) and suggested for TSeF-TCNQ (34, 35) cannot be ruled out with this preliminar analysis. The important point in the present stage is that it shows that the different stacks are modulated independtly from each other which demonstrates the 1-d character of the scattering of TTF-TCNQ above 60°K ;

(iii) in KCP, the polarization (direction of atomic motion or displacements) was strictly longitudinal to the chain axis. In TTF-TCNQ, the polarization is more complex. The $2k_F$ anomaly is found to have both a longitudinal component and a transverse component along c^* (13, 14). The $4k_F$ anomaly is in contrast mainly longitudinal to the chain direction which adds another difference between the two precursor effects.

It is at first puzzling to understand why charge density waves could have a transverse component. One has to recal here that charge density waves arise from the modulation of the intermolecular distances. In KCP, the modulation involves the $Pt(CN)_4$ groups (20) which are perpendicular to the chain direction ; in this case, as for strings of atoms, a longitudinal polarization only, can modulate the intermolecular spacing and give rise to charge density waves. In TTF-TCNQ, the molecular stacking is different (figure 1), the molecules are tilted around the a axis of the unit cell, therefore there are two polarization components which can modulate the intermolecular spacing : the longitudinal component along b, and the transverse component along the perpendicular to the axis of rotation a, which coincides with the reciprocal c^* axis. The comparison of the two cases is shown schematically in figure 6. It is precisely these two polarizations which are mainly observed experimentally, showing that we are really dealing with charge density waves.

B.   The onset of 3-d order   54°K < T < 60°K

Around 60°K, the intensity in the $\pm$ $2k_F$ (0.295**) satellite sheets starts to show additional features : maxima of intensity build up slowly around the wave vectors (0.5 a^*, 0.295 b^*, 0 c^*) as shown in figure 7. These maxima increase in intensity with decreasing temperature, and reveal the progressive coupling between the charge density waves on different molecular stacks, towards a (2a x 3,40b x c) modulated lattice.

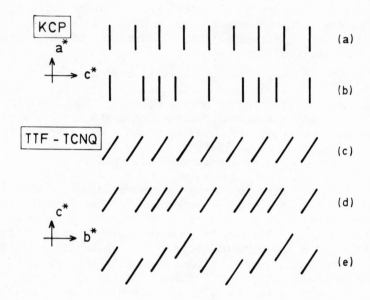

Figure 6 : Polarization of translational modulations, which can modify the intermolecular spacing and give rise to charge density waves.

a) Unmodulated stacking of KCP.

b) The longitudinal polarization only modulates the intermolecular spacing in KCP.

c) Unmodulated stacking in TTF-TCNQ.

d) A longitudinal polarization can modulate the inter-molecular spacing as in KCP, but here a transverse polarization (c^{*}) can also produce a charge density wave (e).

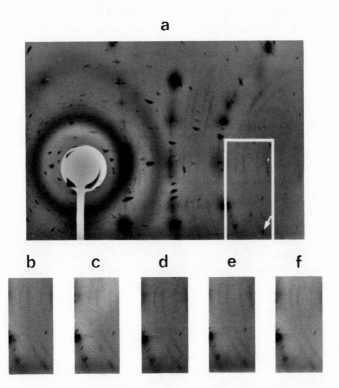

Figure 7 : X-Ray diffuse scattering pattern showing the
           building up of 3-d order below 60°K.

   a) 50°K : modulated phase 1, arrays of well defined
      satellites with wave vector (0.5 a^x, 0.295 b^x, 0 c^x)
      are clearly observable. Note that 1-d 2k$_F$ scattering
      is still visible and that no 4k$_F$ satellites can be
      found.
   b) To (f) progressive smearing out of the satellites
      between 54°K (b) and 60°K (f).

At 54°K, within the resolution limits (about 0.04 $\overset{\circ}{A}^{-1}$ for X-Rays and 0.02 $\overset{\circ}{A}^{-1}$ for neutrons) long range 3-d order is achieved. This temperature coincides with the abrupt drop in conductivity of the metal-insulator transition, and in particular a clearly observable specific heat anomaly (24). No remarkable change is observed in this temperature range on the 4k$_F$ scattering (0.59 b*).

It is this coïncidence between the condensation of the first type of scattering at the wave vector 0.295 b* and the metal insulator transition, which justifies the assignment of 2k$_F$ to this type of scattering. This assignment implies a value of $k_F = 0.1475$ b$^{\ast} = 0.1475 \frac{2\Pi}{b} = 0.295 \frac{\Pi}{b}$, and as there are 2 electrons per electronic band determines the charge transfer to be 0.59 electrons, figure which is very close to the independent estimation made by Coppens (37).

The fact that TTF-TCNQ appears as long range ordered in the modulated low temperature phases constitutes another noticeable difference with the earlier studied KCP (38) TTF-TCNQ undergoes a real phase transition at 54°K while KCP only shows a tendency toward 3-d ordering ; TTF-TCNQ is therefore probably the first compound in which the successive structural stages of a Peierls transition can be investigated.

III.   -   THE LOW TEMPERATURE MODULATED PHASES

Three different modulated phases are observed in TTF-TCNQ, with phase transitions at 54 K (the Peierls transition), 49 K and 38 K as shown in figure 8.

A. Modulated phase I

Modulated phase 1 is stable between 54 and 49 K with satellite reflexions observed at the wave vectors (0.5 a*, 0.295b*, 0 c*), which corresponds to a modulated lattice with a superstructure 2a x 3.40b x c. Satellites of this phase are extremely weak, and only one well defined such a satellite could be followed up to 54 K quantitatively by neutron measurements : namely the satellite with components (0.5a*, 1.295b*, 3c*) which is close to the strongest (0 1 3) reflexion of the unmodulated lattice ; its temperature dependent intensity is shown figure 9. Photographic X-Ray patterns however (figure 7) show series of such weak but well defined satellites confirming the reality of this phase. An estimation of the intensity of these superlattice reflexions relative to the closest Bragg peaks gives a figure of the order of 10^{-5} at 50°K. No satellites are observed in this temperature range with a 4k$_F$

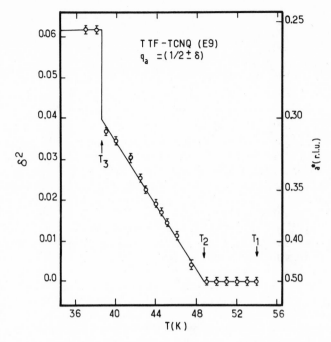

Figure 8 : (from Ellenson et al (11)) - The occurence of 3
phase transitions in TTF-TCNQ is best shown from
the temperature dependence of the satellite peak
position in reciprocal space when plotted as a
function of $S^2=(\frac{1}{2} - q_a)^2$, where q_a is the satellite wave vector along a^*. This was suggested by
BAK and EMERY (40) who first found the 49°K phase
transition.

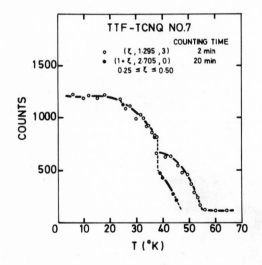

Figure 9 (from Comes et
al (8)).

Temperature dependence of
two satellite peak inten-
sities. Note that the
intensity of the $(1 + \zeta,$
$2,705,0)$ satellite which
has no c^* scattering vector component extrapolates
to zero around 49°K.

wave vector component.

Noticeable is the fact that all satellites observed between 54 and 49 K have a non zero scattering vector component along c^x ; no such satellite could be observed in this temperature range in the (h k o) scattering plane. Since the intensity of satellites of this type (condensed phonons) is proportional to $(\vec{Q}.\vec{u})^2$ where $\vec{Q} = ha^x + kb^x + lc^x$ is the scattering vector and u the polarization of the atomic displacements, this observation seems to imply that only the transverse component along c^x from the higher temperature 1-d scattering (T > 54 K) is condensed in this phase, in consistence with the inelastic neutron measurements which show the sharp phonon anomaly developping around the transverse c^x phonon branch (10, 39). It is also consistent with the subsistence below 54 K of relatively intense 1-d scattering with $2k_F$ wave vector component which could be due to the not condensed longitudinal component of the polarization. The full significance of such a statement must however be considered very carefully since the experimental observations are limited to a few zones.

B.   Modulated phase II   38°K < T < 49°K

1)   The $2k_F$ satellites

At 49 K, a rather subtle and unique type of phase transition takes place, the wave vector component along a^x of the condensed satellite starts to decrease continuously from 0.5 a^x towards about 0.3 a^x and then locks discontinuously in the 0.25 a^x value at 38°K. In terms of modulations in real space, the modulation changes from its simply incommensurate (along b) superstructure 2a x 3,4b x c above 49°K, to a temperature dependent doubly incommensurate (along a and along b) superstructure x(T)a x3,40b x c, before its locks in a simply commensurate superstructure again at 38°K with the modulated lattice 4a x 3,40b x c. The 49°K phase transition was in fact only discovered after a reanalysis by Bak and Emery (40) of the initial data (7) and fully established by the high resolution measurements of Ellenson et al (11) (see figure 8). It is completely confirmed by the independent investigation of Kagoshima et al (13).

Noticeable is the fact that well defined satellites with zero scattering vector component along c^x are clearly observable in this phase 2 and their intensity extrapolates to zero around 49°K (see the (1+ζ, 2 705, 0) satellite intensity plotted in figure 9). With the same restrictions as above (phase I), this clearly shows that the longitudinal component of the polarization of the $2k_F$ scattering is now also condensed, and seems to imply that the 49°K phase transition is driven by a longitudinal $2k_F$

anomaly. There is no available inelastic neutron scattering data around 49°K, but the existence of the shallow $2k_F$ anomaly observed at room temperature and 200°K (10, 39) in the longitudinal phonon branch gives some plausibility to such a suggestion.

2)   The $4k_F$ satellites

Satellites with $4k_F$ wave vector component along $b^{\times}$ were observed on X-Ray photographic patterns (12) as well as by neutron measurements (11), but the temperature dependence of the intensity and position of these satellites was only investigated in detail in the X-Ray counter work of Kagoshima et al (13). This work concludes that the $4k_F$ satellites are clearly visible at low temperature at wave vectors which are always twice those of the $2k_F$ satellites ; with an intensity which extrapolates to zero at 49 K. Their weaker intensity as well as their position, are in this temperature range consistent with a higher order effect. The relative and absolute intensities of the $2k_F$ and $4k_F$ satellites suggest a harmonic of $2k_F$ indicating that below 49 K the modulation along the chain direction is probably not strictly sinusoïdal.

The $4k_F$ satellite therefore does not modify the earlier reported modulated lattices. Figure 10 reproduces the temperature dependent positions of both the $2k_F$ and $4k_F$ satellites.

C.   Modulated phase III   T < 38 K

At 38°K, a first order phase transition with a hysterisis of about 1° takes place (8) and the superlattice locks in its 4a x 3,40b x c modulation and no further structural change is observed below this temperature (figures 7 to 10). The intensity of the $2k_F$ satellites reflexions relative to the closest Bragg reflexions of the main lattice has reached in this phase the order 10^{-4}, from this value one can estimate the distortion amplitude to be about 1 % of the lattice spacing. This amplitude is slightly larger but quite comparable with the amplitude of the charge density waves in KCP.

Figure 11 shows a photographic pattern in this phase, on which both $2k_F$ and $4k_F$ satellites with respective wave vectors

$$0.25a^{\times} \; , \; 0.295b^{\times}, \; 0c^{\times}$$

and

$$0.5a^{\times} \; , \; 0.59b^{\times} \; (=0.41b^{\times}), \; 0 \; c^{\times}$$

are clearly visible.

Figure 12 summarizes the satellite positions in the 3 lower temperature phases. We have omitted in figure 12 the addi-

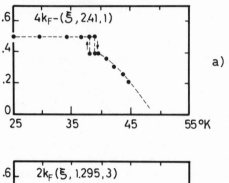

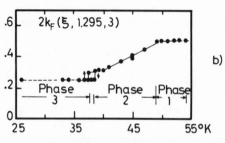

a)

b)

Figure 10
_____

Temperature dependence of
the position (in a^* units).

a) of the $(\zeta, 2.41, 1)$ $4k_F$
   satellite (from Kagoshima
   et al (13)).

b) of the $(\zeta, 1.295, 3)$ $2k_F$
   satellite (from Comes
   et al (7)).

Figure 11 : X-Ray diffuse scattering pattern at 30°K showing
_____   both $2k_F$ satellites with wave vectors (0.25 a^*,
            $1.295b^*$, 0 c^*) and $4k_F$ satellites with wave
            vector (0.5 a^*, 0.41 b^*, 0 c^*).

tional $4k_F$ satellites observed by Kagoshima et al (13) at the wave vector 0 a^x, 0.59 b^x (0.41 b^x), 0 c^x, these satellites could not be observed on any photographic pattern and may need further study.

III - DISCUSSION

In the course of this paper, we have not mentionned the initial neutron scattering measurements on TTF-TCNQ performed by Mook et al (9) who reported the observation of a giant Kohn anomaly at room temperature and at the wave vector 0.290 b^x, which could not be observed in our subsequent neutron measurements (10). The existence of such an important <u>phonon</u> anomaly requires a noticeably enhanced phonon amplitude at the wave vector $2k_F$ (0.295 b^x), which should give rise to strong X-Ray scattering at room temperature. This is completely ruled out by both the photographic and counter X-Ray investigations. The weak cross-sections observed by Mook et al (9) are therefore now suggested by Torrance, Mook and Watson (41) to arise from $2k_F$ spin waves. This brings us to the first physical problem which arises from the structural studies of TTF-TCNQ, namely the observation of both $2k_F$ and $4k_F$ 1-d precursor effects.

Independently and simultaneously to the first experimental observation (12), Torrance (42) had predicted the possibility of a $4k_F$ anomaly in TTF-TCNQ. In his picture, the $2k_F$ <u>phonon</u> anomaly is driven by spin waves, while the $4k_F$ <u>phonon</u> anomaly is driven by charge density waves ; both the existence of spin waves and the occurence of an anomaly at $4k_F$ are due to strongly repulsive interactions between electrons.

Since the first experimental report (12), Emery (41) and subsequently others (44, 45) have worked out theoretical descriptions which at least qualitatively agree reasonnably well with the existence and the different temperature dependence of the $2k_F$ and $4k_F$ anomalies.

Simplest for the present purpose is to use the strong coupling limit which directly leads to a $4k_F$ Kohn anomaly as shown much earlier by Ovchinnikov (46) and Bernasconi et al (47) even if it seems to be only a schematic case.

In the normal case, there are two electrons of opposite spins per momentum state, if a metallic band countains n electrons, the Fermi wave vector is $k_F = \frac{n}{2}\frac{\Pi}{b} = \frac{n}{4} b^x$, and the opening of a gap at k_F requires a lattice distortion with a wave vector in chain direction of $2k_F = n\frac{\Pi}{b} = \frac{n}{2} b^x$.

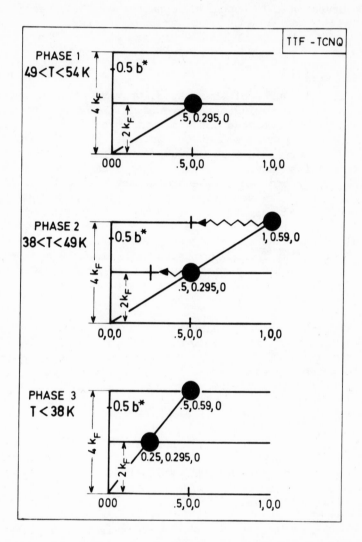

Figure 12 : The satellite positions in reciprocal space in the 3 lower temperature modulated phases of TTF-TCNQ. The position of the $4k_F$ satellite at 49°K in phase II is the extrapolated position deduced from the measurements of Kagoshima et al (13) also shown in figure 10.

In the case of repulsive interaction in the strong coupling limit there is only one electron per momentum state and for the same number of electrons n this requires twice as many momentum states doubling the values of the Fermi wave vector, and of the wave vector of the lattice distortion required in order to open a gap. If we still refer to the values of the normal situation, this produces a $4k_F$ Kohn anomaly.

Another way to describe this schematic case is to use localized electrons in real space (figure 13), for a 1/4 filled band. The normal situation with 2 electrons of opposite spins per site gives a charge density wave period of $4 b = 2\Pi/2k_F$ (fig 13a). The case of repulsive interactions in the strong coupling limit produces a charge density wave of period $2b = 2\Pi/4k_F$ (fig 13b), and as can be easily seen also a spin wave of period $4b = 2\Pi/2k_F$.

The difficulty of a real description is to leave the schematic case of the strong coupling limit and to produce simultaneously the two anomalies but this seems to have been successfully overcome (43-45). These models seem moreover very similar, they all rely on strong coupling (but not infinite) and imply the existence of spin waves.

The slight shift from $0.41 b^{\times}$ to $0.45 b^{\times}$ (or from $0.59 b^{\times}$ to $0.55 b^{\times}$ in the extended zone) of the $4k_F$ scattering creates nevertheless some difficulties. In its simplest interpretation, it suggests a modification of the charge transfer. If this is the case, the $2k_F$ value at room temperature is $0.55b^{\times}/2 = 0.275 b^{\times}$ and spin waves should be observed at this wave vector and not at $0.290 b^{\times}$ as claimed by Torrance, Mook and Watson. The $4k_F$ scattering at room temperature is however already substantially broadened and its position could therefore be slightly different from the value deduced from the exact band filling which could remain constant over the investigated temperature range in spite of the observed shift.

The second and important question raised by the structural investigations is to understand the origin of the sequence of 3 modulated phases below 54°K and in particular phase II with its temperature dependent modulation. This has been remarkably worked out by Back and Emery (40). In their Guizburg-Landau description, the Peierls transition on one type of molecular stack occurs at 54°K, while the transition on the second molecular stack takes place at 49°K, and drives the transverse modulation along $\underline{a}$ from $2\underline{a}$ to larger values, before it locks in 4a at 38°K (a similar suggestion was very briefly mentioned in an earlier paper by Saub et al (49) mostly devoted to KCP).

Here again it is possible to give a very simple qualitative description. In KCP, the low temperature 3-d order in which
the charge density waves on successive platinum chains are simply
in opposition of phase (18, 19) can be easily understood because
this type of ordering minimizes the coulomb interaction (49). A
similar argument can be used for TTF-TCNQ.

As there are two stacks of identical molecules for one
c spacing in TTF-TCNQ (fig 1), if only one type of molecular kind
orders at 54°K (regardless whether TTF or TCNQ), the 2 a x c
ordering of the phase of the charge density waves, observed
between 49 and 54°K, is very similar to the 3-d order in KCP :
successive stacks in a and c directions can order in opposition
of phase (figure 14a).

When the charge density waves on the second type of
molecular stack starts to order at 49°K there is a competition
between the interaction on identical stacks and the interaction
on unlike stacks. From simple Coulomb interaction argument successive identical stacks along a, have a tendency to be in opposition
of phase, while sucessive unlike stacks tend to be in phase because
of the opposite charges carried by the different molecules, this
drives the modulation in a direction to larger values (figure 14b),
until a satisfactory balance is achieved at 38°K (figure 14c).

The two chain model for TTF-TCNQ has some experimental support (48, 49) from which the TCNQ stack is suggested to
order first at 54°K and the TTF stack only at lower temperature.
Up to now, it was impossible to analyse the limited structural
data in order to confirm this sequence, but there is no striking
inconsistency with this model. What the purely structural data
adds to this interpretation is that the mode condensing at 54°K
is mainly transverse (c*) while the mode condensing at 49°K is
probably longitudinal.

In conclusion, it is clear that the structural data
obtained in the last 18 months has considerably contributed to
the understanding of TTF-TCNQ (50). Further X-Ray and neutron
scattering studies should

- determine the atomic motions involved in the distortions,
- provide detailed inelastic studies of the $2k_F$ anomaly close
  to the 54°K phase transition,
- investigate the $4k_F$ scattering (phonon anomaly or central peak?),
- and further characterize the spin density waves.

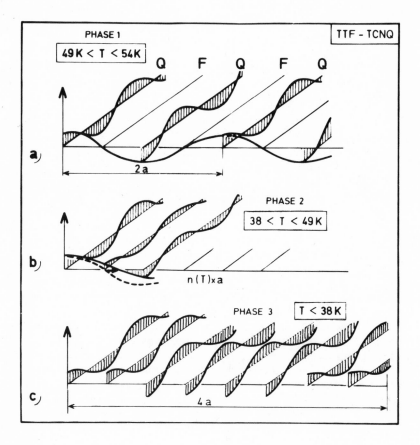

Figure 14 : Schematic reprensentation of the charge density
waves on the successive TTF and TCNQ stacks in
a direction.

    a)   The condensation of one kind of stack accounts for
the 2a transverse multiplicity.

    b)   The condensation of the second type of stack drives
the transverse multiplicity to larger values.

    c)   The 4a transverse multiplicity.

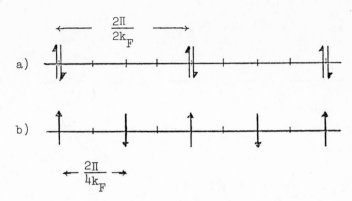

Figure 13   : localization of electrons in the case of
a 1/4 filled band.

a) normal situation : two electrons of opposite spin per site gives rise to a $2k_F$ CDW.

b) strong coupling limit : only one electron
per site gives rise to a $4k_F$ CDW and a
$2k_F$ SDW.

ACKNOWLEDGEMENTS

This work has benefited from innumerable discussions
with most of the touchy but passionate physicists of the one-
dimensional world. We are in particular grateful to V.J. EMERY
and M. LAMBERT who enlightened our physical understanding of
the structural features of TTF-TCNQ.

BIBLIOGRAPHY

(1) A.M. AFANASEV and Yu KAGAN, Sov. Phys. JETP, _16_, 1030 (1963).

(2) R.E. PEIERLS, Quantum Theory of Solids (Clarendon, Oxford),
(1964).

(3) S. BARISIC, Phys. Rev.,_B 5_, 941 (1972) and Ann. Phys.(Paris),
_7_, 23 (1972).

L. GORKOV, in Collective Properties of Physical Systems,
edited by B. Lundquist and S. Lundquist (Academic, New-York,
1973), p. 122.

(4) B. HOROWITZ, H. GUTFREUND and M. Weger, Phys. Rev., _B 9_,
1246 (1974).

(5)   F. DENOYER, R. COMES, A.F. GARITO and A.J. HEEGER, Phys.
      Rev. Lett., 35, 445 (1975).

(6)   S. KAGOSHIMA, H. ANZAI, K. KAJIMURA and R. ISHIGORO, J. Phys.
      Soc. Jap., 39, 1143 (1975).

(7)   R. COMES, S.M. SHAPIRO, G. SHIRANE, A.F. GARITO and
      A.J. HEEGER, Phys. Rev. Lett., 35, 1518 (1975).

(8)   R. COMES, S.M. SHAPIRO, G. SHIRANE, A.F. GARITO and
      A.J. HEEGER - to be published : Phys. Rev. (1976).

(9)   H.A. MOOK and C.R. WATSON, Phys. Rev. Lett., 36, 801 (1976).

(10) G. SHIRANE, S.M. SHAPIRO, R. COMES, A.F. GARITO and
      A.J. HEEGER, Phys. Rev., (1976). A preliminary report of
      these results was reviewed by G. SHIRANE at the Conference
      on Low Lying Vibrational Modes and Their Relationship to
      Superconductivity and Ferroelectricity, SAN JUAN,
      December 1-4, 1975 - to be published in Ferroelectrics.

(11) W.D. ELLENSON, R. COMES, S.M. SHAPIRO, G. SHIRANE,
      A.F. GARITO and A.J. HEEGER - to be published in Sol. St.
      Comm., (1976).

(12) J.P. POUGET, S.K. KHANNA, F. DENOYER, R. COMES, A.F. GARITO
      and A.J. HEEGER, Phys. Rev. Lett., 37, 437 (1976).

(13) S. KAGOSHIMA, T. ISHIGURO and H. ANZAI - to be published.

(14) S.K. KHANNA, J.P. POUGET, R. COMES, A.F. GARITO and
      A.J. HEEGER - to be published.

(15) R. COMES, M. LAMBERT, H. LAUNOIS and H.R. ZELLER, Phys. Rev.,
      B 8, 571 (1973).

(16) H. NIEDOBA, H. LAUNOIS, D.B. BRINKMANN, R. BRUGGER and
      H.R. ZELLER, Phys. Stat. Sol., 58, 309 (1973).

      H. NIEDOBA, H. LAUNOIS, D. BRINKMANN, H.U. KELLER, J. Phys.
      Lett. (France), 35, L 251 (1974).

(17) B. RENKER, H. RIETSCHEL, L. PINTSCHOVIUS, W. GLÄSER,
      P. BRUESCH, D. KUSE and M.J. RICE, Phys. Rev. Lett., 30,
      1144 (1973).

(18) R. COMES, M. LAMBERT and H.R. ZELLER, Phys. Stat. Sol. (b),
      58, 587 (1973).

(19) B. RENKER, L. PINTSCHOVIUS, W. GLÄSER, H. RIETSCHEL, R. COMES,
      L. LIEBERT and W. DREXEL, Phys. Rev. Lett., 32, 836 (1975).

(20) J.W. LYNN, M.I. IZUMI, G. SHIRANE, S.A. WERNER and R.B. SAIL-
      LANT, Phys. Rev., B 12, 1154 (1975).

(21) K. CARNEIRO, G. SHIRANE, S.A. WERNER and S. KAISER, Phys.
      Rev., B 13, 4258 (1976).

(22) L.B. COLEMAN, M.J. COHEN, D.J. SANDMAN, F.G. YAMAGISHI,
      A.F. GARITO and A.J. HEEGER, Sol. St. Comm., 12, 1125 (1973).

(23) J. FERRARIS, D.O. COWAN, V. WALATKA Jr., J.H. PERLSTEIN,
      J. Am. Chem. Soc., 95, 948 (1973).

(24) M.B. SALAMON, J.W. BRAY, G. De PASQUALI, R.P. CRAVEN,
      G. STUCKY and A. SCHULTZ, Phys. Rev., B 11, 619 (1975).

      P.M. CHAIKIN, J.F. FWAK, T.E. JONES, A.F. GARITO and
      A.J. HEEGER, Phys. Rev. Lett., 31, 601 (1973).

      M.J. COHEN, L.B. COLEMAN, A.F. GARITO and A.J. HEEGER, Phys.
      Rev., B 10, 1298 (1974).

(25) D. JEROME, W. MÜLLER and M. WEGER, J. Phys. (Paris) Lett.,
      35, L 77 (1974).

(26) A.J. BERLINSKY, T. TIEDJE, J.F. CAROLAN, L. WEITER and
      W. FRIESEN, Bull. Am. Phys. Soc., 20, 465 (1975).

      S. ETEMAD, T. PENNEY and E.M. ENGLER, Bull. Am. Phys. Soc.,
      20, 496 (1975).

      J.R. COOPER, D. JEROME, M. WEGER and S. ETEMAD, J. Phys.
      Lett. (France), 36, L 219 (1975).

(27) R.H. BLESSING and P. COPPENS, Sol. St. Comm., 15, 215 (1974).

(28) A.J. SCHULTZ, G.D. STUCKY, R.H. BLESSING and P. COPPENS -
      unpublished.

(29) For reviews of this status, see :

            . Low dimensional cooperative phenomena
               Edited by H.J. Keller (Plenum, New-York, 1975).

            . Lecture notes in Physics 34 : One dimensional
               conductors
               (Springer, New-York, 1975).

(30) J.M. WILLIAMS, F.K. ROSS, M. IWATA, J.L. PETERSEN,
      S.W. PETERSEN, S.C. LINN and K. KEEFER, Sol. St. Comm., 17,
      45 (1976).

(31) C.F. EAGEN, S.A. WERNER and R.B. SAILLANT, Phys. Rev., B 12,
      2036 (1975).

(32) B. DORNER and R. COMES in Topics in Applied Physics "Dynamics
      of Solids and Crystals by Neutron Scattering" to be edited
      by T. Springer (Springer, 1977).

(33) B.T. KISTENMACHER, T.E. PHILLIPS and D.O. COWAN, Acta Cryst.,
      B 30, 763 (1974).

(34) C.K. JOHNSON and C.R. WATSON Jr, J. Chem. Phys., 64, 2271
      (1976).

(35)   C. WEYL, E.M. ENGLER, S. ETEMAD, K. BECHGAARD, G. JEMANNO, Sol. St. Comm., $\underline{19}$, 925 (1976).

(36)   H. MORAWITZ, Phys. Rev. Lett., $\underline{34}$, 1096 (1975).

(37)   P. COPPENS, Phys. Rev. Lett., $\underline{35}$, 98 (1975).

(38)   W.H.G. MÜLLER and D. JEROME, J. Phys. Lett., $\underline{35}$, L 103 (1974).Under pressure KCP has a phase transition.

(39)   S.M. SHAPIRO, G. SHIRANE, A.F. GARITO, A.J. HEEGER – to be published.

(40)   P. BAK and V.J. EMERY, Phys. Rev. Lett., $\underline{36}$, 978 (1976).

(41)   J.B. TORRANCE, H.A. MOOK and C.R. WATSON – to be published. J.B. TORRANCE (this conference).

(42)   J.B. TORRANCE (Private Communication, January 1976).

(43)   V.J. EMERY, Phys. Rev. Lett., $\underline{37}$, 107 (1976).

(44)   P.A. LEE, T.M. RICE and R.A. KLEMM – to be published.

(45)   H. SUMI – to be published.

(46)   A.A. OVCHINNIKOV, Sov. Phys. JETP, $\underline{37}$, 176 (1973).

(47)   J. BERNASCONI, M.J. RICE, W.R. SCHNEIDER and S. STRÄSSLER, Phys. Rev., $\underline{B\ 12}$, 1090 (1975).

(48)   A. BJELIS and S. BARISIC – to be published.

(49)   K. SAUB, S. BARISIC and J. FRIEDEL, Phys. Lett., $\underline{56\ A}$, 302 (1976).

(50)   Y. TOMKIEWICZ, A.R. TARANKO and J.B. TORRANCE, Phys. Rev. Lett., $\underline{36}$, 751 (1976). Y. TOMKIEWICZ (this conference).

(51)   E.F. RYBACZEWSKI, A.F. GARITO and A.J. HEEGER, Bull. Am. Phys. Rev., $\underline{21}$, 287 (1976) and to be published. A.J. HEEGER (this conference).

(52)   Compare the content of this conference with ref (28).

ELECTRONIC PROPERTIES OF ORGANIC CONDUCTORS : PRESSURE EFFECTS[+]

D.Jérome and M.Weger[::]

Laboratoire de Physique des Solides associé au CNRS
Université Paris-Sud - Bâtiment 510
91405   ORSAY   (France)

INTRODUCTION

The purpose of this lecture is to present a survey of the electronic properties of organic metals which have been the matter of very intensive work for the last 5 years. Already extended review articles have been published in that subject (1,2,3,4).

A systematic investigation of the effect of pressure on the Peierls transition temperature T_p in organic charge transfer complexes had started in Orsay in 1973. The motivation of this research was as follows : Pressure is known to reduce the metal to insulator transition in systems like V_2O_3 (5,6), and to stabilize a metallic state at OK. Pressure is also known to stabilize the undistorted metallic phase at OK whenever a lattice distortion occurs driven by strong electron-phonon interactions in systems like the transition metal dichalcogenides (7). Thus, there is some chance that it will do likewise in the organics.

Moreover, if the Peierls transition is caused by coupling of the electron to an intra-molecular phonon (8), (such as bond vibration, or bending), pressure should reduce T_p(9); such phonons have a chance to bring about superconductivity(8,10). In TTF-TCNQ,it raises T_p (11 - 13)(as is to be expected for coupling to an inter-molecular

[+]Work supported in part by a DGRST contract n° 75-07-0820 and a CNRS International ATP program.
[::]Permanent address : Racah Institute of Physics. The Hebrew University, Jerusalem, Israël.

phonon). In spite of these disappointments, the effort was continued on HMTSF-TCNQ (14).

Strictly speaking one or two dimensional systems cannot exist in nature. The fact that crystals can be grown is a proof for the existence of some three dimensional(3d)coupling in actual systems. A question occurs : how large is the interchain or interlayer coupling compared to the scale of temperature for the electronic phase transitions characteristic of low dimensional compounds.

We expect some enhancement of 1 d effects only in those systems for which the 3 d coupling is significantly weaker than the temperature of the onset of periodic lattice distortions(PLD)(or charge density waves, CDW). The mean-field approximation, which underestimates the fluctuations, is not an adequate approximation for the description of 1 d systems (15) although probably reasonably appropriate for 2 d systems. We shall emphasize in this lecture the properties of a large number of quasi 1 d conductors going from KCP to SN_x. They all exhibit to some extent, some of the quasi 1 d characteristic features. We shall also mention some results on the transition metal dichalcogenide layer compounds whenever they help in the understanding of 1 d systems.

As for this latter class of materials we refer the reader to a review of the pressure work in reference (16).

We present in this introductive section the resistivity data, figure 1, of a large number of 1 d conductors : KCP, NMP-TCNQ, TTF-TCNQ, TSeF-TCNQ, HMTSF-TCNQ, HMTTF-TCNQ and the polymer SN_x. The conductivity for all compounds at 300 K is presented in Table I. At low temperature the 1-d compounds exhibit an insulating state except for SN_x which remains metallic down to helium temperature. The very interesting case of HMTSF-TCNQ will be discussed later on in this lecture.

A common feature to all compounds undergoing a metal to insulator transition is the existence of an activated conductivity at low temperature as shown by the linear dependence of $\log \rho$ vs $1/T$ down to $\sim$ 10K The corresponding activation energies are listed in Table I.

The organics exhibit a metal to insulator transition more or less sharp at low temperature whereas KCP shows a very broad transition starting at room temperature. The insulating ground state of NMP-TCNQ is that of a magnetic insulator(17), but no sign of intrinsic spin paramagnetism has been noticed in all other materials at low temperature (18).

There are 3 parameters which govern the behaviour of 1 d metals:
i)   The intra-site Coulomb repulsion.
ii)   The electron-phonon coupling.
iii)   The 3 d coupling.
Energies are measured in units of the bandwidth 4 $t_{//}$.

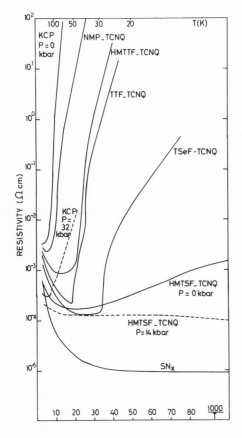

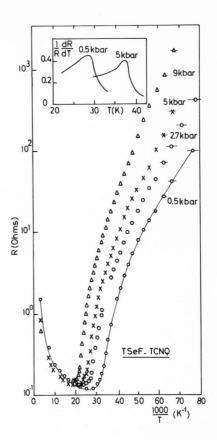

Fig.1. Temperature dependence of the resistivity along the high conductivity axis for several 1d conductors. We thank R.L.Greene for communicating the data of HMTTF-TCNQ prior publication.

Fig.2. Temperature dependence of $\rho(T)$ in TSeF-TCNQ under pressure.
Inset: $(1/R)dR/dT$ in the vicinity of the transition after reference 13.

The intrasite Coulomb repulsion drives a metal to insulator transition of the Mott-Hubbard type and is probably the mechanism for the transition in NMP-TCNQ and related compounds (19). An electron-phonon interaction dominant over the electron-electron repulsion leads to a Peierls-type metal to insulator transition, (20). But, in both cases the sharpness of the transition depends on the amplitude of the 3 d coupling (21). The 3 d coupling may have two different origins which lead to different approaches in the theory of the Peierls transition. i) the lattice distortion characterized by the wave vector $2 k_F$ in 1 d creates a redistribution of the electron density $\delta n/n_o \sim \Delta/E_F$ if the amplitude of the CDW is taken proportionnal to the energy gap 2Δ occuring in a band of Fermi energy E_F(15,22). The redistribution of the

	KCP	NMP–TCNQ	TTF–TCNQ	TSeF–TCNQ	HMTTF TCNQ	HMTSF TCNQ	SN$_x$
$\sigma_{\parallel}$ (RT) (Ωcm)$^{-1}$	300 ±100	380	500 ± 100	800 +100	400 ± 50	1800	1800
Δ(T=0) (K)	682	407	230 ± 20	115 ± 25	240		
T_{M-I} (K)	115	200	53	29	50	$\sim$ 30	

Table I : Conduction at ambient temperature, low temperature gap and Metal to Insulator transition temperature of several quasi one dimensional conductors.

electrons has been measured directly for the case of "weak" PLD(CDW) transitions such as in NbSe$_2$ by NMR on Nb[93] (23). In that system we derived at 6 K, $\delta n/n_o \sim 10\%$. For KCP, the interaction energy for CDW's of neighbouring chains is of the order $E \sim e^2/R \, \varepsilon_o \quad \Delta^2/E_F^2$. With R = 9.8 A, ε_o = 3, Δ = 660 K and $T_F \sim 4000$–6000 K, we calculate E_c = 2–4 meV. In TTF–TCNQ a similar value is to be expected. When all 4 neighbouring chains are taken into account this number may have to be multiplied by 4 (24). ii) the direct overlap between wave functions belonging to different chains leads to a deviation from planar Fermi surfaces(25,26). In TTF–TCNQ the transverse electron energy dispersion can reach values of 5–10 meV. In HMTSF–TCNQ it should be at least 3 times larger.

II – THE PEIERLS TRANSITION IN TSeF–TCNQ, TTF–TCNQ AND HMTSF–TCNQ.

The resistivity behaviour of TSeF–TCNQ under pressure is presented on figure 2 (13). The activation energy can be derived at every pressure and related to the corresponding metal to insulator transition temperature defined by the maximum of (1/R) dR/dT(27). As pressure increases both 2 Δ and T_p increase but $2\Delta/kT_c \sim 6.5$ from 0 to 9 kbar. The common extrapolation of log R when $1/T \rightarrow 0$ at all pressures is consistent with an energy gap varying appreciably but not the carrier mobility. This is an indication for an only weak or even no deformation of the band shape in the 0–10 kbar pressure domain. No other transitions have been detected by resistivity measurements below 29 K in TSeF–TCNQ. Conductivity and X-ray diffuse scattering experiments (28) tend to prove that in TSeF–TCNQ a Peierls transition occurs at 29 K for P = 0 kbar on both TSeF and TCNQ chains. At the same temperature the distortion orders 3 dimensionally with a modulation period 2 a along the a–axis. In TTF–TCNQ the resistivity plot (12), shows two transitions; one at T_H = 53 K, a second at 38 K observed for the first time by low pressure investigations (1.5–3kbar) at T_L = 32 K(9).

A third one has been discovered by neutron scattering experiment at 47-49 K (29) and could produce a very weak anomaly of the resistivity at the same temperature. Under pressure, T_H increases and T_L decreases and the low temperature energy gap follows the pressure dependence of T_L (12).

HMTSF-TCNQ was first synthesized at the John Hopkins University back in the end of 1974 (30). It was found to have $d\rho/dT > 0$ down to about 30 K, and even at lower temperature the conductivity was rather high. The resistivity of HMTSF-TCNQ along the needle axis at various pressures (14) was measured, figure 3. For pressures of 4 kbar or higher, $d\rho/dT > 0$ down to 0.19 K at least (14). The conductivity at helium temperatures exceeds 10^4 $(\Omega-cm)^{-1}$. For temperature 0.2 K<T<2 K, $\rho = A+B\ T^2$, with A = 90 $\mu\Omega$cm; B = 1 $\mu\Omega$cm/K^2 . For the region 2 K<T<30 K, $\rho = a+b\ T$, with a = 86 $\mu\Omega$cm; , b = 1.7 $\mu\Omega$cm/K. Above 30 K, the increase of ρ with T is slower. The change of ρ between 100 K and 30 K is less than 20%, at all pressures. Above 100 K, ρ rises again with T, somewhat like T^2 at P = 0 and more slowly under pressure. Thus, pressure is seen to supress the increase in resistivity occuring at helium temperature.

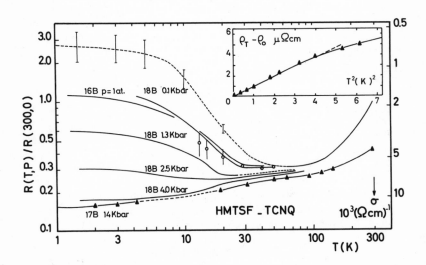

Fig.3. Normalized resistance curves R(T,P)/R(300,0) for pressure runs up to 14 kbar.
The RH scale shows the conductivities obtained by normalizing to a RT value of 1800 $(\Omega$cm$)^{-1}$ at ambient pressure. The dashed line shows the results of reference (30) at ambient pressure. Open circles with error bars represent the average of 4 samples from batch B (Dr.C.S. Jacobsen, private communication). The inset shows the T^2 law obtained for a sample at 14 kbar down to 0.19 K.

The resistivity transverse to the needle axis was also measured as function of temperature and pressure (31). We denote the needle axis the high-conductivity(hc) axis, the direction along the N-Se bonds the intermediate-conductivity (ic) axis, and the direction where the molecular layers are separated by the tri-methylene bridges of the HMTSF molecules, the low-conductivity (ℓc) axis. $\sigma_{hc}/\sigma_{ic} \sim$ 33 and $\sigma_{hc}/\sigma_{ℓc} \sim$ 450 at ambient. In TTF-TCNQ, the hc, ic and ℓc directions are denoted by b, a, c respectively; according to Phillips et al (47) the ic direction is an axis of monoclinic symmetry, and they denote it by c. $\rho_{ℓc}/\rho_{hc}$ increases at low temperatures by a factor of 2 or so, and the effect of pressure is small. ρ_{ic}/ρ_{hc} is an order of magnitude smaller than in TTF-TCNQ at ambient, but it becomes large below 20 K at P = 0. The effect of pressure on it at helium temperature is enormous. Thus, the "transition" occuring below 20 K at P = 0 is seen to be predominantly one of ρ_{ic}, rather than ρ_{hc}. The wide divergence between values of ρ_{hc} of different samples in this temperature range may be due to slight misplacement of contacts (32), which should affect the measured resistance greatly as ρ_{ic}/ρ_{hc} becomes large. A special discussion on the low temperature state of HMTSF-TCNQ will be given in Sect (III). We may point out that a similar pressure removal of a lattice instability has been firmly established in 2H-NbSe$_2$ (7) and in all the other metallic layer compounds (16).

The inorganic quasi 1 d conductor KCP has also been investigated under pressure, up to 32 Kbar (21). Above 25 Kbar, there is a high-temperature domain for which $d\rho/dT > 0$ separated from an activated conductivity low-temperature region by a sharp transition. The logarithmic slope $S(T)=d \ln S/d(1/T)$ has very sharp peaks at 134 K and 144 K at 25 and 32 Kbar, respectively. This transition is similar indeed to what is observed on the resistivity of TTF-TCNQ at 54 K or TSeF-TCNQ at 29 K. We attribute the critical temperature defined by $S(T)$ to sharp metal to semiconductor transitions occuring in KCP under pressure. At low pressures the transition which is probably strongly smeared out by the fluctuations does not show up in the resistivity data. We believe that the critical temperature revealed by the high pressure resistivity data is the extension of the phenomena reported by neutron scattering at P = 0 Kbar (33). A peak in $d\xi/dT$ and in the $(\pi/a, \pi/a, 2 k_F)$ scattering occurs at 115 K (34). A specific heat anomaly has been reported at the same temperature (35). This critical temperature called T_p in this present lecture may correspond to a strongly fluctuation depressed Peierls transition. A resistivity minimum occurs at a temperature T_m; both T_p and T_m are plotted vs pressure in figure 4.

III - ELECTRONIC PROPERTIES OF HMTSF-TCNQ

As it has been already presented earlier HMTSF-TCNQ is the only charge-transfer salt for which the metal to semiconductor transition

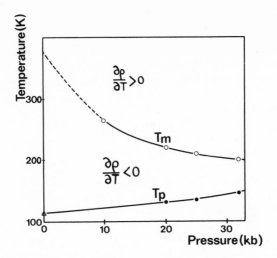

Fig.4. Pressure dependence of T_p (actual phase transition) and T_m (conductivity maximum) in KCP.

can be removed by application of hydrostatic pressure.

With the use of the possibility for stabilization of the metallic state down to practically 0 K a detailed analysis of the galvanomagnetic properties has been performed in HMTSF-TCNQ.

Magnetoresistance, Susceptibility and Hall Effect

The resistivity curves under pressure(fig.3) immediately suggest similarity to Yb,(36) with a reversed effect of pressure(making the material more insulating, rather than more conducting). Thus,they suggest semi-metallic behaviour. A simple test for semi-metallic behaviour is the observation of a large magnetoresistance, since for a given conductivity σ, a semi-metallic state is characterised by a small n, and consequently a large mobility ($\sigma = n\ e\ \mu$). Magnetoresistance is expected to be large when $\omega_c \tau = \mu H/c$ is of order 1, thus a large μ implies a small H to satisfy this condition.

Consequently, the magnetoresistance was measured, and found indeed to be very large (fig.5); a more "professional" presentation of the data in the form of Kohler plots is given in ref.31. Under pressure,$\Delta\rho/\rho$ increases by about 60% for H = 30 kOe in the ic direction and 100% for H in the ℓc direction at helium temperatures. For P = 0, $\Delta\rho/\rho$ is smaller, but $\Delta\rho$ is roughly of the same order (i.e.$\Delta\rho/\rho$ decreases because ρ increases). Under pressure,$\Delta\rho/\rho$ is roughly temperature independent up to 50 K, where it starts to fall, becoming unobservable at 100 K. At P = 0, the fall with increasing temperature is slightly faster. The magnetoresistance is anisotropic. The effect for H along the ic axis is about 1.8 times smaller than for H along the ℓc axis, nearly independent of temperature and pressure. For H along the hc axis (the direction of current flow), the magnetoresistance is at least an order of magnitude smaller and the observed small effect may be due to misalignment of the crystal, since it is hard to align the crystal to better than 5° or so in the pressure bomb. This behaviour is in marked contrast with TTF-TCNQ where there is a small isotropic magnetoresistance at T_p(37). It verifies our hypothesis that HMTSF-TCNQ is semi-metallic at low temperatures.

Independently of the magnetoresistance work, the susceptibility was measured in Copenhagen and found to be diamagnetic below 100 K (38). Diamagnetism is also characteristic of a semimetallic state; the paramagnetic Pauli susceptibility χ_P is proportional to m^*/m_o, while the diamagnetic Landau-Peierls susceptibility χ_L is proportional to m_o/m^*. In semimetals, $\mu = e \tau/m^*$ is large, and since τ cannot be expected to be very large (because of the large number of mechanical defects), m^* must be small. Therefore, χ_L should overcome χ_P. The temperature dependence of χ(fig.6) shows that the effective mass must change drastically, being large at ambient temperature. Thus, the susceptibility measurements show that we have a continuous metal-to-semimetal transition as the temperature is lowered.

For a cylindrical Fermi surface, χ_L (for a field along the axis of the cylinder) is given by :

$$\chi_L^{LT} = - 4 \ \mu_B^2/(3\pi\hbar^2 c) \quad m_e^2/(m_{hc}^* \ m_{ic}^*)^{1/2}$$

for a parabolic band with effective masses m_{hc}^*, m_{ic}^*.

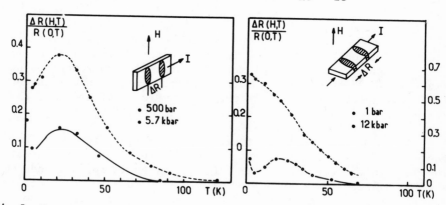

Fig.5. Temperature dependence of the magnetoresistance in HMTSF-TCNQ for H along ℓc direction (left), and ic direction (right) .

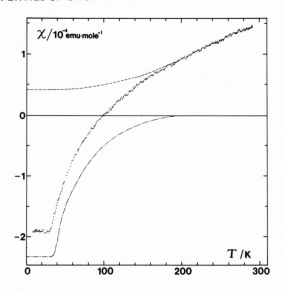

Fig.6. The temperature variation of the intrinsic susceptibility in HMTSF–TCNQ. Expected temperature variations of the Pauli and Landau – Peierls contributions are shown by —————— and —— - —— respectively.

This applies to low temperatures (T< 30 K). For a planar Fermi surface (applicable for T >100 K), obviously χ_L vanishes(there are no closed orbits), while χ_P is given by :

$$\chi_P^{HT} = \frac{4\mu_B^2 \, m^{\ast}}{\pi ac\hbar^2 k_F} \quad thus, \quad \frac{\chi_L^{LT}}{\chi_P^{HT}} = -\frac{1}{3} \frac{m_e}{m^{\ast}} \frac{m_e}{\sqrt{m_{hc}^{\ast} m_{ic}^{\ast}}} \, a \, k_F^{ic}$$

where $m^{\ast}$ is the effective mass in the hc direction at high temperatures, and k_F^{ic} the Fermi vector in that direction (fig.7).

Note that δk_F^{LT} (in the hc direction) cancels out in this ratio. This is a very important property, since if this ratio would depend on $\delta k_F^{LT}/k_F^{HT}$, which is of order $10^{-2} - 10^{-3}$, the ratio χ_L^{LT}/χ_P^{HT} would not be of order unity. This cancellation does not take place for an ellipsoidal Fermi surface. However, for a FS not deviating too strongly from this cylindrical shape, this is probably a reasonable approximation(a decrease in $\chi_L^{LT}(H_{\ell c})$ should be somewhat compensated by an increase in $\chi_L^{LT}(H_{ic})$, for a slightly ellipsoidal surface).

One of the most basic measurements (if not the most basic one) of semi–metals and semi–conductors is the Hall effect.

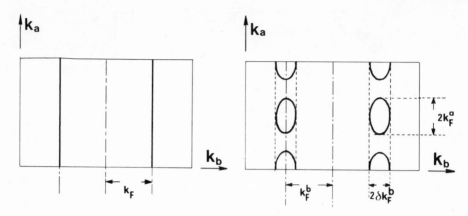

Fig.7. Very schematic picture of the Fermi surface in HMTSF-TCNQ. High-temperature planar surfaces (left). Low-temperature 3d surfaces (right).

The Hall constant(31) is shown in fig.8, as function of temperature, at P = 0 and at about 5 Kbar. It can give an estimate of the carrier density, $n \simeq 1/R_H ec$. A somewhat more elaborate estimate is provided by the use of the resistivity and magnetoresistance, using the formulas (39) :

$$\rho^{-1} = ne(\mu_h + \mu_e) \qquad R_H = 1/nec \quad (\mu_h - \mu_e)/(\mu_h + \mu_e) \qquad \Delta\rho/\rho H^2 = \mu_h \mu_e / c^2$$

They yield $n \approx 1.2.10^{18}$ carriers/cm$^3 \approx 1/500$ per formula unit, $\mu_e \approx 4.10^4$ cm^2/V sec; $\mu_h \approx 1.2.10^4$ cm^2/V sec at helium temperature under pressure. These formulas assume isotropic bands, and therefore do not apply here precisely, but probably the estimates of n and of μ hold within a factor of 2 or so(for a planar Fermi surface(39)$R_H \approx 1/nec$). At ambient temperature, the small mean free path makes the use of the Boltzmann equation somewhat questionable; but we are interested in the low temperature region mainly. Non-linear Hall voltages(as function of the magnetic field) characteristic of semi-metals and semi-conductors are also observed (31).

The Electronic Band Structure

While attempts have been made to apply electronic band theory to TTF-TCNQ(40,41,42) the large value of $U/4t_{//}$ ($\sim$1)makes the applicability of this theory somewhat dubious.

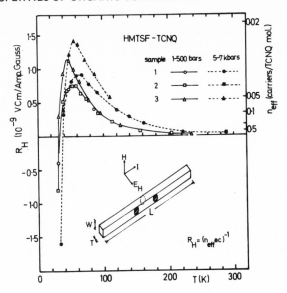

Fig.8. Behaviour of the low field Hall constant in HMTSF-TCNQ vs temperature.

The experimental striking features of TTF-TCNQ are : (a) The existence of several phase transitions at 54 K (43), 48 K (44), 38 K (9,12).(b) The change in the transverse period of the CDW from 2a at 48 K to 4a at 38 K(29). (c) The existence of very strong 4 k_F reflections(45);and (d) : The very strong pressure dependence of the electronic properties ρ, χ, and $T_\perp$, without a strong pressure dependence of the plasma frequency ω_p (46) nor the anisotropy in the resistivity. We cannot account for these phenomena by one-electron band theory in a convincing way, and feel that Coulomb interactions have to be invoked, as will be discussed later. HMTSF-TCNQ has a much smaller value of U, so band theory here should be applicable. Anisotropic galvanomagnetic properties, in particular, call for such a theory.

The essentials of the electronic band structure of TTF-TCNQ like systems are shown in table II. Due to the transfer integrals between chains the Fermi surface is no longer planar. The integrals between like chains(donor→donor and acceptor→acceptor) are "good" in a sense that they result in a large FS and a metallic density of states at the Fermi level. The transfer integrals between unlike chains (donor→acceptor) are "bad" in a sense that they cause a covalency gap, i.e. a vanishing Fermi surface(degenerating to a line at the zone boundary), and a vanishing density of states at the Fermi level(41). In HMTSF-TCNQ the crystal structure is such that in both direction perpendicular to the chain, the neighbours are unlike(47). Therefore, we could naively expect a vanishing Fermi surface(in form of points at the corner of the Brillouin Zone), a quadratically vanishing density

Expression for Energy   Fermi Surface

Uncoupled chains

$$E_Q(k_y) = 2t_Q^b \cos k_y b$$

$$E_F(k_y) = E_o - 2t_F^b \cos k_y b$$

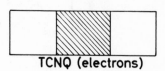

TCNQ (electrons)

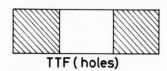

TTF (holes)

Coupling between chains
of same type

$$E_Q(k_y, k_z) = 2t_Q^b \cos k_y b \pm 2t_Q^c \cos \frac{k_z c}{2}$$

$$E_F(k_y, k_z) = E_o - 2t_F^b \cos k_y b \mp 2t_F^c \cos \frac{k_z c}{2}$$

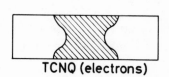

TCNQ (electrons)

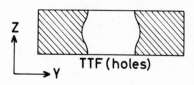

TTF (holes)

Coupling between chains
of opposite type

$$\mathcal{H} = \begin{pmatrix} E_Q(k_y, k_z) & t_{QF}^a \cos \frac{k_x a}{2} \\ t_{QF}^a \cos \frac{k_x a}{2} & E_F(k_y, k_z) \end{pmatrix}$$

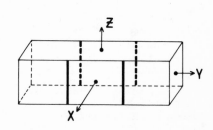

Table II Effect of interchain transfer integrals on the band

Energy Level Diagram   Density of States

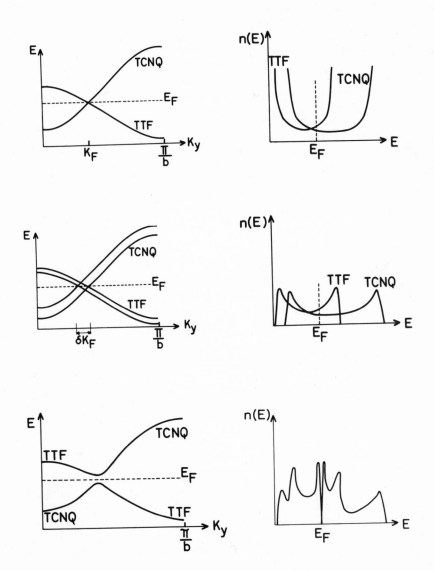

structure of TTF-TCNQ like systems (very schematic).

of states at the Fermi level and a zero-gap semiconductor, somewhat like α-Sn(48) or Hg Te(49), which are Anderson-Insulators when pure. This is in contrast with experiment which shows that HMTSF-TCNQ is highly conducting at low temperatures down to 0.02 K (50) under pressure. The explanation of this paradox is the main task of the band-structure model for this compound.

A possible origin for the semimetallic state is as follows : Consider a TCNQ and a HMTSF molecule side by side (fig.9). The wavefunction on the TCNQ is antisymmetric (between top and bottom) and that of HMTSF is symmetric. Thus, for a symmetric orientation of the molecules, the matrix element for electron tunneling vanishes by symmetry. This has already been pointed out by Berlinsky (40) for TTF-TCNQ. However, if the Se-N bonds are short (47), so that the two-center integral is large (of order 0.1 eV), this integral can connect to the excited state of HMTSF, which is antisymmetric, and this interaction displaces the TCNQ energy level. The matrix element is proportional to $\cos^2 k_{ic} d_{ic}/2$, where d_{ic} is the lattice constant in the ic direction (12.57 A).

Such excited states exist in TTF, TSeF less than 2 eV below the ground state(51). (no data is available on HMTSF). Thus, shifts of order 10 meV are possible. Together with the matrix elements responsible for the covalency gap, they give rise to a band structure, with "light" electrons and "heavy" holes. The Fermi surface is shown in fig.10.

This model is discussed in detail in ref.52. The band structure can be described by an effective 2x2 hamiltonian, with

$$\mathcal{H}_{11} = t_Q \cos k_{hc} d_{hc} + t_{shift} \cos^2 k_{ic} d_{ic}/2$$
$$\mathcal{H}_{22} = E_{HQ} - t_F \cos k_{hc} d_{hc}$$
$$\mathcal{H}_{12} = t_{split} \cos k_{ic} d_{ic}/2 + t_{\ell c} \sin k_{\ell c} d_{\ell c}/2$$

where E_{HQ} is the difference between the HMTSF and TCNQ energy levels at $k = 0$, t_Q, t_H are the integrals along the stacking axis, t_{shift} is the term causing the shift (a similar term should be present in $\mathcal{H}_{22}$ due to the TCNQ excited states), t_{split} is the term causing the covalency gap, and $t_{\ell c}$ is the small integral in the ℓc direction. This effective hamiltonian describes all experimental properties of HMTSF-TCNQ extremely well.

The semimetallic state at 0 K is immediately accounted for. The T^2 law of the resistivity should apply for temperatures small compared with E_F(hole). A T law resistivity should be present for E_F(hole)$<k_B T<E_F$(electron), and the values E_F(hole)$\cong 8$ K, E_F(electron)≈ 40K follow from the parameters of the hamiltonian. The small cross-section of the Fermi surface should give rise to the extreme quantum limit $E_{Landau} \cong E_F$ around 30 kOe; the magnetoresistance is large because of

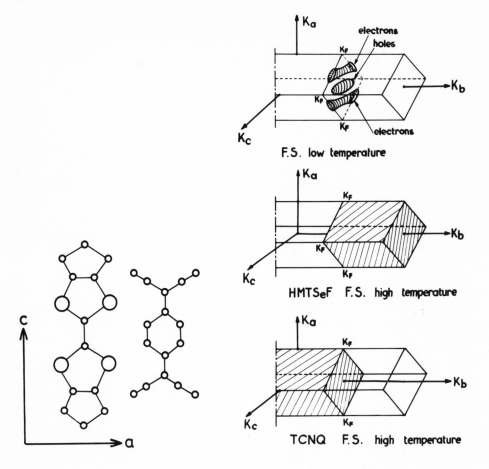

Fig.9. Molecular arrangement of HMTSF and TCNQ molecules in the c a plane, assumed in the simplified calculation.

Fig.10. Fermi surfaces in "high temperature"(T>100 K)and "low temperature"(T<30 K)states. The BZ is centered around(0,0,π/c) since there is no inversion symmetry around (0,0,0).

the compensation of the semimetal(equal number of electrons and holes); the effective mass, particularly of the electrons, is small, giving rise to a large diamagnetism at low temperatures. The relatively small value of ρ_{ic} is due to the strong Se-N bonds; the large mobility is due to the large Fermi velocity in spite of the small value of δk_{hc}. The flat plateau in resistivity is due to the near constancy of n/k_F between the high-temperature 1-D state and the low-temperature 2-D state. (More quantitative estimates are given in ref.52). The large crystalline disorder limits the mean free path so that the lifetime for scattering is very short, and $\tau E_F \simeq h$. This should make it difficult to apply more powerful galvanomagnetic techniques, such as search for de Haas Van Alphen oscillations.

The value of the Hall constant at helium temperatures agrees extremely well with the estimates from the band model. The value at ambient agrees with our belief that HMTSF-TCNQ is a metal there, with about 1.5 carriers per formula unit(28). In particular, the large value demonstrates the asymmetry between electrons and holes, the electrons possessing the higher mobility at helium temperatures, (and the holes above 38 K). The sign above 38 K is in accord with the thermoelectric voltage there (30). Since at high temperatures TCNQ carriers behave like electrons and HMTSF carriers like holes, this suggests a higher mobility on HMTSF chains.

While the success of this band structure as a phenomenological model is extremely good, the preliminary report(47) of the structure of HMTSF-TCNQ suggests that the four nitrogen neighbours of the four seleniums of a given HMTSF molecule belong to FOUR different TCNQ's (rather than 2). For this structure, there is only partial cancellation of the N-Se integrals (by a factor $\sin 1/2 \ k_F d_{hc} \simeq \sin 0.18\pi \simeq 1/2$) and it is hard to see how we get a t_{shift} term large enough to overcome t_{split}. The problem is, that in a tight-binding model, the phases of the wavefunction ψ on neighbouring Se and N atoms do not match. If instead of a TB model we had conducting tubes(where the phase of ψ depends only upon the distance along the tube), this difficulty would not exist. Transfer integrals that make things go in this direction do exist(52), but the question is whether they are strong enough. If t_{shift} is small, we have to postulate the presence of impurities to produce a conducting state at 0 K; moreover, the band structure should be rather symmetrical between electrons and holes. Thorough ab-initio calculations(53) may be required to settle this question decisively.

IV - FREQUENCY DEPENDENCE OF THE NUCLEAR RELAXATION RATE

We present in this section the features of the nuclear spin lattice relaxation particular to quasi one dimensional conductors.

Quite generally nuclear relaxation in conductors is induced by the modulation of the hyperfine coupling and is given by the relation

$$1/T_1 = A_\pm \ g^\pm (\omega_e) + A_z \ g^z(\omega_n) \tag{1}$$

where $A_\pm$, A_z and $g^\pm(\omega)$, $g^z(\omega)$ are geometrical factors and spin correlation functions associated with the scalar $(I^+ S^-)$ and the dipolar $(I^+ S^z)$ parts of the hyperfine hamiltonian(54) and ω_e, ω_n are the electron and nuclear Larmor frequencies.

For the case of itinerant electron spins constrained to move on a chain the electron spin correlation function $g(t)=<S^+(R_i,0)S^-(R_i,t)>$ can be approximated by the probability of return of the electron to the origin after a time t.

For a one dimensional random walk this probability goes like $t^{-1/2}$ (55,56). The nuclear relaxation rate given by the Fourier transform of the spin correlation function diverges as $\omega_{e,n} \to 0$(56,57). However, this divergence is cut-off either by the spin-orbit coupling [thought to be a relatively inefficient mechanism in light-atom one dimensional conductors,(58)] or the finite life time τ of a carrier on the chain (55). The life time effect can be taken into consideration as a $e^{-t/\tau}$ factor in the spin correlation function. Therefore, $1/T_1 \alpha \int_0^\infty \cos(\omega_{e,n} t) t^{-1/2}\exp-t/\tau$ dt becomes field independent if the condition $\omega_{e,n} \tau<1$ is fulfilled but acquires a $1/\sqrt{H_o}$ field dependence for fields such as $\omega_{e,n} \tau>1$.

In the absence of parasitic nuclear relaxation(such as by magnetic impurities) we expect T_1 versus magnetic field to remain constant at low fields up to a cut-off field given by $\omega_e\tau \simeq 1$ and to exhibit a $H_o^{1/2}$ field dependence above, provided the scalar contribution to the nuclear relaxation is dominant upon the dipolar contribution (59).

The frequency dependence of T_1 in TTF-TCNQ(D_4) has been summarized in figure 11. A $H_o^{1/2}$ field dependence is very well observed above 10 kOe. The 64 kOe result (ν_n=276 MHz) deviates significantly from the square root law. This can be attributed to the observation at high field of the dipolar contribution to $1/T_1$ for which $\omega_n\tau$ at 64 kOe is still smaller than unity. The estimate$(1/T_1)_{dipolar} \approx 1/3 (1/T_1)_{exp}$ at H<10 kOe is in very good agreement with the dynamic nuclear polarization results (59).

Two important remarks can be made from the frequency dependence of T_1 in quasi one dimensional conductors.
a) $1/T_1$ can be strongly enhanced over the Korringa value. In fact, as seen on figure (10) the value of the enhancement is influenced by the transverse hopping rate.

An estimation of the diffusion constant of the spin correlation function can be performed using the Hubbard model(60,61). Although, strictly speaking Devreux's approach (61) is valid when kT>$t_{//}$ it was concluded in that paper that an order of magnitude estimation can be performed for kT<$t_{//}$,(the situation in TTF-TCNQ at room temperature). Then, the relaxation rate in low fields is given by :

$$(1/T_1)/(1/T_1)_o \sim (U/\omega_c)^{1/2}.$$

Instead of the Hubbard hamiltonian, the collisions responsible for the electrical resistivity $\rho= m^*/ne^2\tau$ can be invoked, and $(U/\omega_c)^{1/2}$ is replaced by $(3/4 \pi^2\tau\omega_c)^{1/2}$. In this picture the observed near-temperature independence of $\chi_s^2 T_1 T$ at low fields is attributed to accidental cancellation between various temperature-dependent factors.

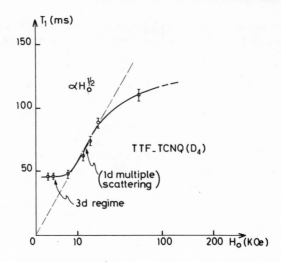

Fig.11. A plot of T_1 vs $H_o^{1/2}$ in TTF-TCNQ(D_4) demonstrating the existence of 3 relaxation regimes.

b) A basic approximation consists in describing the motion of the carriers along the chains as coherent(a questionable assumption at room temperature since the mean free path is not much larger than the intermolecular spacing). At high temperature $kT > t_\perp$ the electron wave functions built from the linear combination of incoherent atomic wave functions on each chain lead to localized electron states. Therefore, the motion of the carriers from chain to chain is diffusive, due to a hopping mechanism. The correlation frequency associated with this hopping mechanism is given by the golden rule :

$$\frac{1}{\tau} = \frac{2\pi}{\hbar} \, t_\perp^2 \, n(E_F) \tag{3}$$

with $\tau = 3.7 \times 10^{-12}$ s, n = 0.55 el/molecule and $t_{//}$ = 0.23 eV the relation (3) yields $t_\perp$ = 5.8 meV, in agreement with the assumption of transverse hopping conduction. For the parameters of TTF-TCNQ namely $t_\perp$ = 5.8 meV T_P = 53 K and T_F = 1800 K we may use the treatment of Horovitz et al (25) to derive the cross over temperature Θ separating the region $T > \Theta$ where fluctuations are 1D from the region $T < \Theta$ where fluctuations become 3 D. By (4.10) of ref 25, $\ln(\Theta/T_P^{MF}) = t_\perp^2 E_F^2 / \pi t_{//}^2 \Theta^2$ as long as $t_\perp$ is large enough to suppress fluctuations, so that $T_P \propto T_P^{MF}$ (this requires $t_\perp E_F/t_{//} T_P^{MF} \gtrsim 4$, by (4.18) there). Thus, when $t_\perp E_F/t_{//} T_P^{MF} < 4$, MF theory does not work, and T_P is strongly depressed by fluctuations, while when $t_\perp E_F/t_{//} T_P^{MF} > 4$, MF theory applies and Θ (or rather $\Theta - T_P$) is of order $t_\perp$, and the correlations near T_P are 3-d, in good agreement with the 3-d critical behaviour of the resistivity there(62).

The T_1 field dependence has been studied for TTF(D_4)-TCNQ and TTF-TCNQ(D_4), figure 12 giving larger ω_c on the TCNQ chains than on the TTF chains (63).The result is in good agreement with a coupling along the c direction being stronger between TCNQ chains than between TTF chains (40;table II) and a larger density of states on the TTF chains (64,24).

The properties of T_1 in HMTSF-TCNQ are as follows (38): (a) T_1 is extremely long at ambient temperature, $\sim$1 sec vs. $\sim$ 0.1 sec in TTF-TCNQ. (b) The Korringa law $\chi_s^2 T_1 T$ is approximately valid down to 140 K; below it, T_1 becomes shorter. (c) T_1 is field independent (up to 64 kOe); at low temperatures(where the Korringa law breaks down), there is a weak field dependence ($T_1 \propto H^{1/4}$, roughly). (d) The values of $\chi_s^2 T_1 T$ (as well as χ_p) are not very far away from the free electron values (in contrast with TTF-TCNQ). (e) The pressure dependence is also weaker than in TTF-TCNQ[a factor 2 at 8 kbar,(65) vs. a factor of 10 in TTF-TCNQ (66)]·Regarding (a), we do not attribute the short T_1 in TTF-TCNQ to dipolar coupling (67); such a coupling should be present in HMTSF-TCNQ as well. Rather, we attribute the longer value of T_1 to a smaller U, due to better screening by the short Se-N distances. As for (b), we attribute the shorter relaxation time at low temperatures to impurities (possibly, Fe, or broken chains containing an odd number of electrons, which may act as paramagnetic centers relaxing the protons and causing Curie-tails in the low temperature susceptibility). Regarding (c), the frequency independence is in accord with the stronger interchain coupling, by the short Se-N bonds. (d) is also accounted for by the smaller value of U. More quantitative estimates are given in ref.38·(e)is also accounted for by a smaller U.

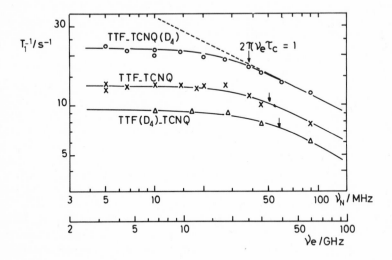

Fig.12.Frequency dependence of the proton relaxation rate in TTF-TCNQ TTF-TCNQ(D_4), TTF(D_4)-TCNQ.

SUMMARY

We tried to demonstrate in this lecture the validity and the limits of applicability of a band picture to organic conductors. We have seen that the non-interacting electron band picture which gives a
reasonable view of the selenium compound HMTSF-TCNQ fails to explain
coherently the experimental data in TTF-TCNQ. Regarding these compounds, there are a number of still not completely settled questions,
namely :
(1) The role of 1-d fluctuations, and the crossover from 1-d to 3-d
regimes.

The parameter characterising the importance of fluctuations is
(25) $\eta E_F/T_P^{MF}$, where $\eta = t_\perp/t_{//}$ for coupling between chains due to tun-
neling(and a similar expression applies to Coulomb coupling between
chains). If this dimensionless parameter is large compared with unity,
mean-field(MF)theory applies, and $T_P \sim T_P^{MF}$, while if it is small,
fluctuations are large and $T_P << T_P^{MF}$. In TTF-TCNQ, η is small(of order
10^{-1}). If $T_P^{MF} \sim T_P = 54$ K, $\eta E_F/T_P^{MF} > 1$ and the model is self-consistent,
but if $T_P^{MF} >> T_P$ then $\eta E_F/T_P^{MF} < 1$ and the model is self-consistent as
well, because in this case T_P is strongly depressed. Another aspect of
this question is the relationship between the interchain coupling(here
represented by $t_\perp$)and the transition temperature Θ between the one-
dimensional structure at high temperatures and the 3-dimensional
structure at low temperatures. Several theories about the relationship between $t_\perp$ and $k_B\Theta$ have been proposed(Lee Rice Anderson,(68)
Bulayevski(69), Scalapino, Imry, Pincus(15), Horovitz, Gutfreund,
Weger(25)). Experimentally, from the magnetoresistance and Hall effect, HMTSF-TCNQ appears to be 3-dimensional up to about 40 K, and
1-dimensional above 200 K (figs.5,8), while $t_\perp$ is estimated to be
about 200 K. In TTF-TCNQ, less data are available, but the deviation
from 1-dimensional behaviour starts at lower temperatures, and the
Peierls transition occurs before the band structure has become 3-di-
mensional.($t_\perp$ was measured(24) to be about 50 K). This agrees with
the mean field approximation(25), but a clear-cut experimental determination of the fluctuations is wanting.(A contribution to $1/T_1$ near the
transition temperature in quasi-1-d systems has been suggested[1] (70)
and observed in Nb_3Al (71) and perhaps KCP(72) and $2H-NbSe_2$(16), but
not yet in TTF-TCNQ).
(2) The relative contribution of tunneling and Coulomb interaction to
the inter-chain coupling.

In this work, we emphasize the frequency dependence of T_1, the
1d-3d cross over of the Hall effect, the magnetoresistance, and the
diamagnetic susceptibility, which demonstrate unambigously the tunneling mechanism. It is likely that compounds such as TTF-TCNQ or TMTTF-
TCNQ exhibit a situation with $T_P > t_\perp/k$ which precludes the observation
of Landau-Peierls contribution to the magnetic susceptibility. This
is also evidenced by the equivalence in temperature dependence between χ_s measured by a Faraday method and the spin contribution to the
susceptibility measured by EPR in TTF-TCNQ or TMTTF-TCNQ,(66,73).Thus,
here the inter-chain Coulomb interaction may play a more

important role in suppressing the 1-d fluctuations(22). But a quantitative determination of these interactions is still wanting.

(3) The value of the intra-site Coulomb repulsion U in the various
compounds, and the role it plays (74).

We suggest that $U/4t_{//} \sim 1$ in TTF-TCNQ. Moreover this quantity is
considerably smaller in HMTSF-TCNQ, and is strongly pressure-dependent.
This provides a plausible explanation for the pressure dependence of
ρ, χ_s, and $1/T_1$ (75,76).
    A phenomenon of key importance missing from our discussion, is the
observation by X-rays of strong reflections of wavevector 4 k_F(45).
One way to account for these reflections is by the spinless fermion
model(77), in which the intra-site Coulomb interaction prevents two
electrons from occupying the same site, and thus has the effect of an
extra "Pauli"principle, halving the number of available states and
therefore doubling k_F. Such an explanation requires a large U, and is
therefore in accord with the present large-U picture for TTF-TCNQ.(Actually, it is not yet clear whether a U of order 4 $t_{//}$ is large enough
for the spinless fermion picture to apply). An alternative explanation
for the 4 k_F reflections has also been proposed (78), treating TTF-
TCNQ as a soft molecular crystal with strong Coulomb forces between the
negative nitrogens and positive sulphurs. This model also accounts for
the change in transverse period of the periodic lattice distortion from
2a to 4a between 48 K and 38 K (29).

(4) The mechanism responsible for the resistivity, and its temperature
dependence.

    This question is highly controversial (79). In this work, we claim
that the long spin-memory time($4x10^{-12}$ sec in TTF-TCNQ at ambient (24),
compared with the collision time of $5x10^{-15}$sec), and the strong pressure dependence, are key factors that any theory, whether phonon(80),or
intra-chain electron-electron collision (81), or inter-chain electron-
electron collision(82)or paramagnon (83), or collective-soliton(1,84)
must account for.

(5) The P = 0 Low-Temperature state of HMTSF-TCNQ, and the question
whether in some other charge transfer compound the P = 0, T = 0 state
may be highly conducting (85), or a larger concentration of carriers may
exist.

(6) The role of the ratio of the $\Delta k = 0$ and $\Delta k = 2$ k_F matrix ele-
ments("geology", 86,87,77).

(7) The role of soft phonons(such as librations (78)),which should be
particularly sensitive to pressure in this soft molecular crystal.

    The picture we have proposed in this lecture can be extended to
the other 1d conductors and summarized on a diagram where compounds are
specified by the parameters $t_\perp/t_{//}$ and $U/4t_{//}$, (Fig.13). In this respect

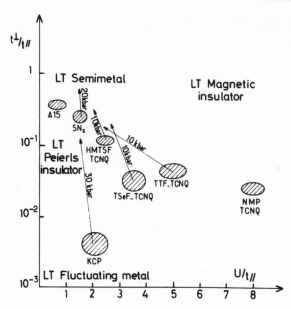

Fig.13. Summary of the effect of pressure on interchain coupling and Coulomb interaction for various 1d conductors.

the organic compounds no longer remain isolated from the mixed valency salts. In particular pressure effects indicated by arrows on fig.13 indicate an increase of interchain coupling(3d behaviour) and a decrease of the electron-electron interactions (88). In this respect pressure appears as a very rewarding tool. Admittedly this new picture of the organic charge transfer salts needs to be confirmed by more experimental work and a further development of the theory. We consider our contribution as a step towards "demystification" of the 1d conductors.

ACKNOWLEDGMENTS

This report is a summary of more detailed articles which are being published and will be published in the near future. We wish to thank our colleagues at Orsay who have participated in the joint effort, particularly J.R.Cooper, G.Soda, C.Weyl, C.Berthier and L. Zuppiroli.

We greatly acknowledge the help given by G.Delplanque and G. Malfait in our high pressure group. The early stage of these researchs benefited greatly from cooperation with S.Etemad, Ed.Engler from IBM and M.Thielemans and R.Deltour from Université Libre, Bruxelles and H.Gutfreund and B.Horovitz from the Hebrew University, Jerusalem.

This research would not have been possible without the very close and efficient cooperation between physicists and the chemistry group of K.Bechgaard in Copenhagen. We thank L.Giral, E.Torreilles, J.M. Fabre and P.Calas who performed part of the chemistry work at Montpellier.

We always have had very useful discussions with N.F.Mott, J. Friedel and S.Barisic.

REFERENCES

1. A.J.Heeger, A.F.Garito, in "Low Dimensional Cooperative Phenomena" ed.H.J.Keller,NATO-ASI Series B7, 89(1975),Plenum Press,New York

2. I.F.Shchegolev, Phys.Stat.Sol.(a) 12, 9 (1972).

3. H.R.Zeller, Festkörper Probleme XIII, 31,(1973).

4. M.Weger and I.B.Goldberg, Solid State Physics 28, 1-177 (1973), Ed.Seitz, Turnbull, Ehrenreich, Academic Press, New York.

5. D.B.Mc Whan and I.P.Remeika, Phys.Rev. B2, 3734 (1970).

6. D.Jérome et al (unpublished); G.Lesino, Thesis, Orsay 1974.

7. C.Berthier, P.Molinié, D.Jérome, Solid State Comm.18, 1393 (1976).

8. H.Gutfreund, B.Horovitz, M.Weger, J.Phys. C 7, 383 (1974); Solid State Comm. 15, 849 (1974); Phys Rev.B 12,1086 (1975).

9. D.Jérome, W.Müller, M.Weger, J.Physique Lett.35, L-77 (1974).

10. B.Horovitz, A.Birenboim, Solid State Comm.19, 91 (1976).

11. C.W.Chu, J.M.E.Harper, T.H.Geballe, R.L.Greene, Phys.Rev.Lett.31, 1491 (1973).

12. J.R.Cooper, D.Jérome, M.Weger, S.Etemad, J.Physique Lett.36, L-219 (1975).

13. J.R.Cooper, D.Jérome, S.Etemad, E.Engler (to be published).

14. J.R.Cooper, M.Weger, D.Jérome, D.Le Fur, K.Bechgaard, A.N.Bloch, D.O.Cowan, Solid State Comm.19, 749 (1976).

15. D.J.Scalapino, J.Imry, P.Pincus, Phys.Rev.B 11, 2042 (1975).

16. D.Jérome, C.Berthier, P.Molinié and J.Rouxel,J.Physique 37 C 4, 105, (1976).

17.   A.J.Epstein, S.Etemad, A.F.Garito, A.J.Heeger, Solid State
       Comm. 9, 1803 (1971).

18.   J.C.Scott, A.F.Garito, A.J.Heeger, Phys.Rev. B10, 3131 (1974).

19.   A.J.Epstein, S.Etemad,A.F.Garito, A.J.Heeger, Phys.Rev.B 5,
       952 (1972).

20.   R.E.Peierls, "Quantum Theory of Solids", Clarendon Press, Ox-
       ford, 1955; H.Fröhlich, Proc.Roy.Soc. A 223, 296 (1954).

21.   M.Thielmans, R.Deltour, D.Jérome, J.R.Cooper, Solid State Comm.
       19, 21 (1976).

22.   K.Saub, S.Barisic, J.Friedel, Phys.Lett.56 A, 302 (1976).

23.   C.Berthier, D.Jérome, P.Molinie, J.Rouxel, Solid State Comm.19,
       131 (1976).

24.   G.Soda, D.Jérome, M.Weger, J.M.Fabre, L.Giral, Solid State Comm.
       18, 1417 (1976).

25.   B.Horovitz, H.Gutfreund, M.Weger, Phys.Rev.B 12, 3174 (1975).

26.   M.Weger, J.Phys.Chem.Solids 31, 1621 (1970).

27.   S.Etemad, Phys.Rev. B 13, 2254 (1976).

28.   C.Weyl, E.M.Engler, S.Etemad, K.Bechgaard, G.Jehanno, Solid Sta-
       te Comm. 19, 925, (1976).

29.   R.Comès, S.M.Shapiro, G.Shirane, A.F.Garito, A.J.Heeger, Phys.
       Rev.Lett. 35, 1518 (1975).

30.   A.N.Bloch, D.O.Cowan, K.Bechgaard, R.E.Pyle, R.H.Bands, T.O.
       Poehler Phys.Rev.Lett. 34, 1561 (1975).

31.   J.R.Cooper, M.Weger, G.Delplanque, D.Jérome, K.Bechgaard, J.
       Physique Lett., Dec.1976.

32.   D.E.Schafer, F.Wudl, G.A.Thomas, J.P.Ferraris, D.O.Cowan, Solid
       State Comm. 14, 347 (1974).

33.   B.Renker,R.Comes,in "Low-Dimensional Cooperative Phenomena"
       ed.H.J.Keller, NATO-ASI Series B7, 235 (1975),Plenum Press,N.Y.

34.   B.Renker, L.Pintschovius, W.Gläser, H.Rietschel, R.Comes, L.
       Liebert, W.Drexel, Phys.Rev.Lett.32, 836 (1974).

35.   K.Franulovic, D.Djurek, Phys.Lett. 51 A, 91 (1975).

36. D.B.Mc Whan, T.M.Rice, P.H.Schmidt, Phys.Rev. 177, 1063 (1969).

37. T.Tiedje, J.F.Carolan, A.J.Berlinsky, C.Weiler , Can.J.Phys.53, 1593 (1975).

38. G.Soda, D.Jérome, M.Weger, K.Bechgaard, E.Pedersen, Solid State Comm. 20, 107 (1976).

39. A.C.Beer, "Galvanomagnetic Effects in Semiconductors", Academic Press, New York.

40. A.J.Berlinsky, J.F.Carolan, L.Weiler, Solid State Comm.15, 795 (1974).

41. U.Bernstein, P.M.Chaikin, P.Pincus, Phys.Rev.Lett. 34, 271 (1975).

42. S.Shitzkovsky, M.Weger, H.Gutfreund, Proc.Siofok Conf.1976,V.K. S.Shante, A.N.Bloch, D.O.Cowan, W.M.Lee, S.Choi, M.H.Cohen, Bull. APS 21, 287 (1976).

43. J.Ferraris, D.O.Cowan, V.Walatka, J.H.Perlstein, J.Am.Chem.Soc. 95, 948 (1973).

44. R.Comes, G.Shirane, S.M.Shapiro, A.F.Garito and A.J.Heeger,Phys. Rev.Lett. 1976. P.Bak and V.E.Emery,Phys. Rev.Lett. 36, 978(1976)

45. J.P.Pouget, S.K.Khanna, F.Denoyer, R.Comes, A.F.Garito and A.J. Heeger, Phys.Rev.Lett.37, 437, 1976. S.Kagoshima, T.Ishiguro, H.Anzai, to be published J.Phys.Soc.Japan.

46. B.Welber, E.M.Engler, P.M.Grant, P.E.Seiden, Bull.APS 21, 311 (1976).

47. T.E.Phillips, T.J.Kistenmacher, A.N.Bloch, D.O.Cowan,JCS Chem. Comm. 334   (1976).

48. S.Groves and W.Paul,Phys.Rev.Lett.11, 194 (1963).

49. S.Groves, R.N.Brown, C.R.Pidgeon, Phys.Rev.161, 779(1967).

50. M.Ribault, private communication.

51. P.Gleiter, M.Kobayashi, J.Spanget-Larsen, J.P.Ferraris, A.N. Bloch, K.Bechgaard, D.O.Cowan, (to be published).

52. M.Weger, Solid State Comm.19, 1149   (1976).

53. D.R.Salahub, R.P.Messmer and F.Herman, Phys.Rev.B 13,4252(1976).

54.   A.Abragam, Nuclear Magnetism, page 264, Clarendon Press, Oxford 1961.

55.   G.Soda, D.Jérome, M.Weger, J.M.Fabre and L.Giral, Solid State Comm. $\underline{18}$, 1417 (1976).

56.   D.W.Hone and P.M.Richards, Ann-Rev.Materials Sci $\underline{4}$, 337 (1974).

57.   F.Devreux and M.Nechtschein, Solid State Comm.$\underline{16}$, 275 (1975).

58.   Y.Tomkiewicz, E.M.Engler and T.D.Schultz, Phys.Rev.Lett $\underline{35}$,456 (1975).

59.   The smallness of the dipolar contribution to T_1^{-1} is confirmed by the observation of a dynamic proton nuclear polarization enhancement in TTF-TCNQ(D_4) and TTF(D_4)-TCNQ reaching + 200 at infinite saturating power.
       J.Gallice, J.P.Blanc, H.Robert,J.Alizon, private communication and report at the"Siofok Conference 1976."

60.   J.Villain, J.Physique Lett.$\underline{36}$,L-173 (1975).

61.   F.Devreux, Phys.Rev.B, $\underline{13}$, 4651 (1976).

62.   P.M. Horn and D.Rimai, Phys.Rev.Lett. 36, 809 (1976).

63.   It appears from the latest results of G.Soda et al(to be published) that in new samples $1/T_1$(TTF(D_4)-TCNQ) is slightly smaller than the data of figure (12), leading consequently to a larger cut-off frequency on the TCNQ chain. This discrepancy may be attributed to an incomplete deuteration of early samples.

64.   Y.Tomkiewicz, B.A.Scott, L.J.Tao and R.S.Title, Phys.Rev.Lett $\underline{32}$, 1363 (1974).

65.   G.Soda,Proc. "Conference on Organic Conductors and Semiconductors", Siofok/Hungary, 1976.

66.   C.Berthier,J.R.Cooper, D.Jérome, G.Soda.C.Weyl, J.M.Fabre and L.Giral, Mol.Cryst Liq Cryst $\underline{32}$, 267 (1976).

67.   E.F.Rybaczevski, L.S.Smith, A.F.Garito, A.J.Heeger, B.G.Silbernagel (to be published).

68.   P,A.Lee, T.M.Rice and P.W.Anderson, Phys.Rev.Lett.$\underline{31}$, 462 (1973)

69.   L.N.Bulaevskii, Sov.Phys.Usp.$\underline{18}$, 131 (1975).

70.   M.Weger, T.Maniv,A.Ron, K.H.Benneman, Phys.Rev.Lett.$\underline{29}$, 584(1972

71.   E.Ehrenfreund, A.C.Gossard, J.H.Wernick, Phys.Rev.B $\underline{4}$, 2906
        (1971) and F.Y.Fradin (private communication).

72.   H.Niedoba, H.Launois, D.Brinkman,H .V.Keller, J.Physique Lett.
        $\underline{35}$, L-251 (1974).

73.   C.Berthier, D.Jérome, G.Soda, C.Weyl,L.Zuppiroli, J.M.Fabre and
        L.Giral, Mol Cryst Liq Cryst $\underline{32}$, 261 (1976).

74.   J.B.Torrance, B.A.Scott and F.B.Kaufman, Solid State Comm $\underline{17}$,
        1369 (1975).

75.   D.Jérome and L.Giral, Proc. "Conference on Organic Conductors
        and Semiconductors", Siofok/Hungary, 1976
76.   D.Jérome, M.Weger, to be published.

77.   A.Blandin, Lecture Notes, Orsay. V.E.Emery (in this book).

78.   M.Weger and J.Friedel (to be published).

79.   G.A.Thomas et al, Phys.Rev.B $\underline{13}$, 5105, (1976). M.J.Cohen et al
        Phys.Rev.B $\underline{13}$, 5111 (1976).

80.   D.E.Schafer, G.A.Thomas, F.Wudl, Phys. Rev.B $\underline{12}$, 5532   (1975).

81.   P.E.Seiden and D.Cabib, Phys.Rev.B $\underline{13}$, 1846 (1976).

82.   P.A.Lee, T.M.Rice and R.E.Klemm, preprint 1976.

83.   D.L.Mills and P.Lederer, J.Phys.Chem.Solids $\underline{27}$, 1805 (1966.).

84.   M.J.Rice, A.R.Bishop, J.A.Krumhansl and S.E.Trullinger, Phys.
        Rev.Lett $\underline{36}$ 432   (1976).

85.   G.J.Ashwell, D.D.Eley, M.R.Willis, Nature $\underline{259}$, 201 (1976).

86.   N.Menyhard and J.Solyom, J.Low Temp Phys.$\underline{12}$, 529 (1973).

87.   B.Horovitz, Solid State Comm.$\underline{18}$, 445(1976); $\underline{19}$, 1001 (1976).

88.   For SN_x, see the pressure investigation performed by R.H.Friend
        D.Jérome, S.Rehmatullah and A.D.Yoffe, J.Phys.C. to be published
        1977.

CHARGE DENSITY WAVES IN LAYERED COMPOUNDS

F. J. DiSalvo

Bell Laboratories

Murray Hill, New Jersey 07974 USA

This conference concerns the properties of "One Dimensional" metals. However, I'm going to talk about Charge Density Wave (CDW) states in two dimensional metals. The motive for doing that at this conference is that there are a number of similarities in the kinds of phenomena that occur in the one and two dimensional systems and in the theories proposed to explain them. In some ways the two dimensional systems are easier to understand, in that their normal properties are not too dissimilar from normal 3 dimensional metals, the mechanisms for instabilities that occur are at least qualitatively understood, and the CDW properties are reasonably well described in terms of Landau models.

One should be careful in applying the results on these two dimensional systems or the ideas proposed to explain them directly to the one dimensional systems. The reason is that the magnitude of the parameters used to characterize the different systems are not the same. These parameters might be for example: t (overlap integral-related to bandwidth), U_{ii} (onsite Coulomb interaction), U_{ij} (the Coulomb interaction between sites) and λ (electron phonon coupling). It is at least clear that the calculated transfer integral (t_{2d}) for the layer compounds is approximately an order of magnitude larger than for the organic linear conductors (t_{1d}). In the 2D compounds it is also clear that λ is large and it seems likely that $t_{2D} \gg U_{2D}$. In contrast, in the organic 1D conductors it is possible that all or many of these parameters are within small factors of each other (in particular $U_{ii} \approx t_{1D}$). Finally the conductivity anisotropy in the metallic layered compounds is only on the order of 10. Thus, we can look on the layered systems

as unusually anisotropic 3D metals - in that we do not expect
fluctuations of the order parameter to play a dominant role in
determining the physical properties. This is in contrast to the
1D systems where the conductivity anisotropy is several orders of
magnitude larger and fluctuation effects are thought to play a
dominant role in the properties over a large temperature interval.

The one dimensional conductors are comprised of a large
variety of compounds that are chemically and physically quite
different. On one side we have the inorganic linear chain
compounds such as KCP and $NbSe_3$, and on the other the organic
linear stacks such as TTF-TCNQ, with $(SN)_x$ somewhere in the middle.
I'm going to talk about a more restricted class of compounds, the
metallic transition metal dichalcogenides. The work I will
describe was performed in collaboration with many others including
J. D. Axe, S. Mahajan, D. E. Moncton, D. W. Murphy, J. A. Wilson
and J. V. Waszczak. While it is impossible to cover all the
interesting phenomena observed in layered compounds, I will
reference a large number of publications in this field for those
who wish to follow it up further.

Let me outline what we will talk about. We'll start by
studying the structure of those layered compounds which are
expected to be metallic conductors. The physical properties, such
as electrical transport and magnetic susceptibility show clear
anomalies that I will assert are associated with CDW formation
and/or changes in the CDW structure. At this point I will describe
using a simple model what a CDW is and how it comes about, thus
introducing the parameters that characterize the CDW. Then we'll
come back to the layered compounds to get a feel for the magnitude
and temperature dependence of the CDW parameters. Finally we'll
discuss the many effects of impurities (or disorder) on the CDW
and properties of layered compounds.

1.   STRUCTURE AND PROPERTIES

The layered transition metal dichalcogenides (1) have the
chemical formula MX_2, where X = S, Se, or Te and M can be any one
of a large number of metals from the periodic table, but here we
will discuss primarily M = V, Nb, Ta (group Vb) and Ti (from
group IVb). Since the anions are divalent, the electron configuration of the group Vb metals is d^1 and of IVb is d^0. The
structure of the compounds, schematically illustrated in Fig. 1,
is formed from 3 atom thick sandwiches. The top and bottom sheet
of the sandwich is comprised of close packed chalcogenide (X)
atoms, while the middle sheet is comprised of metal atoms. The
bonding within a sandwich is strong (covalent or ionic), but
between sandwiches (between adjacent X sheets) it is weak -

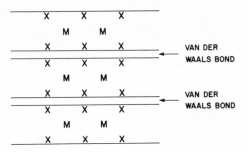

Figure 1. A schematic of the structure of the MX_2 layered
compounds shows the three atom thick sandwiches held together by
relatively weak forces between adjacent sheets of X atoms.

usually labeled van der Waal's bonding. Consequently the physical
properties of these compounds are anisotropic or "quasi-two
dimensional". For example, these materials cleave easily parallel
to the sandwiches (or layers) much like graphite or mica. Many of
the compounds are polymorphic (1), for two reasons: (a) The M
atoms in a given sandwich are either all octahedrally coordinated
(0) by X atoms or all trigonal prismatically coordinated (TP),
(b) the layers can be stacked on top of one another in several
different ways due to the weak interlayer forces. The unit cells,
however, can all be described in the hexagonal system with the
a-axis equal to the intralayer M-M distance and the c axis some
multiple of the layer thickness. We will concern ourselves
primarily with the two simplest polytypes: 1T - in which all the
M atoms are 0 coordinated, and 2H - in which all M atoms are TP
coordinated.

Since the group Vb compounds are d^1, we expect them to be
metallic because of the moderately close M-M distance (a $\sim$ 3.3A)
and the largely covalent nature of the bonds. More sophisticated
theory, such as the APW band calculations of L. F. Mattheiss (2),
leads to the same conclusion. The uppermost unfilled bands are
primarily based on M d states, the density of states at the Fermi
level for the group Vb compounds being 5 to 20 times that of Cu,
for 1T and 2H polytypes respectively. Consequently we expect
these materials to be metallic conductors with conductivities that
are approximately one order of magnitude smaller than that of Cu
metal.

The resistivities (current parallel to the layers) of the 1T
and 2H polymorphs of TaS_2 and $TaSe_2$ are shown in Fig. 2. The
original investigators of the transport properties are given in
the references: $1T-TaS_2$ (3), $1T-TaSe_2$ (4), $2H-TaS_2$ (5), and
$2H-TaSe_2$ (6). While the resistivity (ρ) of the 2H-polytypes has

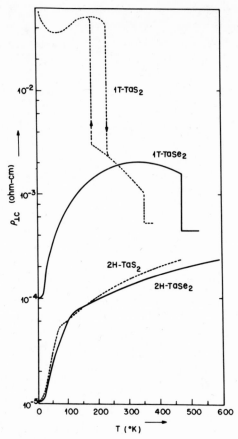

Figure 2. The electrical resistivity parallel to the layers of
several layered compounds vs. temperature.

a metallic like slope, there is a sharp decrease in ρ at low
temperatures. The ρ of the 1T polytypes, however, does not look
like that of a simple metal, and there are sharp discontinuities
at first order transitions.

Anomalies are observed in other transport properties, thermal
properties, and magnetic susceptibility. The magnetic susceptibility of the same compounds is shown in Fig. 3. The 2H-polymorphs
are Pauli paramagnetic, the susceptibility shows a sharp decrease
at the same temperature as the decrease in ρ. The 1T polytypes are
diamagnetic, but the APW calculated density of states should
produce a large enough Pauli paramagnetism to overcome the
diamagnetic contribution of the atom cores.

At this point let me just assert that these anomalies in the
physical properties are due to CDW formation and proceed to discuss

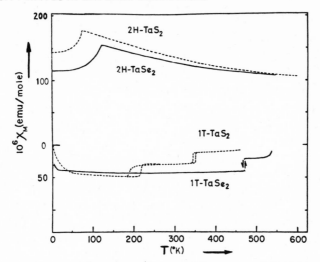

Figure 3. The powder (average) magnetic susceptibility of several layered compounds vs. temperature.

what a CDW is, and how it occurs, before attempting to understand these measurements. Several review articles (7,8) are available on this subject and I will only try to pass on the main ideas to you.

Charge Density Wave instabilities were theoretically proposed by A. W. Overhauser in 1968 (9) where he placed an emphasis on the correlation energy as being the source of the instability. The source of the instability in these compounds appears to be somewhat different as outlined below.

A CDW is a static, coupled, periodic distortion of both the conduction electron density and the lattice. One is not the consequence of the other, but they are intimately tied together. We can see why, using a simple one dimensional model of a metal (Fig. 4). We consider a row of uniformly spaced positive ions and a uniform conduction electron density to preserve overall charge neutrality. If a sinusoidal perturbation of the conduction electrons occurs, the net charge, including ions, oscillates from negative to positive at maxima and minima in the wave. The Coulomb energy of such a state is large and it will not likely be the stable ground state of the system. In fact such an excitation is a plasmon, one quantum of which in normal metals costs on the order of 10 eV. However, if the positive ions move toward the maxima and away from the minima, the Coulomb energy can be greatly reduced and such a coupled distortion may become the stable ground state of the system. Consequently, a CDW is more likely to occur in systems with large electron-phonon (electron-lattice) coupling. Below the onset temperature of the CDW, T_O, the charge

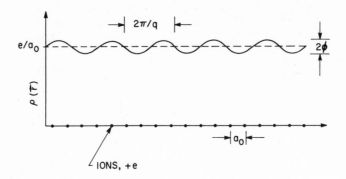

Figure 4. One dimensional model of a metal - used to show why electron phonon coupling is important and the CDW is a <u>coupled</u> distortion of the lattice and the conduction electron density.

density and atomic displacements would be given by (for this simple model)

$$\rho(r) = \rho_o(r)[1+\phi \cos \vec{q}\cdot\vec{r}] \qquad \text{eq. (1)}$$

$$\Delta\vec{x} = \vec{A} \sin \vec{q}\cdot\vec{r} \qquad \text{eq. (2)}$$

where ϕ and $\vec{A}$ are the amplitude in charge density and atomic displacement respectively.

We further expect that lattice waves with this wave vector $\vec{q}$ and displacements parallel to $\vec{q}$ (longitudinal phonon) will be lowered in energy <u>above</u> T_o, because of the effective screening of the ionic charge by the electrons. Thus a dip, called a Kohn anomaly, will appear in the phonon dispersion curve (phonon energy vs. wave vector). Simple one dimensional models for this effect are published (10); some examples for the layered compounds will be given below.

Such a CDW instability can occur if the Fermi surface (F.S.) has the proper shape. Further, the F.S. determines the $\vec{q}$ of the distortions. We can see what kind of F.S. is needed by considering a simple linear response model of the conduction electrons to a perturbation with wave vector $\vec{q}$. If the response to an infinitesimally small perturbation, becomes macroscopic, the system will spontaneously move to a distorted state. The response of a non-interacting electron gas to an electric potential is porportional to the second order perturbation energy:

$$\chi^{O}(\vec{q}) = \sum_{\vec{k}} \frac{f_{\vec{k}}(1-f_{\vec{k}+\vec{q}})}{\varepsilon_{\vec{k}+\vec{q}}-\varepsilon_{\vec{k}}} \qquad \text{eq. (3)}$$

where ε_k = energy of state with wave vector k
  f_k = Fermi occupation factor

We can approximately include the effect of the electron phonon interaction in the following way. The response of the real system to an externally applied field is given by the generalized susceptibility χ^g, so that $\Delta\rho = \chi^g E_{ext}$. From Eq. (3) we also have $\Delta\rho = \chi^O E_{total}$, with $E_{total} = E_{ext} + E_{ions}$. But the electric field of the ions is proportional to the ion displacement $\vec{A}$ which through the electron phonon coupling is proportional to $\Delta\rho$, therefore $E_{ions} = g\Delta\rho$. Thus $\chi^O(E_{ext}+g\Delta\rho) = \Delta\rho$ and $\Delta\rho(1-g\chi_o) = \chi^O E_{ext}$

$$\frac{\Delta\rho}{E_{ext}} = \chi^g = \frac{\chi_o}{1-g\chi_o} \qquad \text{eq. (4)}$$

where g is proportional to the electron phonon coupling. Thus we see that when $g\chi^O$ equals 1, the susceptibility of the interacting system diverges. (The system is unstable toward a macroscopic deformation $\Delta\rho$ with wave vector q). In normal metals χ^O is small and the product $g\chi^O$ is less than one. With the proper F.S. we can make χ^O large. From the form of Eq. (3) it is clear that if we sum over many states k where the denominator vanishes or is very small, while the numerator is nonzero, χ^O will be large. This will occur if many occupied states at the F.S. are connected by the vector $\vec{q}$ to unoccupied states at the F.S. One possibility is an F.S. with two plain parallel sections, or equivalently nesting sections (11) (pieces with the same curvature that can be translated by $\vec{q}$ to be coincident with each other). For the parallel or nesting section, the peak in $\chi^O(\vec{q})$ when $\vec{q}$ just reaches or "spans" from one side of the F.S. to the other is proportional to $\ln(E_F/kT)$. Thus as T is decreased, $\chi^O(q)$ increases and at some low enough temperature, T_o, the instability occurs. Other possibilities exist, such as having saddle points in the band structure near the F.S. (12). I should point out that for some interactions g itself is q dependent, peaking when $\vec{q}$ "spans" the F.S. as for $\chi^O(\vec{q})$. In fact Overhauser originally suggested that the CDW instability arose from a peak not in $\chi^O(q)$ but in the interaction itself. A more rigorous calculation for the instability condition within the Hartree-Fock approximation is given by Chan and Heine (13).

What happens below T_o? The instability produces a lattice distortion of wave vector $\vec{q}$, producing a new lattice potential. This potential connects the states at k and k+q in first order perturbation, splitting the states away from the Fermi level. That is, a gap is produced at the F.S., just over those regions spanned by $\vec{q}$. Consequently these states contribute to χ^0 with the gap energy Δ as a denominator, not zero as above T_0. Consequently, χ^0 (and χ^g) decrease in magnitude below T_0. Within simple models, the gap is expected to increase with decreasing temperature in exactly the same way as the BCS superconducting gap (10).

Since $\vec{q}$ is determined by the F.S., the wavelength, $2\pi/|q|$, of the CDW will usually be incommensurate with the lattice; that is, the wavelength will not equal a lattice translation. However, in many of the layered compounds a first order transition to the commensurate state occurs at $T_d < T_0$. We comment later on the driving force for this transition.

The CDW state below T_0 can best be observed by diffraction techniques (7,8). Elastic scattering will occur at points in reciprocal space given by $\vec{k} = \vec{G} \pm n\vec{q}$ where $\vec{G}$ is any one of the reciprocal lattice vectors of the undistorted lattice that exists above T_0 and n is any integer. Thus each main Bragg peak will have a series of satellite peaks about it with an intensity given from simple kinematic scattering theory by $I_n \sim (\vec{k}\cdot\vec{A})^{2n}/n!$ (Eq. (5)) for $\vec{k}\cdot\vec{A} \ll 1$. In fact in the layered compounds $\vec{A}$ is usually quite small and the satellite peaks have only 10^{-3} or less intensity than the strongest main Bragg peak. This made the CDW's very difficult to find by simple, powder X-ray diffraction techniques. They were first discovered by electron diffraction, since the intensity of the satellite peaks can be greatly enhanced over the kinematic formula by multiple scattering. While the position of the satellite peaks is easy to obtain by electron diffraction (thus obtaining $\vec{q}$), the intensity cannot be used to obtain lattice displacements. These latter are obtained from X-ray or neutron diffraction measurements.

This completes our discussion of the origin of CDW's. We have introduced the parameters T_0, T_d, $\vec{q}$, ϕ, $\vec{A}$, and Δ; the last four of which are expected to be temperature dependent. We now return to the layered compounds and consider some of these parameters in more detail.

Table I lists a number of layered compounds with the CDW onset temperature, T_0, the lockin temperature, T_d, and CDW wavelength in multiples of the a-axis. In all cases the CDW consists of a super-position of three CDW's with q vectors that are related by a $120°$ rotation about the c-axis (the normal to the layers).

A large variation of T_0 is seen <u>both</u> with a change in polytype and between different compounds. Of the metallic compounds, $2H\text{-}NbS_2$ is the only one in which a CDW has not been detected. Note also that the CDW wavelength is moderately short - between 3 and 4 lattice parameters.

Table I

Material	$T_o(k)$	$T_d(k)$	λ_{CDW}	Ref.
1T-TaSe$_2$	$\approx$600	473	$\sim$3.5a	7, 14
2H-TaSe$_2$	122	$\sim$95	$\sim$3.0a	7, 15
1T-TaS$_2$	$\approx$600	$\approx$200	$\sim$3.5a	7, 16
2H-TaS$_2$	$\sim$80	?	$\sim$3.0a	7, 17
2H-NbSe$_2$	32k	no	$\sim$3.0a	15
1T-VSe$_2$	112k	$\sim$80k	$\sim$4.0a	8, 18

The CDW in 2H-NbSe$_2$ remains incommensurate down to 4.2K. $\vec{q} = (1-\delta)a^*/3$, where a^* is a reciprocal lattice vector in the plane and δ decreases smoothly from 0.02 at 32K to approximately 0.01 at 4.2K.

The intensity of a first order satellite peak (n=1) in 2H-NbSe$_2$ as measured by neutron diffraction is shown in Fig. 5 vs. temperature (15). The transition at T_o is seen to be second order, since the intensity smoothly drops to zero. The atomic displacements at 4.2K calculated from this data are approximately 0.05 Å for Nb (parallel to the layers) and half that for the Se atoms (some component perpendicular to the layers.) The amplitude in charge density ϕ at 4.2K is estimated from early NMR measurements to be $\approx$5% of the conduction electron density (19).

Next we consider the CDW transition in 2H-TaSe$_2$. The results of neutron scattering measurements are shown in Fig. 6. The inset in the upper left shows the scattering peaks in the (HKO) plane. The open circles are the main Bragg peaks. The dark circles are the satellite peaks. If we look in the main part of the figure about the position labeled 4 (position 4 is at $4a^*/3$) we see just below $T_o = 122K$ that $\vec{q} = (1-\delta)a^*/3$ and $\delta \sim 0.02$. As T decreases, δ decreases until it discontinuously goes to zero at $T_d \sim 95K$. Note also that a weak secondary peak appears on the other side of $4a^*/3$ by 2δ. This peak occurs at the same position as the second order satellite coming from the Bragg peak at 6 (i.e. $2\vec{a}^*-2\vec{q} = \frac{a^*}{3}(4+2\delta)$). However, the intensity calculated from Eq. (5) is much too small to be the source of this peak. Consequently, this is due to a second periodic distortion at $q_{2\delta} = a^*/3(1+2\delta)$. A Landau free energy model has been developed by D. E. Moncton et al. (15) that shows how this secondary distortion develops and its role in "pulling" $\vec{q}$ toward the commensurate value of $a^*_o/3$. The displacements grow to $\vec{A} \sim 0.1$ A for Ta and $\sim$0.05 A for Se at 4.2K.

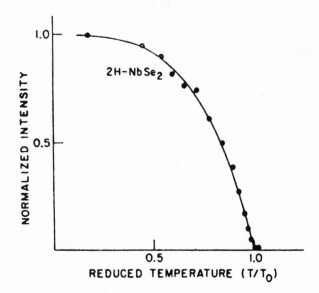

Figure 5. The satellite intensity vs. temperature for 2H-NbSe$_2$
below T$_O$ = 32K. The intensity is proportional to the lattice displacement squared. The lattice displacement may be used as an
order parameter in a Landau theory.

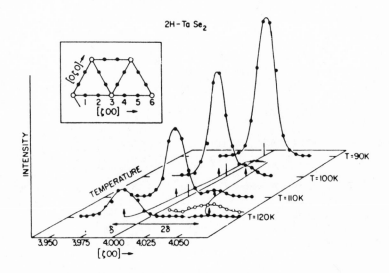

Figure 6. Elastic neutron scattering measurements of the
satellite peak intensity and position in 2H-TaSe$_2$ vs. temperature.

If we scale by the ratio of atomic displacements from the $2H-NbSe_2$ data, we estimate ϕ (4.2K) $\sim$ 0.1 (e/a). Little data exists on the behavior of Δ as a ftn. of T. Infrared reflectivity measurements in $2H-TaSe_2$ show weak gap like features below T_0 (20). However, these lead to the prediction that Δ (4.2K) $\approx$ 25 kT_0. This number seems too large, since simple theories predict $\Delta \sim$ 3 to 5 kT_0. Further measurements at lower photon energies seem appropriate before this reflectivity data can be adequately interpreted.

The CDW behavior of the 1T polytypes is quite different. Electron diffraction patterns of the (HKO) plane show a commensurate superlattice at room temperature (Fig. 7a) and an incommensurate CDW (Fig. 7b) above the first order transition apparent in the resistivity (Fig. 2) at T_d = 473K. Many of the main Bragg peaks appear as the brightest spots. In the incommensurate state $\vec{q}$ = $0.285\vec{a}^*$ (i.e. $\vec{q}$ is parallel to the line joining main Bragg peaks).

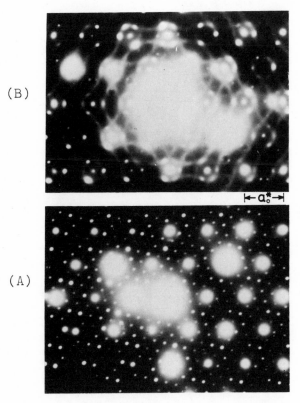

(B)

(A)

$\longmapsto a_o^* \longmapsto$

Figure 7. (a) The basal plane diffraction pattern of the CCDW in $1T-TaSe_2$ shows the $\sqrt{13}$ a superlattice. The bright spots are the main Bragg peaks and the weaker ones are satellite peaks. (b) The basal plane diffraction pattern of the ICDW in $1T-TaSe_2$ above T_d = 473K.

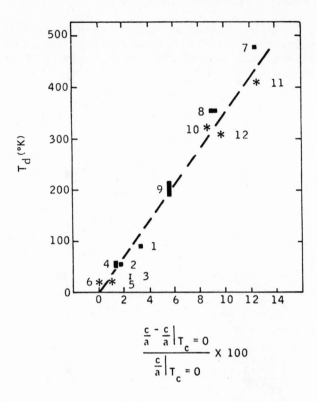

$$\frac{\frac{c}{a} - \frac{c}{a}\big|_{T_c = 0}}{\frac{c}{a}\big|_{T_c = 0}} \times 100$$

Figure 8. T_d is plotted vs. c/a (normalized to a specific value) from Thompson (24). Each point represents a different layered compound or a different polymorph.

At T_d $\vec{q}$ rotates by 13 54' and shrinks slightly (∿2%) to produce the 3×1 superlattice apparent in Fig. 7a. This transition involves primarily a rotation of $\vec{q}$. At room temperature and below the atomic displacements are quite large, ∿0.25 Å for Ta. Further the amplitude ϕ is about 1 e/a! This might have been expected if we scale ϕ from 2H-NbSe$_2$ by the ratio of the onset temperatures T_o. With such a large charge oscillation, the binding of the Ta core electrons shifts enough to be observable in ESCA (21,22). Splittings in the 4f binding energies of ∿0.5 eV are clearly observed at room temperature. This means that at room temperature and below, the CDW is not a weak perturbation of the F.S. Rather the commensurate phase may be thought of as a valence dispro-portionation of Ta^{4+} into Ta^{5+}, Ta^{4+}, Ta^{3+}. Summarizing the behavior of 1T-TaSe$_2$, we see that as the temperature is reduced from above T_o, the material passes through a number of states: a normal undistorted metal, then a second order transition to an incommensurate CDW state where, at least close to T_o, the

amplitude ϕ is small, and finally this state evolves through the
transition at T_d to a "valence disproportionation". As yet there
are no adequate theories to explain the overall behavior of 1T-
TaSe$_2$, although weak coupling Landau models are able to qualitatively predict the sequence of transitions (23).

1T-TaS$_2$ is even more complicated than 1T-TaSe$_2$, as is
apparent from the two first order transitions seen in the resistivity (Fig. 2). Below 200K the CDW shows the same commensurate
state as 1T-TaSe$_2$ (16). Above the transition at $T_d' = 350K$, the
CDW is incommensurate with $\vec{q} = 0.288\vec{a}^*$ (7,16). At 350K, $\vec{q}$ rotates
by $\sim 12°$ but stops short of becoming commensurate! $\vec{q}$ continues to
rotate until at $\sim 200K$ it jumps the last fraction of a degree to
become commensurate. A full explanation of this behavior has not
been offered. Again it is clear from ESCA and X-ray diffraction
that the CDW amplitude is as large in 1T-TaS$_2$ as in 1T-TaSe$_2$.

We have talked about transitions to the CCDW, but
have not tried to indicate why they occur. A hint to their origin
can be obtained from a correlation between T_d and the crystallographic c/a ratio discovered by A. H. Thompson (24). Figure 8
shows the almost linear relation between T_d and c/a for 12 different compounds (or polytypes) that I will not bother to identify in
detail. Previously, F. R. Gamble had shown that the c/a ratio in
these compounds was related to the ionicity difference between the
cation and anion (25). Consequently, we see that T_d is related to
an ionicity difference. This result suggests that the driving
force toward the commensurate state involves local ionic Coulomb
or covalent bonding forces. One might see how these forces arise
by considering a simple case. Suppose an incommensurate CDW exists
in a two dimensional hexagonal packed sheet of metal atoms. Since
the CDW is incommensurate, at some points the charge will be
increased at some nonsymmetrical position. Ionic or covalent
energies will be minimized (or maximized) by placing this charge
at a center of symmetry; such as, (a) on a metal atom, (b) half
way between two atoms to maximize the bonding charge, (c) at a
geometrical center, like the center of a triangular set of three
atoms, to make a bonded metal cluster, (d) etc. In the real
compounds we must also consider the metal-nonmetal bonds, but this
simple model gives the qualitative idea.

So far we have talked about the CDW as a single layer
phenomena. The CDW's interact from one layer to the next producing a given stacking sequence. This sequence is consistent
with that obtained by minimizing interlayer Coulomb interactions
(7,26). For example, if we consider the simplest CDW $\Delta\rho = \frac{\phi}{3}$
$\sum_{i=1}^{3} \cos \vec{q}_i \vec{r}$, the contours of $\Delta\rho$ have hexagonal symmetry as shown in

Fig. 9a. The maxima (or minima, depending upon the sign of ϕ) are

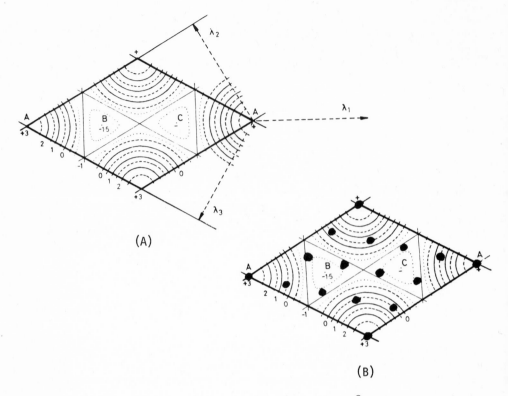

Figure 9. (a) The CDW pattern created by $\rho = \sum\limits_{i=1}^{3} \cos \vec{q}_i \cdot \vec{r}$ with the
3 CDW wavelengths $\lambda_i = 2\pi/q_i$ shown. (b) The CDW pattern and metal
atom positions expected for the CCDW in $1T\text{-}TaS_2/Se_2$.

at the cell edges, with minima at the center of each triangle.
Using hexagonal notation, we label the maxima A and the two minima
B and C as shown. In the incommensurate phase, the origin of the CDW
can be placed arbitrarily at any point in the layer. If we choose
point A in the first layer, then the Coulomb interaction is minimized with the next layer by placing its charge maxima over point
B in the first layer. The third layer minimizes its Coulomb energy
with both the first and second layer by putting its maxima over
point C. This stacking sequence leads to a three layer repeat for
the CDW as is found in the 1T polytypes. The interaction energy
in the 2H polytypes is modified by the screw symmetry between
adjacent layers and the CDW repeat appears to be two layers. In
the commensurate phase, the origin of the CDW cannot be arbitrarily chosen. Rather it appears that the CDW cell origin lies
at a Ta site. Figure 9b shows that the CDW in the second layer

can minimize its interaction energy with the first by translating
the CDW origin by $2\vec{a}$. By continuing this sequence the overall
interaction energy is minimized. The origin of the CDW repeats
itself energy 13 layers. However, the true unit cell is triclinic
with a = $\sqrt{13}a$, b = $\sqrt{13}a$ and c = $|\vec{c}+2\vec{a}|$ (26).

Having discussed in some detail the magnitude of the CDW
parameters, we briefly consider the resistivity. It is apparent
from Fig. 2 that below the onset of the CDW the 1T polytypes
become more resistive and the 2H polytypes less resistive. The
simple picture introduced in discussing the driving force of the
CDW leads one to expect the behavior observed for 1T-TaSe$_2$. That
is, below T_0 the resistivity should increase (compared to the
normal metal) as T decreases due to the formation of gaps at the
F.S. Clearly below 600K ρ (1T) > ρ (2H). At T_d the gaps increase
discontinuously further increasing the resistivity. Finally at
low temperatures we expect metallic like conductivity from those
portions of the F.S. not destroyed by CDW gaps.

The resistivity of the 2H polytypes decreases below T_0. A
natural explanation of this fact arises from the model for the
instability introduced by Rice and Scott (12). In this model the
instability arises from saddle points in the band structure near
the F.S. Carriers in the region of the saddle points have large
effective masses. Consequently, these carriers do not contribute
much to the conductivity by their own motion, rather this high
density of states region behaves as a scattering sink for lighter
electrons on the F.S., reducing their scattering time. Consequently, when the F.S. near the saddle points is removed by gaps,
the conductivity of the light electrons increases. While this
model seems quite plausible, definite proof of its validity is not
yet available.

Finally we come to 1T-TaS$_2$. At each transition toward the
commensurate state we expect the gaps to increase in size and the
resistivity to increase. However, in the commensurate state the
resistivity is very high and not at all metallic like. This state
is not understood. In fact the conductivity is close to the estimates for minimum metallic conductivity - beyond which some kind of
localization of the remaining carriers is expected to occur (27).

In the remainder of this article, we consider the effects of
impurities. In particular, we consider the random substitution of
the cation by other transition metals and the effect of this substitution on the transport properties and on CDW formation. This
substitution causes randomness in the lattice potential and may
also change the average conduction electron density, z. Each of
these effects is related in different ways to changes in the CDW
behavior.

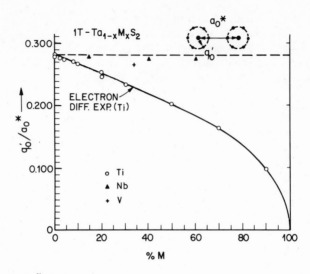

Figure 10. q/a^*, the wavelength of the CDW in the ICDW state
divided by the reciprocal lattice vector, is shown vs. x for
$1T\text{-}Ta_{1-x}M_xS_2$ where M = Ti, Nb, or V.

First consider the effect of changing z. The calculated F.S.
for the undistorted 1T polytypes has the shape of an ellipsoid in
the plane of the layer with near perpendicular walls along C^*.
This F.S. then has sections that are near to parallel, leading to
the CDW instability. If Ti is substituted for Ta z decreases, and
in the rigid band approximation the F.S. will shrink but remain an
ellipsoidal cylinder. Consequently, in the ICDW phase we expect
that q/a^* will decrease with increasing x in $1T\text{-}Ta_{1-x}Ti_xS_2$. The
results of such measurements are shown in Fig. 10 (7,28). The
solid line is a fit to the data which is close to that expected
from the rigid band approximation. Also in Fig. 10 q/a^* is shown
for $1T\text{-}Ta_{1-x}Nb_xS_2$ and $1T\text{-}Ta_{1-x}V_xS_2$. In these two cases q/a^* is
close to constant, as expected, since Nb and V are isoelectronic
with Ta and z is constant. In the 1T polytypes the effect of
changing z is to smoothly change the F.S. and consequently q/a^*.
The data shown in Fig. 9 are obtained at (or above) room temperature, consequently T_0 remains greater than 300K even for $x \gtrsim 0.7$.
It appears from magnetic susceptibility that T_0 is reduced, but
slowly, with increasing x. This is an expected effect of disorder
(23) and occurs with T_i, Nb or V substitution (or indeed any
cation disorder).

The disorder slowly reduces T_0, but rapidly suppresses the
commensurate state. This can be seen in the resistivity of
$1T\text{-}Ta_{1-x}Ti_xSe_2$ (Fig. 11). With increasing x, both the transition

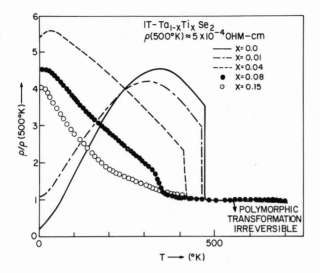

Figure 11. The electrical resistivity of $1T$-$Ta_{1-x}Ti_xSe_2$ shows the decrease in T_d with increasing x.

temperature from the incommensurate to the commensurate state, T_d, and the magnitude of the resistive anomaly at T_d decrease. This data, combined with measurements of the enthalpy of transition (28), show that for $x > x_c \sim 0.10$ the CCDW does not occur. We can see how this occurs with a simple chemical argument. Consider a one dimensional string of atoms A and B that are randomly placed on lattice sites. We wish to compare the free energy of the CCDW and ICDW. Let us consider B to be the dilute species and assume that B is more electronegative than A. (We could assume it is more electropositive, but will obtain the same result). Because the B atoms are more electronegative, the free energy will be a minimum when the CDW charge maxima lie at B sites. If there is a CCDW and the alloy is random, many of the B atoms will not lie at maxima and we must pay some free energy proportional to the ionicity difference $(X_B-X_A)^2$. If there is an ICDW, the CDW can change its phase (or equivalently its local wavevector q) so that each B atom lies at charge maxima. In this case, we must pay some elastic energy to distort the CDW - but this turns out to be small. (The elastic energy to change the wave vector q from its commensurate to incommensurate value, must be relatively small for the CCDW to even exist in the pure material.) Consequently, we expect the increase in free energy with cation substitution to be larger in the CCDW than the ICDW and thus the ICDW becomes more stable (i.e. T_d is suppressed and for $x \gtrsim x_c$ the CCDW does not exist.) A more rigorous and elaborate free energy model reaches the same conclusions (23).

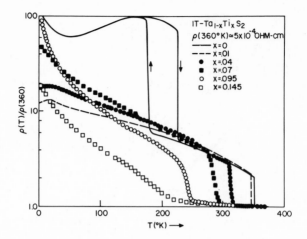

Figure 12. The electrical resistivity of $1T\text{-}Ta_{1-x}Ti_xS_2$ shows that both first order transitions toward the CCDW state are suppressed with increasing x, but the commensurate state is suppressed for $x \gtrsim 0.002$.

Similar effects are seen in $1T\text{-}Ta_{1-x}Ti_xS_2$ (Fig. 12). The transition at 200K to the commensurate state is suppressed for $x \gtrsim 0.002$, while the transition at T_d' (toward the commensurate state or into the "quasi-commensurate" state) is slowly suppressed and is finally lost when $x \gtrsim 0.14$.

Figure 12 shows another interesting phenomena. For samples with $x \approx 0.08$, the low temperature resistivity increases rapidly at low temperatures, appearing to diverge as $T \to 0$. While we believe this behavior is also due to disorder, a clearer example of such behavior is seen in $1T\text{-}Ta_{1-x}Fe_xS_2$ (Fig. 13).

$1T\text{-}Ta_{1-x}Fe_xS_2$ is an alloy of two compounds that both form dichalcogenides, but FeS_2 has the cubic pyrite structure (1). The Fe in FeS_2 is divalent, since singly bonded S_2 molecules exist in this compound. Mossbauer measurements (29) confirm that the Fe is divalent in $1T\text{-}Ta_{1-x}Fe_xS_2$ and electron microscopy and X-ray diffraction indicate that the Fe randomly replaces the Ta. Electron diffraction shows that an ICDW exists at 300K for $x \lesssim 0.15$. For $x \gtrsim 0.15$ the satellite peaks due to the CDW become quite diffuse, the coherence length of the distortions becomes quite small and it is difficult to identify a particular temperature interval to associate with CDW onset. For overall charge compensation some Ta must be pentavalent, producing $Fe_x^{2+}Ta_{1-3x}^{4+}Ta_{2x}^{5+}S_2$. Thus we expect at $x = 1/3$ the concentration of Ta^{4+} and the conduction electron density will be zero. In fact $x = 1/3$ is the limit of Fe substitution possible, as we would guess from the above model.

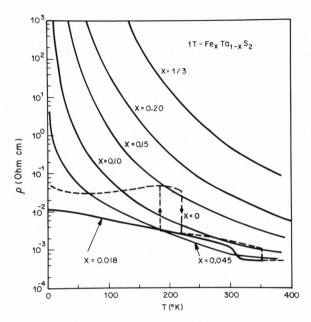

Figure 13. The electrical resistivity of $1T-Ta_{1-x}Fe_xS_2$ shows that not only are the transitions toward the CCDW suppressed with increasing x but also the low temperature resistivity diverges at a large enough doping level.

Consequently, with increasing x there are two effects; (1) the conduction electron concentration, and the width of the occupied conduction band, decreases, (2) the disorder increases. The data of Fig. 13 show that for $x \gtrsim 0.02$, the low temperature resistivity diverges as $T \to 0$. This behavior appears to be due to Anderson localization of the carriers by the disorder (30). In the Anderson model, when the r.m.s. amplitude of the random part of the potential becomes on the order of the occupied bandwidth, the conductivity at T = 0 is zero. The models predict that the conductivity at T = 0 decreases with increasing disorder and reaches a minimum value, σ_{min}, at a critical doping level beyond which $\sigma = 0$ (27). This is precisely the behavior observed here.

Similar behavior is expected in one dimensional systems. The low temperature resistance of KCP appears to be dominated by the random position of the Br atoms (31). It may even be possible that substitution compounds prepared with _isoelectronic_ species will produce a large enough random potential to cause localization. It is at least certain that such "isoelectronic disorder" can produce large changes in T_o and T_d. These points should be considered when attempting to interpret the results of experiments on 1 dim. conductors, such as $TTF_{1-x}TSeF_xTCNQ$.

While we have seen that both T_o and T_d are reduced by disorder (the CCDW eventually being completely eliminated), we also
suggest that the CDW plays some role in producing localization.
The impurity potential seen by the conducting electrons is ampli-
fied by the CDW. In a normal metal the impurity potential is
screened out in a short Thomas-Fermi length of several Angstroms.
In a CDW unstable material, <u>even above T_o</u>, the impurity potential
has a <u>large</u> temperature dependent range and amplitude (23). Thus
the CDW enhances the potential fluctuations over that expected in
normal metals and consequently localization is more likely.

So far, we have seen three major effects that cation disorder
can produce in the CDW unstable layered compounds. First, we can
<u>smoothly</u> change the F.S. size and the wave vector of the ICDW
distortions. Second, the onset temperature is reduced and the
commensurate state can be eliminated beyond a critical doping
level. Finally, some impurities at a high enough concentration
produce Anderson localization of the conduction electrons at low
temperatures.

We have already covered a large variety of physical effects
that are observed in layered compounds, but there is much more
even within the realm of CDW phenomena. I will briefly mention
some of these areas, so that those interested can follow them up.

The Fe^{2+} in $1T-Ta_{1-x}Fe_xS_2$ shows also a low spin to high spin
transition with increasing temperature (29). This appears to be
due to an (accidental) coincidence of the crystal field splitting
10 Dq and the Hunds rule energy difference between the low spin
and high spin state. 10 Dq also appears to have an unusual temperature dependence due to the lattice distortions driven by the CDW.

$4Hb-TaSe_2$, in which the symmetry of the $TaSe_2$ layers
alternates (in the layer stacking sequence) between octahedral and
trigonal prismatic coordination, shows separate CDW transitions in
the different symmetry layers (four transitions in all) (32).
This is similar to the large number of transitions observed in
TTF-TCNQ which apparently occur in different stacks (33).

The 2H polymorphs are superconductors. Further "guest"
molecules or atoms can be inserted between the layers, to form
intercalation compounds (34-36). The superconducting transition
temperature was known to generally increase upon intercalation,
especially with organic molecules. Later it was shown that intercalation usually reduced T_o and/or the CDW amplitude. It is this
interplay between the CDW and superconductivity that explains the
increasing superconducting transition temperature upon intercalation (37).

ENDING REMARKS

I have discussed the CDW behavior of several layered compounds, tried to give you a feeling for the magnitude of the parameters that characterize the CDW, and indicated some of the changes that occur in the CDW with cation substitution and disorder.

This has not been a review in the larger sense - I have not tried to cover the literature in a balanced way - rather I have illustrated what I wanted to say by drawing largely from my own work and a few key works of others. The intent was to familiarize you with the behavior of metallic two dimensional systems; in particular, those aspects which show similarities to the phenomena observed in one dimensional systems.

REFERENCES

1.   J. A. Wilson and A. D. Yoffe, Adv. in Phys. $\underline{18}$, 193 (1969).
2.   L. F. Mattheiss, Phys. Rev. $\underline{B8}$, 3719 (1973).
3.   A. H. Thompson, F. R. Gamble and J. F. Revelli, S.S. Comm. $\underline{9}$, 981 (1971).
4.   F. J. DiSalvo, R. G. Maines, J. V. Waszczak and R. E. Schwall, S.S. Comm. $\underline{14}$, 497 (1974).
5.   A. H. Thompson, F. R. Gamble and R. F. Koehler, Phys. Rev. $\underline{B5}$, 2811 (1972).
6.   H. N. S. Lee, M. Garcia, H. McKinzie and A. Wold, J.S.S. Chem. $\underline{1}$, 190 (1970).
7.   J. A. Wilson, F. J. DiSalvo and S. Mahajan, Adv. in Phys. $\underline{24}$, 117 (1975).
8.   P. M. Williams, "Crystallography and Crystal Chemistry of Materials with Layered Structures", Vol. 2, F. Levy (ed), Reidel Pub.
9.   A. W. Overhauser, Phys. Rev. $\underline{167}$, 691 (1968).
10. M. J. Rice and S. Strassler, S.S. Comm. $\underline{13}$, 125 (1973).
11. L. M. Roth, H. J. Zeigler and T. A. Kaplan, Phys. Rev. $\underline{149}$, 519 (1966).
12. T. M. Rice and G. K. Scott, Phys. Rev. Letts. $\underline{35}$, 120 (1975).
13. S. K. Chan and V. Heine, J. Phys. $\underline{F3}$, 795 (1973).
14. J. A. Wilson, F. J. DiSalvo, S. Mahajan, Phys. Rev. Letts. $\underline{32}$, 882 (1974).
15. D. E. Moncton, J. D. Axe and F. J. DiSalvo, Phys. Rev. Letts. $\underline{34}$, 734 (1975).
16. C. B. Scruby, P. M. Williams, and G. S. Parry, Phil. Mag. $\underline{31}$, 225 (1975).
17. J. P. Tidman, O. Singh, A. E. Curzon and R. F. Frindt, Phil. Mag. $\underline{30}$, 1191 (1974).

18. F. J. DiSalvo and J. V. Waszczak, Proc. of Conf. on Metal-Non Metal Transitions Autrans 1976, to be published in Journal de Physique.
19. E. Eherenfreund, A. C. Gossard and F. R. Gamble, Phys. Rev. B5, 1708 (1972).
20. A. S. Barker, Jr., J. A. Ditzenberger and F. J. DiSalvo, Phys. Rev. B12, 2049 (1975).
21. G. K. Wertheim, F. J. DiSalvo, and S. Chiang, Phys. Rev. B13, 5476 (1976).
22. H. B. Hughes and R. A. Pollak, Comm. Phys. 1, 61 (1976).
23. W. L. McMillan, Phys. Rev. B12, 1187 (1975) and Phys. Rev. B14, 1496 (1976).
24. A. H. Thompson, Phys. Rev. Letts. 34, 520 (1975).
25. F. R. Gamble, J.S.S. Chem. 9, 358 (1974).
26. D. E. Moncton, F. J. DiSalvo, J. D. Axe, L. J. Sham, and B. R. Patton, to be pub., Phys. Rev. B, 1976.
27. N. F. Mott, M. Pepper, S. Pollett, R. H. Wallis and C. J. Adkins, Proc. Roy. Soc. London Ser. A345, 169 (1975).
28. F. J. DiSalvo, J. A. Wilson, B. G. Bagley and J. V. Waszczak, Phys. Rev. B12, 2220 (1975).
29. M. Eibschutz and F. J. DiSalvo, Phys. Rev. Letts. 36, 104 (1976).
30. F. J. DiSalvo, J. A. Wilson and J. V. Waszczak, Phys. Rev. Letts. 36, 885 (1976).
31. A. N. Bloch, R. B. Weismann and C. M. Varma, Phys. Rev. Letts. 28, 753 (1972).
32. F. J. DiSalvo, D. E. Moncton, J. A. Wilson and S. Mahajan, Phys. Rev. B14, 1543 (1976).
33. Many papers in this conference.
34. F. R. Gamble, J. H. Osiecki and F. J. DiSalvo, J. Chem. Phys. 55, 3525 (1971).
35. F. R. Gamble, J. H. Osiecki, M. Cais, R. Pisharody, F. J. DiSalvo and T. H. Geballe, Science 174, 493 (1971).
36. F. J. DiSalvo, G. W. Hull, Jr., L. H. Schwartz, J. M. Voorhoeve and J. V. Waszczak, J. Chem. Phys. 59, 1922 (1973).
37. D. W. Murphy, F. J. DiSalvo, G. W. Hull, Jr., J. V. Waszczak, S. F. Meyer, G. R. Stewart, S. Early, J. V. Acrivos, and T. H. Geballe, J. Chem. Phys. 62, 967 (1975).

COMPARISON OF COLUMNAR ORGANIC AND INORGANIC SOLIDS

Z. G. Soos* and H. J. Keller

Anorganisch-Chemisches Institut der Universität
Heidelberg, D-6900 Heidelberg 1, GFR

*Department of Chemistry, Princeton University,
Princeton N.J. 08540, U.S.A.

I. INTRODUCTION

Shchegolev (1) noted the striking similarities in the physical properties of quasi one-dimensional (1-d) systems (2,3) based on partly oxidized inorganic complexes such as Krogmann (4) salts and ion radical organic crystals such as TCNQ salts (5). These systems are very different chemically; furthermore, there is enormous variety within each group. Nevertheless, the rather detailed picture of the inorganic conductor KCP ($K_2Pt(CN)_4Br_{.3} \cdot 3.2H_2O$) summarized by Zeller (6) has also been proposed (7) for the organic conductor TTF-TCNQ (tetrathiafulvalene-tetracyanoquinodimethane). Thus clear similarities persist, even on the more thorough level of modern solid-state methods, in the electric properties of 1-d inorganic and organic conductors. These similarities undoubtedly reflect the columnar arrangement of planar subunits in both cases.

In addition to a few good conductors, there are many semiconductors and insulators based on either columnar inorganic or organic solids. Soos and Klein (8) discuss the possibilities afforded by organic crystals containing π-electron donors (D) and acceptors (A). By adapting molecular exciton theory (9) to open-shell (radical) subunits, they find that the characteristic optical excitations (charge-transfer bands) and the qualitative features of the magnetic and electric properties follow naturally from unperturbed π-radicals stacked according to the observed structure. Miller and Epstein (10) summarize the many 1-d inorganic solids based on direct metal-metal interactions. Other reviews (11,2,3) treat 1-d magnetic insulators containing ligand-bridged chains of transition metal ions. Such

391

low-dimensional networks do not require planar subunits and will not be emphasized here.

The chemical differences and vast scope of both columnar inorganic and organic systems have led to separate treatments, in spite of the early recognition (1) of important similarities. In our opinion, comparison of 1-d inorganic and organic systems is a challenging and fruitful approach to solid state properties. Such comparisons minimize the differences due to the complicated molecular properties of either transition metal complexes or of organic molecules. Rather, they emphasize the collective magnetic and electric behavior arising from the columnar structure. We present here several general connections between columnar organic and inorganic systems based on planar subunits.

II. ELECTRONIC FACTORS IN COLUMNAR STRUCTURES

There is ample structural evidence, for either planar inorganic complexes (6,10) or organic ion-radicals (8,7), that clearly discernible molecular subunits occur in the solid state. Molecular species in different columns are generally separated by van der Waals distances, while some slightly closer (~ 0.5A) separations are typical within a column. This reduction is of course the basis for quasi 1-d behavior. The electronic interaction t between neighbors in a column is smaller than the energy associated with chemical bonds or with electronic excitations of a subunit. In zeroth-order, the columns consist of essentially unperturbed molecular subunits. Three basic electronic factors are then needed to rationalize a large variety of behavior:

(a) The interaction energy t within a column; several different t may occur, depending on the columnar structure, but a single t describes all crystals with one molecule per unit cell along the column.

(b) The site energy ΔE_S, which describes any change in the energy of the highest occupied MO along the stack; ΔE_S can arise from either chemical or crystallographic inequivalences along the column.

(c) The occupancy of the highest-energy filled MO, which is two for spin-paired (diamagnetic) complexes, one for paramagnetic complexes, and can also assume nonintegral values, for example in TCNQ salts or in mixed-valence complexes.

Various possibilities for these three factors can occur. The currently known good 1-d conductors, for example, all have nonintegral electron occupancy, a single t, and $\Delta E_S \sim 0$. Under special conditions, these criteria may be relaxed and finding such conductors, whether organic or inorganic, would be very interesting. These electronic factors, together with planar subunits arranged in columnar structures, provide general comparisons between organic and inorganic systems. More extensive treatments, in which effects restricted to a few systems are included, can be found in reviews on inorganic complexes (4,10,12), on magnetic insulators (11), on organic charge-transfer complexes (8), or on 1-d conductors (1,6,7,13). Two conference proceedings (2,3) provide further examples of the variety of organic and inorganic systems. We consider here the broad similarities and differences among the majority of columnar organic and inorganic complexes, even at the possible expense of occasional exceptions.

III. MOLECULAR BUILDING BLOCKS

It is an empirical observation that columnar structures are usually based on rigid, planar subunits, although by no means all such molecules crystallize as columns. Fig. 1 contains the representative π-donors D=TTF and TMPD (N,N,N',N'-tetramethyl-p-phenylenediamine) and π-acceptors A=TCNQ and chloranil. These molecules form many 1-d structures, with columns of A^- and/or D^+ ion-radicals (8). Fig. 2 contains representative square-planar d^8 inorganic complexes such as $[Pt(CN)_4]^{2-}$ and $[Ir(CO)_2X_2]^-$; both form columnar structures (10), especially in mixed-valence compounds (14,15). Also shown in Fig. 2 are the metal dithiolates, whose ability to assume multiple oxidation states and to form columnar structures is well estabished (16,10). Many transition metals (M) can be used. We note in Fig. 2 such planar binuclear complexes as $[Cu_2Cl_6]^{2-}$, as found (17) in $K_2Cu_2Cl_6$; the d^9 Cu(II) ion is especially flexible and is important for magnetic studies (18). Other square-planar complexes include (10,12) macrocyclic or multidentate ligands, with a single molecule providing several of the ligands (L) around the metal center. Several bidentate oxalato, (Fig. 2) vic-dioximato, and salicyaldiminato systems form columnar structures. Tetradentate ligands like the phthalocyanines (19) or the porphyrins also display columnar structures, but have generally not provided examples of

π Donors π Acceptors

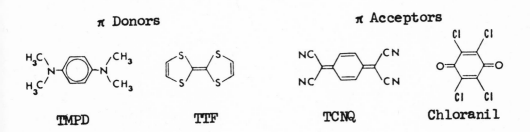

TMPD   TTF   TCNQ   Chloranil

Fig. 1. Representative planar π-electron donors (D) and acceptors (A).

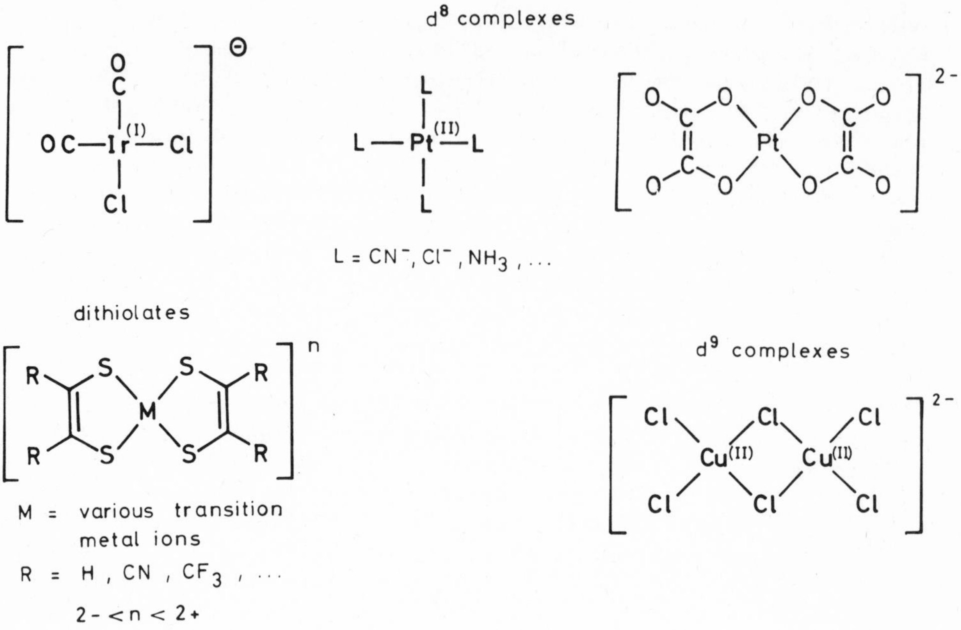

d^8 complexes

L = CN⁻, Cl⁻, NH₃ , ...

dithiolates

d^9 complexes

M = various transition
    metal ions

R = H , CN , CF₃ , ...

2- < n < 2+

Fig. 2. Representative planar four-coordinate transition metal
         complexes.

highly-conducting systems. A surprising recent exception (20) is
the phthalocyaninatolead(II), a good conductor in which the large
Pb^{+2} ion does not lie in the molecular plane of the columnar phthalocyanines. Multidentate ligands promote the formation of square-
planar complexes, although the greater bulk of such rigid ligands
often limits the minimum possible spacing in the column.

The common feature of the molecules in Figs. 1 and 2 is their
planarity. Columnar structures can then be rationalized as fulfil-
ling the condition of maximum space filling, or of maximum nonbonded (eg. van der Waals) interactions.

The lowest electronic excitations are, by contrast, quite different for the columnar solids based on organic molecules (Fig. 1)
and inorganic complexes (Fig. 2). The lowest excitation of organic
crystals based on D and A molecules is a charge-transfer (CT) band
and such bands have been extensively studied (21) for DA complexes
in solution. The lowest electronic excitations of transition metal
complexes are of course the d-d transitions involving ligand-field
splittings. Polarizability terms due to the interaction t for

neighbors in a column involve, in second order, such energy denomi-
nators. The molecular-exciton condition (9) of essentially unperturbed molecules can be guaranteed when t is far smaller than the
lowest electronic excitation and in practice may be a sensible
approximation even for t comparable to the lowest excitation.

It is worth noting that some restrictions arise for either organic or inorganic building blocks. To form columnar structures, and
especially to form ion-radical columns, the planar organic molecule
must also be a reasonably strong π-donor or π-acceptor. Transition
metal complexes, on the other hand, only form the necessary planar
building blocks in special cases such as those represented in Fig.2.
Thus a great many molecules or complexes can be excluded from consideration as building blocks. Furthermore, close spacing along the
column will generally favor a substantial interaction t and such
close spacing cannot be achieved for complicated (hence bulky) ligands. Thus chemical modifications of favorable building blocks are
also limited. These restrictions on the molecular building blocks
are not particularly serious at this early stage, when many possible
classes remain to be examined.

IV. COLUMNAR STRUCTURES

The tendency of planar subunits to stack is consistent with
the maximal nonbonded interactions for a space-filling columnar
structure. On the other hand, such planar aromatic molecules as ben-
zene, naphthalene, or anthracene form space-filling structures not
based on columns; similarly, not all planar transition-metal ions
form stacks. Both planar organic (8) and inorganic (12,18,19) species often show multiple phases, not all of them columnar. Thus
there is presently no possibility of predicting crystal structures,
since many competing and comparable intermolecular interactions cannot be estimated with sufficient accuracy. Columnar structures based
on planar subunits can nevertheless be rationalized by noting that,
in addition to the easy possibility of filling space by stacking
planar units, some weak specific interactions favor such structures.

Such specific electronic interactions are illustrated in Fig.3
and are quite different for organic and inorganic systems. Both
mixed (..DADA... or ...$D^+A^-D^+A^-$...) and segregated (...$A^-A^-A^-A^-$...
or ...$D^+D^+D^+D^+$...)columns of organic donors and acceptors are stabilized by CT interactions (8). The π-electron MO in question is
delocalized over the molecule and permits good overlap even when
the molecular planes are not perpendicular to the column. Tilted
planes are in fact more common (22). In either case, CT excitations
polarized along the column are the lowest optical transition of the
crystal (8,23).

Inorganic complexes in Fig. 3 can form columns by taking advantage of M...M stabilization or of M...L stabilization. The directi-

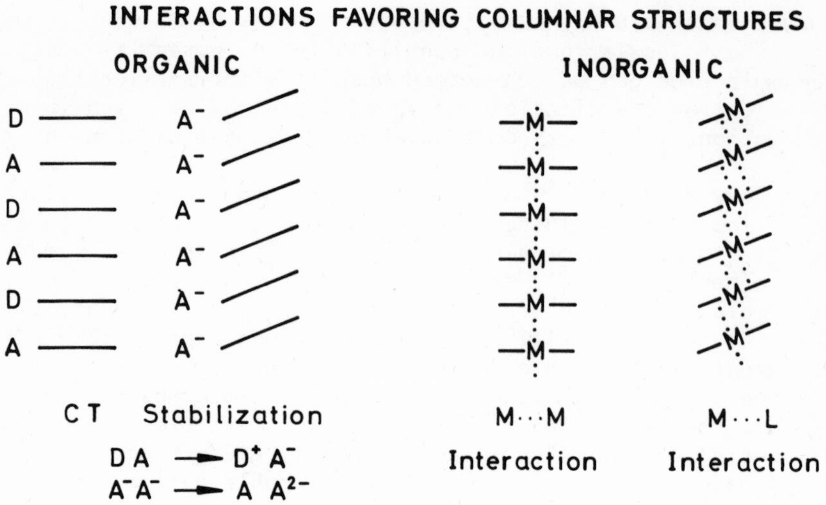

INTERACTIONS FAVORING COLUMNAR STRUCTURES

ORGANIC   **INORGANIC**

CT   Stabilization   M···M   M···L

$$DA \longrightarrow D^+A^-$$
$$A^-A^- \longrightarrow A\ A^{2-}$$

Interaction   Interaction

Fig. 3. Electronic stabilization of columnar structures based on
planar molecular subunits.

onal properties of the metal AO now lead to very different cases on
tilting the complex. Direct M...M interaction requires very nearly
perfect stacking, while any substantial tilting the planes leads to
M...L stabilization. At the present, M...M stacking has been required for conductors, while M...L interactions produce insulators.
The recent description (24) of a zigzag M...M...M chain in $K_{1.75}$·
$Pt(CN)_4 \cdot 1.5H_2O$ shows that a small tilting of the molecular
planes does not destroy M...M coupling.

M...L interactions can also occur without planar complexes.
Fig. 4 shows how octahedral complexes can share a vertex, or an
edge, or a face to produce a 1-d ligand-bridged chain. All three
possibilities are commonly realized and provide important realiza-
tions of low-dimensional magnetic insulators. Such ligand-bridged
networks based on essentially octahedral, and hence definitely non-
planar, complexes do not belong with columnar structures.

V. SOLID STATE PROPERTIES OF MOLECULAR SOLIDS

Molecular solids are held together by relatively weak forces,
which do not significantly alter the shape or the bonding of the
constituent molecules. The reduced symmetry in the crystalline state
produces splittings and shifts of various molecular excitations, as
discussed extensively for closed-shell organic crystals in terms of

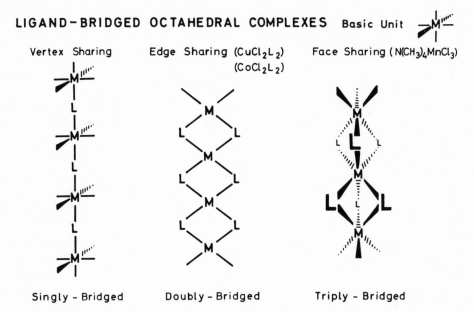

LIGAND–BRIDGED OCTAHEDRAL COMPLEXES Basic Unit

Vertex Sharing   Edge Sharing $(CuCl_2L_2)$ Face Sharing $(N(CH_3)_4MnCl_3)$
$(CoCl_2L_2)$

Singly – Bridged   Doubly – Bridged   Triply – Bridged

Fig. 4. Ligand-bridged networks of octahedral subunits leading to one, two, or three superexchange pathways.

molecular-exciton theory (9). Crystalline anthracene remains the prototype for the physics of such solids (25). The occurrence of molecular ion-radicals D^+ and/or A^- in organic charge transfer or free radical solids indicates that excitations of a single subunit (eg. intermolecular vibrations, electronic excitations) will be slightly perturbed by the crystalline environment. The same holds for transition-metal complexes in which the interaction t is suffi- ciently small. There are many spectroscopic investigations of the effects of the crystalline environment on the excitations of a single molecular building block. These studies include NMR, EPR, ir, visible, u.v., and even higher-energy probes. Aside from the basic integrity of the subunits, there is little resemblence between organic and in- organic systems. Their constituent building blocks represent quite different bonding and thus have different types of excitations. Such molecular excitations will not be discussed here.

    In a closely spaced column, with relatively large t, the ques- tion of essentially unperturbed molecular building blocks is more delicate. For sufficiently large t, the subunits become chemically bonded and a polymer is formed, rather than a molecular solid. Thus the polymer (26,27) $(SN)_x$, for example, does not contain essentially unperturbed molecular subunits. Partly-oxidized inorganic columnar systems, with the smallest M...M spacing, have significant t (~ 1 ev) and may contain somewhat perturbed subunits. Longer M...M spacing (smaller t) poses no problems. Organic systems have small t ($\sim$ 0.1 ev) even for the closest spacing. Even the highest-energy MO, which should be the most sensitive to forming a columnar structure, is found (28, 8) by EPR to be adequately given by the isolated-molecule result.

The solid-state magnetic, electric, and optical properties of columnar solids reflect different arrangements of essentially <u>unperturbed</u> molecular building blocks. In contrast to the small change in molecular excitations, different crystalline arrangements can lead to orders of magnitude changes. For example, the same building blocks can lead to insulators, or semiconductors, or conductors, depending on the interplay of the electronic factors (a), (b), (c) mentioned in Section II. Such collective magnetic and electric phenomena, and also electronic excitations which depend sensitively on the columnar structure, have been of primary interest in quasi 1-d organic and inorganic solids.

The occurrence of essentially unperturbed molecular subunits in these crystals makes it natural to apply molecular-exciton methods, as extended (8) to open-shell molecules. In addition, the sensitivity of the magnetic and electric properties to the crystal structure clearly requires investigations in the solid state, as conventional chemical analyses of the building blocks do not probe such collective behavior.

There are of course many other, perhaps even subtler, collective properties: the crystal structure, the nature of phase transitions, elastic properties, etc. It is not clear whether comparisons between inorganic and organic solids, or even between different types of solids within each class, at this level will prove to be instructive.

Thus we focus on the collective magnetic, electric, and optical excitations arising from a given columnar arrangement of planar subunits. Some rather detailed correspondences can then be developed between organic and inorganic systems, each based on essentially unperturbed building blocks. Such a structural classification (8) of stacked organic systems shows that the "organic metals" naturally form a subclass, while magnetic insulators and triplet-exciton systems form other subclasses. The usefulness of such general considerations is illustrated by the demonstration of partial charge transfer (29) in NMP-TCNQ, a 1:1 conductor which otherwise fell in a subclass of large-gap semiconductors. General comparisons of columnar organic and inorganic crystals also serve as a ready guide for the possible qualitative properties of systems yet to be prepared; indeed, novel systems provide a severe test for classification schemes. Finally, by focusing on gross solid-state electronic properties, rather than on molecular excitations, such general comparisons provide a basis for the striking similarities (1,7) of, for example, quasi 1-d organic and inorganic conductors.

VI. DIAMAGNETIC COMPLEXES: OPTICAL EXCITATIONS

Most molecules are diamagnetic, with two electrons in the highest filled MO. The vast majority of organic (8,21) D and A complexes are neutral, diamagnetic solids. Square-planar d^8 complexes also have spin-paired electrons in the highest AO. Weak-field ligands of Ni(II) might lead to planar high-spin d^8 complexes, with S = 1, but the known paramagnetic Ni(II) complexes are tetrahedral. Neutral organic CT complexes form mixed stacks (...DADA... in Fig. 3). Mixed stacking of $[PtCl_4]^{2-}$ and $[Pt(NH_3)_4]^{2+}$ subunits also occurs in the inorganic complex Magnus Green Salt (12) (MGS). Diamagnetic d^8 complexes such as $[Pt(CN)_4]^{2-}$ also exhibit segregated stacking, as illustrated by $K_2[Pt(CN)_4] \cdot 3H_2O$ in which the column contains identical subunits. Mixed columns have $\Delta E_s \neq 0$ due to chemical differences, which usually are large in comparison with t, while segregated columns only have $\Delta E_s \neq 0$ for crystallographically inequivalent sites.

Weakly interacting diamagnetic molecules are not expected to form paramagnetic or conducting solids. Thus columnar structures primarily affect the optical properties, which for both inorganic and organic complexes become sensitive to the structure. Furthermore, the lowest-energy, structure sensitive transition often falls in the visible region of the spectrum. Such color changes have fascinated researchers for over a century. Finally, the characteristic optical excitations depend on having weakly-interacting subunits, rather than on the electron occupancy, and thus occur in paramagnetic semiconductors and conductors also.

Table I compares neutral organic CT crystals and diamagnetic d^8 complexes. The lowest optical excitation in the organic system is a CT band, $DA \rightarrow D^+A^-$, which clearly depends on the ionization potential of D, on the electron affinity of A, and on various polarization contributions. The CT band is polarized along the column and occurs below any molecular (eg. $\pi \rightarrow \pi^*$) excitations of D or A. It is sensitive to the neighbors and, through polarization and overlap terms, also to the columnar structure. Since $h\nu_{CT} \sim 1$ ev in many organics, these solids are semiconductors (8,30) with an activation energy ΔE_c often in the range of 0.5 ev.

The lowest excitation of inorganic complexes are d-d transitions, E_{dd}, involving a single subunit. Thus the difference between mixed or segregated columns is simply that the former can have two sets of d-d excitations per column. E_{dd} in columnar d^8 structures now reflects the sensitivity of the crystal field splitting to the M...M (or the M...L) separation R. For large t, or strong M...M interaction, the square-planar complex becomes more octahedral and, as indicated in Fig. 5, the filled d_{z^2} and the empty $d_{x^2-y^2}$ AOs are then degenerate, with vanishing d-d splitting. In a segregated (chemically equivalent) column of N complexes, each d-d transition is N-fold degenerate. Such degeneracy is known from molecular-exciton theory (9) to lead

Table 1

Columnar Structures with Closed Shell (Diamagnetic) Building Blocks

	Organic	Inorganic
Molecule/Complex	D, A Weak donors acceptors	Low-spin d^8
Structure	Mixed Stack ...DADA...	mixed: MGS segregated: $K_2[Pt(CN)_4]\cdot 3H_2O$
Optical Excitation Sensitive to Structure	CT band (∥ to Column)	d-d transition
Magnetism Conduction	Diamagnetic Semiconductor $\Delta E_m \sim \Delta E_c \sim 1/2\ \Delta E_{CT}$	Diamagnetic Insulator $\Delta E_c > 1$ ev

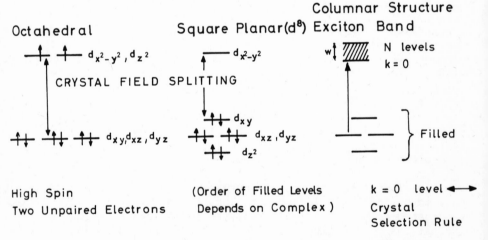

Fig. 5. Schematic energy-level diagram for a d^8 complex in an octahedral, an isolated square-planar, and a columnar square-
planar ligand field. The d-d transition in the solid produces a $k \approx 0$ exciton in a band which is however not a simple
metal AO but a complicated admixture.

to a bandwidth $W \sim 4t^*$, with the crystal selection rule ($k \approx 0$) re-stricting the transition to all complexes having the same phase (Fig. 5). Thus increasing t (or W) also decreases E_{dd}. Day (31) has adopted the molecular-exciton picture which, in the point dipole approximation, leads to $W \propto R^{-3}$ and has demonstrated such behavior in E_{dd} vs R for a number of Pt(II) and Pd(II) complexes in which the M...M spacing R is known from structural data.

In inorganic d^8 complexes, the important variable for the d-d transitions is the M...M separation, while in organic CT complexes the chemical identity of the neighbor is more important. The nature of the structure-sensitive excitations are also quite different. In both cases, however, the columnar arrangement of weakly interacting, diamagnetic subunits leads to characteristic low-energy optical exci-tations. Since the Coulomb interaction of additional electrons in a metal AO is large (~ 10 ev), the activation energy for semiconduction in d^8 systems is rather large, as indicated in Table I. In mixed columns such as MGS, smaller ΔE_c can occur in principle for suffi-ciently unlike subunits (large ΔE_s), but this is not the case (32) for MGS.

VII. PARAMAGNETIC COMPLEXES: UNPAIRED ELECTRONS

The strongest π donors like TMPD or TTF and the strongest accep-tors like TCNQ or Chloranil form solids based on D^+ or A^- ion radi-cals. A few mixed stacks (...$D^+A^-D^+A^-$...) are known (8), but segre-gated stacking is more common. In these free radical (FR) salts of either D^+ or A^-, electric neutrality is maintained by diamagnetic ca-tions or anions that lie outside the column. The magnetic properties of TCNQ salts with segregated columns were from the first associated with the number of FR units per unit cell along the stack (33). Se-veral different t leads to dimerized columns and to triplet spin excitons (33), while equivalent FR spacings yield a single t and a single, temperature-dependent EPR absorption (34).

The corresponding inorganic complexes in Table II are based on d^n ions, with n odd. In terms of planar complexes, the d^9 system (3, 18) of Cu(II) gives both M...M stacking and, more commonly, M...L stacking in Fig. 3. The chain structure (35,12) of $N(CH_3)_4MnCl_3$ is based on face-sharing octahedra and illustrates high-spin d^5 ions of Mn(II). Other possibilities such as d^7 for Co(II) can readily be cited (36), albeit usually for ligand-bridged chains not based on planar complexes.

---

*W depends on the interaction between an excited molecule and its nonexcited neighbor, whereas ground-state interactions are impor-tant for many of the other properties.

Table II

Columnar Structures with Open Shell (Paramagnetic) Building Blocks

	Organic	Inorganic
Molecule/Complex	D^+, A^- (S = 1/2)	d^9 Cu(II) complexes ligand-bridged d^n complexes various S
Structure	Mixed (CT crystals) Segregated (FR crystals)	Segregated, usually with M...L stabilization
Optical Excitation Sensitive to Structure	CT band $\Delta E_{CT} \sim U$ in FR	d-d transition
Magnetism	Exchange in Column $J \sim t^2 / \Delta E_{CT}$	Exchange in Column J from superexchange through bridging ligands
Conductivity	Semiconductors $\Delta E_c < \Delta E_{CT} \sim$ lev	Insulators $U \sim 10$ ev

In all of these, the molecular building blocks contain unpaired electron(s) and are formally open-shelled. As discussed by Mott (37), open-shell systems remain insulators when the correlation energy (U) for putting two electrons on the same site is considerably larger than the bandwidth (4t) for the unpaired-electron MOs. These ideas provide a qualitative picture for the great variety of magnetic insulators such as ligand-bridged transition-metal crystals of any dimension. In organic ion-radical systems, where $h\nu_{CT}$ is the lowest-energy electronic excitation and the bandwidth parameter t is formally associated with the Mulliken CT integral, the model of strong on-site correlations $(U \gg t)$ leads in simple salts to paramagnetic semiconductors (8). Furthermore, the highly anisotropic one-dimensional exchange constant $J \sim t^2/h\nu_{CT}$ plays a central role in the magnetic properties (34).

The low values of $\Delta E_{CT} \sim 1$ ev in organic ion-radicals is related to the reduction of U due to the considerable size of the π-electron MO. While a quantitative analysis of U is still in the stages of identifying the largest contributions, it is evident that U becomes far larger for two electrons in the same transition metal AO. Now values of U of the order of 10 ev are appropriate and, in spite of open-shell ions, such transition-metal salts are insulators $(\Delta E_c > 1$ ev). Furthermore, t can be further reduced in the usual ligand-bridged geometry. The t^2/U contribution to the superexchange is no longer dominant, but is one of several effects (38).

The considerations are summarized in Table II. Columnar paramagnetic complexes are readily understood as molecular building blocks
with integral numbers of unpaired electrons. The characteristic optical excitations(Sec. VI) remain unchanged: CT excitations polarized
along the ion-radical stack for either mixed or segregated organic
systems and d-d transitions for either stacked or ligand-bridged
transition-metal chains. The smaller values of ΔE_{CT} and of ΔE_c in
π-molecular radicals reflect the delocalization of the MO. The magnetic properties in both cases are dominated by exchange or superexchange interactions along the columns. As long as the correlation
energy U is large compared to t, an FR column remains a semiconductor or an insulator.

VIII. FRACTIONAL CHARGES: STRUCTURAL POSSIBILITIES

The picture of essentially unperturbed molecular subunits follows
when the interaction t is small compared to the electronic excitations
of a subunit. The correlation energy U for creating an adjacent electron-hole pair in a segregated column will in general be comparable
to subunit excitation energies and will therefore usually exceed t.
Weakly interacting open-shell systems thus do not become conducting,
at least for integral electron occupancy in the highest MO. There
remains the possibility of nonintegral electron occupancy, as achieved
in complex TCNQ salts or in partly-oxidized transition-metal complexes.

Isolated molecular building blocks clearly have an integral number of electrons, as does a column in a solid of nonoverlapping columns. An integral number (N_e) of electrons in a column of N sites
may in principle produce a nonintegral electron occupancy or a mixed
valence system. However, it is by no means sufficient to have $N_e \neq N$.
The interplay of the electronic energies t and ΔE_s must also be taken into account. The detailed nature of the columnar structure now
becomes important. For example, the most straightforward way of demonstrating (22) fractional charges is through crystallographically
equivalent sites in a salt with complex stoichiometry, as happens in
several 1:2 TCNQ salts.

Crystallographic equivalence demonstrates that $\Delta E_s=0$ and, when
combined with information that $N_e \neq N$, implies fractional charges. The
occurrence of a single subunit per unit cell along the column demonstrates that all the t are equal. It must be emphasized that these conditions on $\Delta E_s=0$ and on a single t are quite different. As an extreme case, ΔE_s is large in a mixed column ...$D^+A^-D^+A^-$..., but the
structures (39) of TMPD-TCNQ and TMPD-Chloranil yield a single t.
Recent structural work (40) on KCP shows that there are two Pt sites
and in principle two Pt...Pt distances that are equal within experimental error. Small variations in t and in ΔE_s may therefore occur.
Structural data (41) in the TTF-TCNQ family imply a single t and
$\Delta E_s=0$, although care must be taken to check for possible different
space groups in closely related systems. The occurrence of several

subunits per unit cell along the column poses a quantitative question: are the sites in sufficiently different environments, as in the segregated $Cs_2(TCNQ)_3$ stack (42), to correspond to essentially neutral TCNQ molecules and essentially ionic $TCNQ^-$ radicals? Such a situation follows (43) for the $\Delta E_s > t$; fractional charges are expected and found (28) for $t < \Delta E_s$. Finally, when several different t are possible, then the ratio $\Delta t/t$ distinguishes between regular ($\Delta t/t \ll 1$) and alternating ($\Delta t/t \gtrsim 1$) columns.

The important parameter for fractional charges is thus the relative magnitude of ΔE_s and of t. Similar general considerations led to the three classes of mixed-valence inorganic systems discussed by Robin and Day (15). Class I corresponds to $\Delta E_s \gg t$ and describes building blocks with different oxidation states or different integral numbers of electrons. The diamagnetic, ligand-bridged network (15) of octahedral Pt(IV) and square-planar Pt(II) in Wolffram's Red Salt (X=Cl, L = ethylamine) and in related (44) $[PtL_4X_2PtL_4]^{4+}$ systems is thus readily understood, as are Au(I)-Au(III) chains. Pressure studies (45) on the former decrease the inequivalence of the Pt^{IV}-X... Pt^{II} bonds, thus decreasing the ratio $\Delta E_s/t$ and increasing the conductivity. Class II includes $\Delta E_s \sim t$ and leads to nonintegral, but still unequal, electron occupancies; many complex TCNQ salts belong in this class. The final case $\Delta E_s \ll t$ (Class III) includes the special case $\Delta E_s = 0$ of equivalent sites and corresponds to the fractional charge N_e/N or to fractional oxidation states. The Robin-Day approach can be applied to any degree of electron occupancy in the highest MO, including integral occupancy, while the present discussion emphasizes the case of fractional charges.

The interplay of t and ΔE_s is illustrated schematically in Fig. 6 for a columnar structure with two equivalent and one inequivalent site per unit cell For t = 0 and $\Delta E_s \neq 0$, two electrons occupy the equivalent sites, as shown for isolated nonoverlapping (t = 0) sites. Electron-electron correlations U here exceed t and lead to adjacent free-radicals and a closed-shell site. CT excitations U produce an electron-hole pair. However, it is clear that conduction can occur, in the case shown of two electrons per three sites, by single occupancy of the empty level at ΔE_s. Thus the high-energy U processes need not be involved in complex ($N_e \neq N$) salts of weakly interacting subunits. The formation of a columnar structure when ΔE_s exceeds t is shown on the right in Fig. 6 and leads to a semiconductor. The condition $U > t$ now corresponds to one electron per k-state in the band. The case $\Delta E_s = 0$ and equal interactions t is shown on the left, again for $U > t$, and now formally represents a partly-filled metallic band.

The additional structural possibilities afforded by having t_1, t_2, ... along the chain lead to more complicated band structures, with energy gaps of the order of $\Delta t = |t_1 - t_2|$. Such alternating interactions are important for describing (34) the collective magnetic

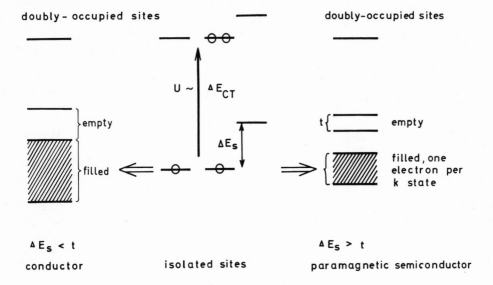

Fig. 6. Schematic energy-level diagram for two electrons on three
         sites per unit cell and large U, showing the possible case
         for $\Delta E_s > t$ and for $\Delta E_s > t$.

properties and can also lead to activation energies for semiconduction. Now the limiting case of large $\Delta t/t$ describes noninteracting
segments containing several interacting subunits, while the opposite
case of $\Delta t/t = 0$ corresponds to a regular system with a single band.
The intermediate case of $\Delta t/t \sim 1$ may be representative (28) of many
tetrameric TCNQ salts and requires further analysis. The simplest
structures have naturally been of primary interest in establishing
the basic trends. The structural possibilities for $\Delta t \neq 0$ and for
$\Delta E_s \neq 0$ in segregated columns with $N_e \neq N$ illustrate the richness
and complexity of the three electronic factors introduced in Section II.

IX. COLUMNAR CONDUCTORS

The preceding discussion clearly indicates that equivalent
($\Delta E_s \ll t$) and regular ($\Delta t \ll t$) columns of fractionally-charged
subunits lead to quasi 1-d conductors. These criteria are collected
in Table III for several organic and inorganic conductors. Many complex TCNQ semiconductors in Table III probably have $\Delta E_s \sim t$ and
$\Delta t \sim t$ and are semiconductors. Complex stoichiometry has long been
known to enhance the conductivity (46) and is another indication that
the correlation energy U, as discussed in Section VI in connection
with CT excitations, exceeds the columnar interaction t. These more
complicated columnar structures provide additional applications for

Table III

Columnar Structures with Fractional Charges

	Organic	Inorganic
Molecule/Complex	$D^{+\gamma}$, $A^{-\gamma}$ ($\gamma \neq 1$)	Partly oxidized d^8
Structure	Segregated	Segregated
Optical Excitation	Several CT Bands	d-d transitions and intraband transitions
Bandwidth	$t \sim 0.1$ ev	$t \sim 1$ ev (for $R \lesssim 3$ A)

Conductors:

(ΔE_s, $\lvert t_1 - t_2 \rvert \ll t$) regular stacks equivalent sites	TTF-TCNQ ($\gamma \simeq 0.6 \pm 0.10$)	KCP ($\gamma' = 2.33$) [*] $K_{1.15}[Pt(CN)_4]1.5H_2O$ ($\gamma' = 2.25$; ref. 24)
	NMP-TCNQ ($\gamma = 0.94$) (disordered phase)	$Ir(CO)_3Cl_{1.1}$ ($\gamma' \simeq 1.10$; ref. 51)
	$Q(TCNQ)_2$ ($\gamma = 0.50$)	$[Ir(CO)_2X_2]^{n-}$ ($1.4 < \gamma' < 1.6$; ref.14)
	TTF X_γ ($\gamma \approx 0.80$; ref.49)	

Nonconductors

alternating stacks and/or inequivalent sites	2:3, 1:2 TCNQ Salts [**] (triplet exciton systems) $Cs_2(TCNQ)_3$ ($\emptyset_3As\ CH_3$)$(TCNQ)_2$ ($\emptyset_3PCH_3$)$(TCNQ)_2$ TEA $(TCNQ)_2$	$Ni(BCD)_2\ 0.5\ I$ ($\gamma' = 2.17$; ref.54) partly oxidized $[Pt(ox)_2]^{2-}$ salts [***]

[*] γ' is the formal (nonintegral) oxidation state

[**] See ref. 8 for a summary

[***] A large number of columnar structures with different superlattices and/or small crystals have hampered conductivity measurements; see refs. 4,10 and 12. Thus some of the oxalates and/or cation deficient compounds may well turn out to be conductors.

the molecular-exciton approach of unperturbed sites.

Most members of the TTF-TCNQ family have $\Delta E_s = 0$ and a single t, even if there are magnetically inequivalent columns. Fractional charges then necessarily occur if $N_e \neq N$ (incomplete transfer). By contrast, KCP has small ΔE_s and possibly some small variations in t, although the relatively strong columnar interaction at the short M...M spacing of 2.89 A yields regular, equivalent sites. The partial oxidation, by either halogen ion or by cation deficiency, estabilshes that $N_e \neq N$. Such nonstoichiometric compounds may well have additional contribution to ΔE_s, as might happen in disordered structures (47). Now many ΔE_s values can occur. The band structure is not very greatly changed when t exceeds such variations (48). Several conducting TTF^{+Y} $(X^-)_\gamma$ (X^-= halogen, SCN) salts are clearly nonstoichiometric (49), with fractionally charged TTF sites. The conducting phase of NMP-TCNQ has disordered NMP cations (50), and partial charge transfer (29), with a regular segregated stack of $TCNQ^-·94$. No differences in the Ir sites of the columnar solid $Ir(CO)_3Cl_{1.1}$ have been observed, which led to the postulate (51) of additional Cl^- ions between the $Ir(CO)_3Cl$ columns, as in KCP. The persistent, but not conclusive, nonstoichiometry of this material is certainly reasonable for its physical properties. To avoid circular arguments, however, it is important to demonstrate independently that $\Delta E_s \ll t$, that $\Delta t \ll t$, and that $N_e \neq N$ in quasi 1-d conductors.

It is perhaps worth emphasizing that the semiconduction in systems based on chemically inequivalent subunits (mixed column) depends on the smallness of t relative to ΔE_s. Increasing t and/or decreasing ΔE_s by modification of the ligands could in principle reverse the magnitudes and lead to $\Delta E_s < t$. Now for $N_e \neq N$ a 1-d conducting alloy would be expected. The smallness of t relative to electronic excitations, and thus relative to ΔE_s in mixed stacks, suggests that such presently unknown conductors should be rare. The best chance is to reduce ΔE_s through variation of the ligands. At least formally, merely increasing t to exceed electronic energies amounts to forming a polymer instead of a columnar solid of essentially unperturbed organic or inorganic subunits.

X. DISCUSSION

An important role of general comparisons is to bypass difficult quantitative problems in molecular quantum mechanics. There is for example no satisfactory quantitative estimate for the correlation energy U of producing a doubly occupied site. Nevertheless, the absence of highly conducting FR systems with integral electron occupancy points to $U > t$ in organic systems. The further observation that quasi 1-d conductors, whether organic or inorganic, have fractional charges, with $N_e \neq N$ and conduction possible without producing doubly-occupied sites, is again consistent with $U > t$. Such conditions

follow for weakly interacting sites and do not require evaluation of
either U or t. Both of these parameters in fact tend to be small and
complicated; they are small compared to the accuracy of approximate
quantum molecular computations and they are complicated by various
types of renormalizations in a polarizable, distortable molecular
solid. Thus it is quite interesting that a survey of columnar inorganic and organic systems, both insulating and conducting, so clearly
indicates that electron-electron correlations exceed the bandwidth
in the available systems.

We have already suggested that chemical modification could in
principle lead to small ΔE_s even in mixed stacks; a normal t could
then lead to a conductor. Since t is far smaller in organic solids
($\sim$.1 ev) than in inorganic solid ($\sim$1 ev) with R $\gtrsim$ 3 Å, a search for
such a mixed conductor is far more likely to succeed with inorganic
building blocks.

The present discussion has focused on weakly interacting molecular building blocks. The charge of the ion-radical subunits has not
been central, and both cationic and anionic systems can readily be
cited. Nevertheless, the available 1-d organic and inorganic conductors are salts, rather than neutral molecular solids. While the early
stable organic radicals such as $\emptyset_3C$ or DPPH are neither small nor
planar, there are small planar radicals (52) such as substituted phe-
nazines and phenothiazines. Such solids, if they form columns, would
not have the repulsive Coulomb interaction of a segregated ion-radical
stack. The physical properties of solids containing small neutral π-
radicals are not presently known.

Comparison of columnar d^8 complexes (10) strongly suggests that
rather strong t is needed for 1-d conductors, which so far have M...M
spacing* of < 3 Å. In addition, nonintegral electron occupancy seems
to be required, a condition which often produces shortened R values.
On the other hand, it has been proposed (53) that the "most favorable"
model for a 1-d superconductor is a partly oxidized t column with
R$\sim$3.4 Å, a minimum requirement unfortunately imposed by large organic ligands. Calculations (53) suggest that the partly-oxidized column would be metallic in the absence of ligands, a conclusion that
has little support from known systems. It is in fact possible to produce (54) partly-oxidized Ni columns with comparable Ni...Ni separations (R = 3.15 Å) and disordered anions between the columns. Such
systems are not metallic. The problem may well be that t has been reduced to less that the disorder in the site energies, but such disordered anions are also part of the proposed (53) Pt column. Each specific system must of course be examined for a conclusive evaluation

---

*A larger R = 3.73 Å is given in ref. (20) for the main group colum-
nar conductor phthalocyaninatolead(II)

of its conducting properties. But highly approximate computations whose conclusions are rather contrary to available data involve very high risks for difficult synthetic work.

The delocalized π-electron MO in organic radicals is responsible for the reduced energy of the CT excitation and the semiconducting behavior of systems with integral numbers of electrons. The smaller U is accompanied by smaller t, again in part due to the delocalized MO. At least two organic conductors remain (55) metallic down to at least .050 K, without however becoming superconducting. The delocalized nature of the MO, which necessarily precludes close contact with polarizable ligands, has been cited (53) as a major shortcoming of TCNQ systems as model superconductors. Such theoretical speculations are supported by the currently available results on delocalized π-electron MOS, which have been quite successful in producing conductors, but not excitonic superconductors.

There are a number of possibilities for combining, at least in principle, some advantageous features of both organic and inorganic systems. Rather than citing current examples, which are naturally preliminary and have yet to produce a really good conductor, we conclude by noting that such efforts presuppose some familiarity with both types of materials. Many of the qualitative aspects of columnar solids follow from a picture of weakly interacting molecular building blocks. Such a simple model can be used for orientation and comparison among both organic and inorganic systems. Qualitative comparisons can focus attention on the important empirical requirements for achieving a desired property such as high conductivity. In view of the increasingly detailed and complex solid-state theories for particular columnar systems, the picture of weakly interacting molecular sites in either organic or inorganic crystals also serves to simplify and to unify a rapidly expanding field.

Z. G. Soos is grateful to the Anorganisch-Chemisches Institut, Heidelberg, for their hospitality as a visiting professor in 1976, where this work was performed.

REFERENCES

1. I. F. Shchegolev, Phys. Status Solidi(a) 12, 4 (1972)
2. H. J. Keller, ed. "Low-Dimensional Cooperative Phenomena", (Plenum, New York, 1975) NATO-ASI Series B, Vol. 7
3. L. V. Interrante, ed. "Extended Interactions Between Metal Ions in Transition Metal Complexes", (ACS Symposium Series No.5, 1974)
4. K. Krogmann, Angew. Chem. Intl. Edt. Enq. 8, 35 (1969)
5. L. R. Melby, R. J. Harder, W. R. Hertler, W. Mahler, R. E. Benson and W. E. Mochel, J. Amer. Chem. Soc. 84, 3374 (1962)

6. H. R. Zeller, pp. 215 - 233 in ref. 2.

7. A. F. Garito and A. J. Heeger, Acct. Chem. Research 7, 232 (1974);
   A. J. Heeger and A. F. Garito, pp. 89 - 123 in ref. 2

8. Z. G. Soos and D. J. Klein, in "Molecular Association" Vol. I.
   (ed. R. Foster, Academic, London, 1975) p. 1 - 109;
   Z. G. Soos, Ann. Rev. Phys. Chem. 25, 121 (1974)

9. D. P. Craig and S. H. Walmsley, "Excitons in Molecular Crystals",
   (Benjamin, New York, 1968);
   A. S. Davydov, "Theory of Molecular Excitons", Plenum, New York,
   1971)

10. J. S. Miller and A. J. Epstein, Prog. Inorg. Chem. 20, 1-154 (1976)

11. L. S. de Jongh and A. R. Miedema, Adv. Phys. 23, 1 (1974);
    D. W. Hone and P. M. Richards, Ann. Rev. Mat. Sci, 4, 337 (1974).
    See also ref. 3

12. T. W. Thomas and A. E. Underhill, Chem. Soc. Rev. 1, 99 (1972)

13. F. Wudl, ed. "Highly Conducting Organic Materials", (Academic
    Press, New York, to be published)

14. A. P. Ginsberg, J. W. Koepke, J. J. Hauser, K. W. West, F. J.
    DiSalvo, C.R.Sprinkle, R. L. Cohen, Inorg. Chem. 15, 514 (1976)

15. M. B. Robin and P. Day, Adv. Inorg. Chem. Radiochem. 10, 247 (1967)
    G. C. Allen and N. S. Hush, Progr. Inorg. Chem. 8, 357 (1967)

16. J. A. McCleverty, Progr. Inorg. Chem. 10, 49 (1968)

17. R. D. Willett, C. Dwiggins, R. F. Kruh and R. E. Rundle, J. Chem.
    Phys. 38, 2429 (1963)

18. W. E. Hatfield and R. Whyman, Trans. Metal. Chem. 5, 47 (1969)
    G. F. Kokoszka and G. Gordon, Trans. Metal. Chem. 5, 181 (1969)

19. A. B. P. Lever, Adv. Inorg. Chem. Radiochem. 7, 27 (1965)

20. K. Ukei, Phys. Letters 55A, 111 (1975);
    J. Phys. Soc. Japan 40, 140 (1975)

21. G. Briegleb "Elektronen-Donor-Akzeptor-Komplexe" (Springer Ver-
    lag, Berlin 1961); R. Foster, "Organic Charge Transfer Complexes"
    (Academic Press, New York 1969)

22. F. H. Herbstein in "Perspectives in Structural Chemistry"
    Vol. IV (eds. J. D. Dunitz and J. A. Ibers, Wiley, New York 1971)
    pp. 166-395

23. J. B. Torrance, B. A. Scott and F. B. Kaufman, Solid State Comm.
    17, 1369 (1975)

24. K. D. Keefer, D. M. Washecheck, N. P. Enright and J. M. Williams,
    J. Amer. Chem. Soc. 98, 233 (1976)
    A. H. Reis Jr., S. W. Petersen, D. M. Washecheck and J. S. Miller,
    J. Amer. Chem. Soc. 98, 234 (1976)

25. R. G. Kepler, in "Treatise on Solid State Chemistry", Vol. III
    (ed. N. B. Hannay), Plenum New York, (1976), pp. 615-678

26. C. M Mikulski, P. J. Russo, M. S. Saran, A. G. MacDiarmid,
    A. F. Garito and A. J. Heeger, J. Amer. Chem. Soc. 97, 6358 (1975)

27. R. L. Greene, G. B. Street and L. J. Suter, Phys. Rev. Letters 34,
    577 (1975)

28. A. J. Silverstein and Z. G. Soos, Chem. Phys. Letters 39, 525
    (1976)

29. M. A. Butler, F. Wudl and Z. G. Soos, Phys.Rev. B 12, 4708 (1975)

30. F. Gutmann and L. E. Lyons, "Organic Semiconductors" (John Wiley, New York, 1967)
31. P. Day, J. Amer. Chem. Soc. 97, 1588 (1975) and in this volume, H. Yersin, G. Gliemann and U. Rössler, Solid State Comm., preprint
32. L. V. Interrante, J.C.S. Chem. Commun. 1972, 302
    L. V. Interrante and R. P. Messmer, Inorg. Chem. 10, 1174 (1971)
33. P. L. Nordio, Z. G. Soos and H. M. McConnell, Ann. Rev. Phys. Chem. 17, 237 (1966)
34. Z. G. Soos, ref. 2, pp. 45 - 64, and J. Chem. Phys. 46, 4284 (1967)
35. B. Morosin and E. J. Graeber, Acta Cryst. 23, 766 (1976)
36. K. Takeda, S. Matsukawa and T. Haseda, J. Phys. Soc. Japan 30, 1330 (1971)
37. N. F. Mott, Proc. Poy. Soc. Ser. A, 62, 416 (1949) and "Metal-Insulator Transitions", Taylor and Francis, London, 1974
38. C. Herring, in "Magnetism", Vol. II B. (eds. G. T. Rado and H. Suhl), Academic Press, New York (1963) pp 1 - 185;
    P. W. Anderson, in "Solid State Physics", Vol. 14 (eds. F. Seitz and D. Turnbull), Academic Press, New York (1963) pp 99 - 214
39. A. W. Hanson, Acta Cryst. 19, 610 (1965);
    J. L. DeBoer and A. Vos, Acta Cryst. B 24, 720 (1967)
40. C. Peters and C. F. Eagen, Inorg. Chem. 15, 782 (1976)
41. A. J. Schultz, G. D. Stucky, R. H. Blessing and P. Coppens, J. Amer. Chem. Soc. 98, 3194 (1976)
42. C. J. Fritchie, Jr., and P. Arthur, Jr., Acta Chryst. 21, 139 (1966)
43. Z. G. Soos and D. J. Klein, J. Chem. Phys. 55, 3284 (1971)
44. Ö. Bekaroglu, H. Breer, H. Endres, H. J. Keller, H. Nam Gung, Inorg. Chim. Acta, in press
45. L. V. Interrante, K. W. Browall and F. P. Bundy, Inorg. Chem. 13, 1158 (1974)
46. W. J. Siemons, P. E. Bierstedt, and R. G. Kepler, J. Chem. Phys. 39, 3523 (1963); see also ref. 8
47. A. N. Bloch, D. O Cowan, and T. O. Pohler, in "Charge and Energy Transfer in Organic Semiconductors" (eds. M. Masuda and M. Silver) Plenum New York (1974) p. 167, 159
48. E. Ehrenfreund, S. Etemad, L. B. Coleman, E. F. Rybaczewski, A. F. Garito, and A. J. Heeger, Phys. Rev. Letters 31, 269 (1972). The role of disorder is not really established here for NMP-TCNQ, however, as the starting point of a 1/2-filled TCNQ⁻ band is inadequate (Ref. 29)
49. R. J. Warmack, T. A. Callcott and C. R. Watson, Phys. Rev. B12, 3336 (1975) and F. Wudl, in this volume
50. C. J. Fritchie, Acta Cryst. 20, 892 (1966); B. Morosin, Phys. Lett. A 53, 455 (1975)
51. A.P.Ginsberg, R. L. Cohen, F. J. DiSalvo and K. W. West, J. Chem. Phys. 60, 2657 (1974)
52. J. Brandt and M. Zander, Chemiker Zeitung 99, 272 (1975)
53. D. Davis, H. Gutfreund and W. A. Little, Phys. Rev. B 13, 4766 (1976)

54. H. Endres, H. J. Keller, M. Mégnamisi-Bélombé, W. Moroni,
    H. Pritzkow, J. Weiss and R. Comès, Acta Cryst. $\underline{32\ A}$, (1976)
    in press
55. G. J. Ashwell, D. D. Eley and M. R. Willis, Nature $\underline{259}$, 201
    (1976); A. N. Bloch, D. O. Cowan, K. Bechgaard, R. E. Pyle,
    R. H. Banks, and T. O. Poehler, Phys. Rev. Letters $\underline{34}$, 1561
    (1975)

APPENDIX

List of contributors to the discussion sessions

J.R.Andersen et al., Risø, Denmark	One-dimensional organic conductors with low-symmetric acceptors
K.Bechgaard et al., Copenhagen, Denmark	Electrochemistry of TCNQ and related compounds
K.Carneiro, Copenhagen, Denmark	Lattice dynamics of $K_2Pt(CN)_4$ $Br_{0.3} \cdot 3.2D_2O$ amd $K_{1.75}Pt(CN)_4$ $1.5H_2O$
W.G.Clark, Los Angeles, USA	Specific heat of $Qn(TCNQ)_2$ at very low temperatures
H.P.Geserich et al., Karlsruhe GFR	Optical investigation of the anisotropic electrical conductivity of $(SN)_x$ single crystals
W.D.Gill et al., San Jose, USA	Intrinsic transport properties of polymeric sulfur nitride $(SN)_x$
W.E.Hatfield, Chapel Hill, USA	New synthetic approaches to the preparation of one-dimensional compounds
H.W.Helberg, Göttingen, GFR	Indicatrix orientation in complex salts of TCNQ
F.Herman, San Jose, USA	Electronic structure of TTF, TCNQ and related molecules
L.V.Interrante et al., Schenectady, USA	Synthesis and solid state properties of π-donor-acceptor compounds based on bis-dithiolene metal complex acceptors
H.Kahlert et al., Vienna, Austria	Galvanomagnetic effects in poly-sulfur nitride, $(SN)_x$
P.Mengel, Karlsruhe, GFR	X-ray and ultraviolet photo-emission properties of $(SN)_x$
H.Morawitz, San Jose, USA	Chiral charge density waves in quasi one-dimensional organic conductors

413

H.Naarmann, Ludwigshafen, GFR   Peryleneimides (PI) a new type
                                       of semiconducting material

H.Niedoba, Orsay, France   NMR study of the 1d conductor KCP

B.Renker et al., Karlsruhe, GFR   Inelastic neutron scattering
                                       study of the $2k_F$ instability in
                                       $K_2Pt(CN)_4Br_{.3} \cdot 3D_2O$, (KCP)

T.D.Schultz, Yorktown Heights,   Contrasts in the behavior of
    USA   TSeF-TCNQ and TTF-TCNQ and what
                                       they imply about the differences
                                       in underlying models

G.Soda, Orsay, France   Nuclear spin relaxation studies
                                       of the inter-chain coupling in
                                       one-dimensional organic conduc-
                                       tors, TTF-TCNQ and HMTSF-TCNQ

K.Seeger, Vienna, Austria   Thermistor-effect in semiconduc-
                                       ting TTF-acceptor complexes

G.B.Street, San Jose, USA   The chemistry of $(SN)_x$ - a poly-
                                       meric superconducting metal

H.Temkin, Ithaca, USA   Raman scattering in metallic
                                       polymers: $(SN)_x$

Y.Tomkiewicz, Yorktown Heights,   Organic alloys
    USA

A.E.Underhill, Bangor, England   Effect of lattice alterations on
                                       the electrical conduction proper-
                                       ties of "Krogmann compounds"

C.Weyl-Bertinotti, Orsay, France   X-ray scattering in the metallic
                                       state of TSeF-TCNQ and HMTSF-TCNQ
                                       compounds

J.M.Williams, Argonne, USA   Synthesis and characterization of
                                       new metallic Pt-Pt chain forming
                                       compounds

H.Yersin, Regensburg, GFR   Spectroscopic properties of
                                       $M_x[Pt(CN)_4] \cdot yH_2O$ single crystals

SUBJECT INDEX

Absorption
   edge, 310
   optical, 137, 139, 161, 177,
   313
   spectra, 30, 306
Acceptors, 25, 28, 31, 244 -
   246, 391, 395
Acetylacetonate
   complexes of, 205
Acridinium-TCNQ, 51
Activation energy, 76, 77, 123,
   140, 150, 342, 344, 399, 401
Alloys, linear, 236, 240
Anderson-insulator, 354
Anisotropic properties
   1, 87, 189, 201, 234 - 236,
   274 - 275, 348, 351
Annihilation operator
   5, 93, 259
Anomaly
   electronic, 152
   $2k_F$, 320-323, 329, 334
   $4k_F$, 20, 98, 323, 331, 333
   of resistivity, 345
Anthracene, 395, 397
Antiferromagnetism, 3
Antisoliton, 13
A-15 superconductors
   186, 217, 280
Aza-TCNQ (ATCNQ), 249, 252
Azide (N=N=N), 234
Backscattering
   11, 63, 98, 105, 269-272
Band
   edge, 77
   energies, 283
   gap, 207, 214, 217

Band
   filling, 10, 30, 95, 144, 246,
   282, 336, 404
   shape, 208, 344, 348
Band structure
   calculations, 70, 172-186,
   350-356
   of $(SN)_x$, 172
   one-dimensional, 87, 172
Band theory, 350-356
Bandwidth, 11, 15, 25, 30, 48, 103,
   137, 146, 157, 290, 369, 387,
   401, 402, 406, 408
BCS
   energy gap, 157, 376
   equation, 280, 283, 291, 293
   state, 260
   theory, 260
Binary compounds, 199
Binding energy, 133
Bipyrimidine complexes, 204
Bloch wave, 105
Boltzmann equation, 350
Born-Oppenheimer approximation, 92
Boson, 3, 7, 13
   representation, 11
Bragg peak, 376-379
Brillouin zone, 50, 79, 92, 173,
   275, 288, 318, 351, 355
Bulky ligands, 199, 281
   substituents, 55
Carbenes, 301, 302
Carrier
   concentration, 179, 356, 361
   density, 350
   lifetime, 357
   mobility, 344, 358

415

Carrier
   propagation, 70
Cation-anion pair, 58
Cation stack, 26, 28
Chain
   anion bridged, 201, 207
   antiferromagnetic, 203
   end effects, 186
   ferromagnetic, 203
   Heisenberg, 153
   imperfection, 176
   ligand-bridged, 391-403
   magnetic, 236
   materials, 1-21, 252, 297
   mixed valence, 197, 211, 223,
   236, 281
   platinum, 205-212, 225-231,
   280-282, 334, 393, 399
   regular, 405
   single metallic, 268-276
   structure, 88, 189, 401
   susceptibility, 126
   TCNQ, 25-166, 233-256, 315-
   339
CDW
   excitation, 369-376
   gaps, 371
   incommensurate, 98, 379 ff.,
   388
   instabilities, 16, 49, 373-
   375, 384
   phase, 105
   phenomena, 87-135
   pinned mode, 116, 125
   spectrum, 13
   states, 1-23, 49, 50, 152-158,
   269-275, 315-339, 342-344
   transition, 16, 377
Charge
   distribution, 25, 26, 143
   excitation, 139-160
   fluctuations, 213-215
   oscillation, 380
   transfer, 54, 215, 326, 333
   transfer bands, 391, 399
   transfer complex, 30, 48, 259,
   341, 361, 362
   transfer excitations, 395, 403-
   405, 409

Charge
   transfer integral, 47, 48
   transfer interaction, 395, 396
   transfer partial, 407
   transfer solids, 248
   transfer state, 213
Chemical synthesis of CT salts
   25-45
Chloranil, 25, 401
Chromophore, 281, 282, 294
Coherence length, 68, 73, 94-106
Coherency, 358
Collective
   behaviour, 126, 200
   electronic states, 204
   electron transport, 127
   mechanism, 1, 22
   modes, 10, 21, 70, 99-127
   mode life time, 106
   properties, excitations, 398
Columnar
   arrangement, 391
   compounds, 225, 391, 409
   conductors, 405
   inorganic systems, 392, 398,
   408
   organic systems, 392, 398, 408
Columnarstructure, 392-406
Commensurability, 94, 102, 125
Commensurate
   CDW, 381, 385, 387
   phase, 380
   state, 383, 381, 386
   value, 17
Complex
   order parameter, 104, 105, 115,
   126
   salts, 404
   TCNQ salts, 30, 403, 404
Condensation, 326, 328, 335
Conduction
   electron density, 383-387
   electrons, 2, 53, 138, 279,
   373, 374, 377, 383, 386 ff.
Conducting chain, 28, 48, 73, 238,
   279, 281, 290
Conductivity
   1, 22, 25, 29, 47, 51, 62ff.,
   67ff., 88, 98ff., 140ff., 157ff.,
   176 ff. 186, 214, 219, 233ff.,

Conductivity
    255, 258, 309, 326, 342ff.,
    369ff., 383, 398, 402, 409
Conductivity
    activated, 342, 346
    infrared, 101
    optical, 177
    single crystal, 236
Cooper chanel, 269
Cooper pairs, 10, 270
Correlation
    energy, 48, 373, 402ff.
    frequency, 358
Coulomb
    coupling, 50, 55, 94, 102,
    113, 360
    energy, 15, 98, 113, 138, 145,
    284, 373
    interaction, 2, 17, 60, 91,
    113, 126, 137-166, 212, 248,
    268, 283, 334, 351, 360-369,
    381, 401, 408
    potential, 274
    repulsion, 28, 91, 137, 198,
    280, 288, 342, 361
Coupling
    dipolar, 213, 356-359
    electron-electron, 92
    electron-exciton, 233, 290
    electron-phonon, 2, 21, 92,
    116, 161, 267, 275, 292, 369,
    375
    interaction, 3, 15, 30, 49,
    87, 111-115, 126, 172, 184,
    187, 255, 274, 293, 315, 323
    lattice-electron, 316
    phonon, 159
Creation operator, 93, 259
Crossover temperature, 358
Crystal field splitting, 399
Crystal field transition, 213
Curie tails, 359
    behaviour, 77
Cut-off field, 357
Cyanine dye, 281, 286, 294
Debye
    frequency, 258
    temperature, 70, 258-260

Defects, 68, 78, 106, 121, 127,
    168, 170, 177, 185, 189, 226,
    230-233, 299, 303, 310, 348
Density
    oscillations, 271
    of states, 54, 70, 80, 90, 98,
    175, 184, 275, 280, 290, 351,
    353, 359, 371, 383
Diacetylenes, 298-314
Dichalcogenides, transition me-
    tal, 341 - 386
Dicyanamide, 234
Dicyanoargentate(I), 234
Dielectric
    constant, 109, 286, 288, 310
    response function, 89, 117, 176
    regime, 126
Diffuse scattering, 68, 95, 104,
    109, 119, 126, 137, 148, 318,
    344
Diffusion constant, 357
Diffusive motion, 358
Diffusive streaking, 60
Diimine complexes, 207, 211
Dioxalatoplatinates, structure,
    225-231
Dioximato complexes, 205, 207, 393
    partially oxidized, 406
Depinning, 121, 125
Dipolar shifts, 210, 219
Direct process, 272
1,3-diselenole-2-selone, 37, 39
Dislocation, 310
Disorder, 28, 31, 48-53, 226, 234,
    240, 355, 384-388, 407
Dispersion
    curves, 202
    of spinwaves, 202
    relation, 139, 148
Distortion
    19, 21, 91-94, 229, 236, 329,
    334, 344, 373-377
    dynamical, 98-101, 110
1,3-dithiole-2-thiones, 32, 33
Dithiolato metal complexes, 236,
    393
DMTTF-TCNQ, 27, 33, 62, 66
Donor, 27, 30, 211, 212, 240-244,
    391, 395, 401

Drude analysis, 105, 109, 179
DSDTF-TCNQ, 30, 60, 62, 76, 118, 120
Dyes, 280, 286
Dynamical effects, 22
Dynamic nuclear polarization, 357
Elastic deformation energy, 20, 213
Electric dipole transition, 203, 206, 213, 214
Electron
  affinity, 399
  correlation, 214
  creation operator, 93
  density, 115, 343
  diffraction, 170, 171, 376, 386
Electron-electron
  collision, 361
  correlation, 408
  interaction, 2, 17, 19, 30, 49, 64, 79, 92, 93, 136, 158, 253, 260, 267-269, 276, 290, 343, 390, 404
  exchange, 198
Electron-exciton interaction, 282-290
Electron gas, 19, 20, 49, 267, 276, 315
Electron-hole
  excitation, 142, 143, 152, 212, 268
  pairs, 19, 63, 179, 270, 404
  pockets, 19, 174
  scattering, 70
Electronic
  cutoff, 272
  excitation, 20, 100, 106, 202, 215, 219, 308, 392, 394, 397, 398, 402, 403, 407
  polarizability, 49, 71
  stabilization, 396
  transition, 96, 206
Electron-phonon
  collision, 275, 361
  interaction, 16, 50, 61, 98, 116, 138, 142, 156-159, 182, 198, 216-219, 267, 276, 284, 286, 341-343, 375

Electron-phonon
  scattering, 177
Electron spin resonance, 61-65, 121, 246, 303, 360
Electron transfer, 211
Electron tunneling, 354, 360
Elementary particle excitations, 13,
Energy
  gap, 10, 77, 100, 102, 109, 119, 157, 260, 343-345
  level diagram, 353, 405
  loss experiments, 180
  spectrum, 6, 10, 142
Enthalpy of transition, 385
Entropy, 4
ESCA measurements, 152, 380, 381
Etch-pit technique, 310
Exchange
  ferromagnetic, 201
  integral, 201, 203, 204
Exchange integral, nearest neighbour, 202, 219
Exchange interaction, 145, 280, 403
  antiferromagnetic, 10, 140, 158, 201
Exchange of electronic excitation, 279
Excitation
  magnetic, 202
  spectrum, 140, 272
Excited state, 8, 10, 52, 197, 207, 219, 354
Exciton
  156, 202, 203, 210, 212, 215, 219, 279, 284
  attraction, 293
  band, 283-287
  conduction electron coupling, 37
  exchange interaction, 280, 283
  interaction, 293
Excitonic
  interaction, 207
  model, 257, 280
  polarizability, 156, 161
  screening, 145, 156
  superconductor, 277, 279-296, 409

Exciton
  magnon combination band, 204
  mechanism, 279, 280, 290
  system, 290
  wavefunction, 212
Extended Huckel tight binding
  methods, 172
Far infrared spectra, 109, 110,
  116, 126
Faraday method, 360
Fermi energy, 49, 51, 52, 55, 146,
  172, 174, 182, 226, 260, 275,
  280, 343, 351, 354, 371, 375
Fermion model, 361
Fermions, 5, 13, 15, 21, 140, 142
Fermi
  screening, 287
  sea, 8, 10
  statistics, 13
  surface, 5, 10, 28, 49, 70, 80,
  90, 94, 98, 140, 152, 154, 174,
  176, 184, 199, 268, 270, 279,
  283, 290, 344-355, 374, 375,
  383, 384
  velocity, 10, 100, 271, 292,
  355
  wave-vector, 5, 98, 139-142,
  316, 331, 333, 349
Ferromagnet, linear, 204
Fibrous morphology, 168-189
Field-theory model, 9-16
Filamentary compound, 286
Fluctuating currents, 262
Fluctuations
  3, 4, 7, 17, 48, 76, 79, 94-
  127, 187, 257-266, 268, 274,
  276, 293, 342, 358-361, 370,
  388
  amplitude, 102-104, 264-265
  one-dimensional, 94, 99, 100,
  104
  thermodynamical, 267, 274
Fourier transform, 5, 271, 357
Fractional
  band filling, 30, 31, 103
  charges, 403-408
Franck-Condon principle, 213
Free electron, 91, 359

Free
  energy model, 4, 17, 102, 112,
  113, 261, 262, 377, 385
  radicals, 25, 397, 404, 407
Frenkel excitons, 203, 209-217,
  219
Fröhlich state, 102-126
Gadolinium chloride, 217
Galvanomagnetic techniques, 356
Gap
  13, 105, 145, 150, 157, 331,
  333, 344, 351, 375, 376, 383
  covalency, 351, 354
  function, 283, 289
  insulating, 49, 51
"g-ology", 268-277, 361
g_2-process, 272
Granular
  metals, 187
  superconducting films, 186
Graphite, 252
Guinier powder pattern, 229
g-value, 61, 77, 126, 138
Hall effect, 179, 261, 347-351,
  360
Halocarbonyliridates, 225, 229,
  230
Hartree-Fock approximation, 375
Hartree potential, 54
Heat capacity, 29, 67, 76
Heisenberg chain, 149, 159
Heitler-London configuration, 202
Heterogeneous growth, 304
High-spin
  complexes, 399
  state, 388, 401
High-temperature
  conductivity, 30, 80
  resistivity, 70-76, 297
  superconductor, 1, 2, 259, 279
Hexamethyltetrathioflulvalenium-
  TCNQ, 27, 30, 38, 47, 85, 160,
  189, 255, 342-361
Hopping
  2, 7, 8, 15, 20, 48, 88, 268,
  275, 358
  transverse, 357
  virtual, 10

Hubbard model, 5, 51, 138-162, 198, 200, 208, 216, 268, 357
Hybridization, 7, 19, 145
Hysteresis, 112, 240
Imperfections, 187, 203
Impurities, 3, 17, 78, 102, 106, 121, 126, 189, 276, 356-359, 383, 388
Inhomogeneities, 187
Inorganic
  chain compounds, 197-231, 391-409
  complexes, 197-231, 391-409
Instabilities
  49, 79, 89, 199, 257, 269, 280, 290, 375, 383
  inherent, 269
Insulating oxides, 200
Insulating state, 172, 198, 207, 342
Insulator
  20, 25, 28, 47, 109, 213, 215, 393, 402
  antiferromagnetic localized, 199
  magnetic, 198, 201, 219, 342, 393-402
  transition, 235
Interacting metal atoms, 197
Interaction
  magnetic exchange, 201, 213
  operator, 209
  interchain, 50, 53, 170, 172, 189, 274
  interionic, 197, 207, 219
  π-π, 48
  parameters, 270
  processes, 269
Interband transition, 105, 117, 179
Intercalation, 388
Interchain
  bond lengths, 57, 80, 113, 170, 176, 189
  charge-transfer integrals, 80
  contacts, 71
  Coulomb interaction, 55, 275
  force constants, 184
  lattice coupling energy, 182
  scattering, 63
  separation, 204

Interlayer forces, 371
Intermolecular
  electron exchange, 208
  spacing, 323, 332
Intersite repulsion, 7, 8, 10
Intraband absorption, 144
Intrachain
  effects, 58, 64, 67, 204
  interaction energy, 204, 392
  optical phonon modes, 182
Intramolecular excitation, 144
Intrinsic anisotropy, 182, 186, 189
Intrinsic bandgap, 177, 207
Ionization potentials, 200, 399
Ionicity difference, 381
Iridium
  acetylacetonates, 205, 207
  carbonyl halides, 229-231, 236, 238, 393, 406, 407
$K_2Pt(CN)_4Br_{0.3}\cdot3.2H_2O$ (KCP)
  19, 88, 176, 217, 220, 225, 238, 274, 277, 282, 286, 315, 342, 370, 387, 391, 403, 406
Kernel, 283-286
Knight shift, 100, 104, 112
Kohn anomaly, 18, 88, 148, 157, 184, 315, 319, 331, 333, 374
Korringa value, 357, 359
Kramers-Kronig analysis, 109, 120, 310
Landau damping, 79
Landau-Peierls susceptibility, 348, 349, 360
Lattice binding forces, 189
Lattice vibration, 19, 20, 374
Layered compounds
  3, 343, 346, 369-390
  van der Waal's bonding in, 371
Layered chalcogenides,
  polymorphic forms, 371
  polytypes, 381
Layered transition metal dichalcogenides, 369-390
Ligand-bridged
  network, 396, 397, 404
  transition metal crystals, 402

Ligand field
    excited states, 197, 202, 203, 206
    spectra, 203
Linear
    antiferromagnet, 204
    phase phonon, 103
    polymers, 297-314
Lindhard function, 90, 91, 95
Little model compound, 156, 161, 298
Localization, one-dimensional, 88
Localized electrons, 333, 336
Long chain molecule, 297
Long range
    attraction, 295
    forces, 4
    order, 4, 17, 50, 96, 111, 127, 262-264, 270, 326
    triplet correlations, 8
Lorentzian oscillator, 110, 119
Lorentz transformation, 11
Low frequency behaviour, 271, 272
Low spin state, 338
Low temperature
    effects, 323-331, 342-360
    phase, 3, 199, 329, 361
Macromolecules, 305
Magnetic circular dichroism spectroscopy, 24
Magnetic
    properties, 99, 145, 234, 347, 351, 403
    susceptibility, 21, 64-80, 98, 100, 119, 137-145, 233-240, 370, 372, 384
Magnetoresistance, 180, 181, 347-350, 360
Magnons, 203, 213
Magnus' Green Salt, 197, 198, 205-207, 399
Mean field theory, 16, 50, 94, 99, 102, 114, 119, 263, 267-275, 293, 342, 358, 360
Mean free path, 28, 68, 88, 127, 350, 355, 358
Meissner effect, 187
Merocyanine, 294

Metal chain compounds, 197, 223, 225-231, 391-409
Metal-metal
    direct contact, 200, 201
    interaction, 197, 216-219, 280, 294, 391, 395
    spacing, 407, 408
Microwave studies, 126
Migdal approximation, 292
Mixed valence compounds, 30, 197-223, 244, 361, 392-404
Mobility, 77, 123, 143, 347, 355
Modulation, 7, 19, 316, 334, 344, 356
Molecular crystals, 208, 209, 391-401
Molecular excitation, 391-401
Momentum transfer, 150, 268, 285, 291
Mößbauer measurements, 386
Multipole expansion, 209
Naphthalene, 395
Néel ground state, 158
Neutron scattering, 7, 8, 17, 19, 29, 71, 78, 88, 95, 97, 110, 118, 137, 140, 148-158, 182, 184, 202, 216, 217, 225, 315-339, 345, 346, 376, 377, 378
Niobium sulfide, 371
Niobium selenide, 376-380
NMR measurements, 126, 137, 161, 344, 377
NMP-TCNQ, 28, 51, 342-345, 398, 406, 407
Nonintegral
    electron occupancy, 392, 408
    oxidation states (NOS), 392, 403, 404
Non-interacting electrons, 6, 137, 139, 374
Nonlinear interaction, 11, 13
Nonlinear Bragg scattering, 17
Nonlinearity, 122, 123
Nonstoichiometric compounds, 197-231, 391-409
Nuclear Relaxation rate, 98, 126, 147, 356, 357
Nucleation, 303, 305

OMTSF, 27, 31, 38
OMTSF-TNAP, 31
One-dimensional
    correlations, 216, 319
    effects, 15, 342
    electron gas, 1, 3, 268, 271
    Fermi system, 268
    "g-ology", 269
    inorganic solids, 197-231, 391-409
    instability, 88
    magnet, 202, 391
    regime, 318, 323
    superconductivity, 263, 279-296
One-electron band theory, 92, 139
Optical
    excitation, 137, 391, 399, 401-406
    modes, 70
    properties, 176
    phonon screening, 156
    scattering time, 126
    spectra, 177, 180, 185, 203, 214
    transition, 139, 141, 395
Orbital diamagnetism, 67, 76
Ordering, 3, 22, 270, 329, 334
Order parameter, 3, 4, 71, 101, 112, 113, 263-265, 274, 342
Ordering temperature, 114
Ordering transition, 96
Organic
    CT-salts, 25-45, 47-85, 87, 137, 233-248, 393, 397, 399, 401
    metals, chemistry of, 25-45
    radicals, 248, 409
OPW (Orthogonalized-plane-wave)
    calculation, 172-176
Oscillator strength, 106-126
Overhauser instabilities, 269
Overlap, 147, 211, 369
Oxidation potentials, 30
Oxidation waves, 244-246
Packing properties in diacetylenes, 299-301
Paraconductivity, 267
Paramagnetism in complexes, 342, 399-403
Paramagnon, 361

Parquet diagrams, 269
Partially oxidized Pt-complexes, 225-231, 280, 391-407
Particle-hole channel, 269, 270
Particle-particle channel, 269, 270
Pascal's constants, 64
Peierls distortion, 19, 20, 49, 50, 87-135, 142-159, 167, 172, 174, 199, 267-276, 315, 326, 333, 341-346
Peierls-Fröhlich CDW, 87-135, 157
Perovskites, 201, 203, 205
Phase
    fluctuation, 102, 115, 262, 264
    locking, 16, 21
    transformation, mechanism of, 304
    transition, 1-23, 29, 30, 60-85, 94, 121, 126, 138-161, 199, 263, 264, 268, 270, 280, 303, 306, 310, 315, 326-328, 342, 351
    variation, 264
    velocity, 292
1,10-phenanthroline complex, 281
Phonon
    2-22, 63
    acoustic, 148-150, 182, 318
    anomaly, 19, 20, 68, 97, 152, 160, 319, 328, 334
    branch, 14, 328
    condensed, 328
    dispersion, 63, 319, 321, 374
    frequency, 50, 78, 94, 116, 118, 275
    longitudinal, 374
    low frequency, 79, 279, 292, 293, 341
    mode, optical, 79, 150, 156, 159
    modulation, 20, 61, 98
    optical, 79
    scattering, 29, 127, 149, 150
    side bands, 310
    softening, 63, 64, 95, 267
    spectrum, 18, 88-95

Photoconductivity, 208, 212, 312, 313
Photoemission response function, 176
Phthalocyaninatolead(II), 394
Pinning, 94-127
Pinning frequency, 94-127
Planar complexes of transition metals, 225-231, 394-395
Plasma edge, 87, 88, 107, 109, 147, 156, 179, 219, 229
Plasma frequency, 88, 116, 161, 179, 351
Plasmon, 8, 91, 373
Plasma tensor, 176
Point-dipole-approximation, 401
Polarizability, 28, 89, 90, 293, 307, 394
Polarization, 2, 19, 323, 328,399
Polarizable dye, 281, 290
Polaron binding energy, 156
Polaron theory, 79
Polydiacetylenes, 297-314
Polyformaldehyde, 303
Polymer
    backbone, 167-192, 297, 397
    crystal, 303, 305
    single crystal, 167, 297-313
Polymeric compounds, 244, 252
Polymeric solid, 189, 297-314
Polymerization, 253
Polymethyleneditelluride$(CH_2Te_2)_n$, 253
Porphyrins, 393
Potential, random, 48
Precursor effect, 96, 98, 318,331
Pressure effect, 187, 188, 210, 341-367
Prussian blue, 215, 216
Pseudo-gap, 99-106, 145
Pseudohalide salts, 234
Pyrazine complexes, 204, 214
Pyrene analogues, 244-246
Quinolinium-TCNQ, 28, 51, 406
Quinoxaline complexes, 204
Radical
    ion salts, 25-45, 233-252
    neutral, 248
Raman spectra, 182, 187, 308

Random potential, 48, 49
Random walk, 357
Reflectivity
    87, 107, 117, 179, 310
    infrared, 182, 379
    polarized, 87-108, 211
Resistance fluctuations, 73,74
Resistivity, 29, 68-78, 121, 167, 177, 259, 264, 303, 340-361, 381-387
Response function, 49, 89, 90, 142, 271-273, 315
Retardation effects, 276
Rhodium(I) complexes, 207
Rotons, 20, 21
Rutile lattice, 198, 205
Salicylaldiminato complexes, 393
Superlattice reflexions, 95, 96, 316-332
Satellites, $2k_F$, 95-100, 126, 322-330, 346
Satellites, $4k_F$, 97, 126, 319-334, 351, 361
Scale temperature, 112, 115
Scanning electron microscope pictures, 168
Scattering processes, 88, 179, 186, 268, 269, 272
Screening, 91, 94, 155-166, 256, 286, 287, 359
    metallic, 92, 156
Screw symmetry degeneracy, 172, 173
Segregated stacks, 25, 28, 31, 399, 401-408
Selenium derivatives of TTF, 30, 33-40, 60, 61
Selones, 37
Single particle excitation, 105
Single particle oscillator strength, 106, 115
Singlet pairing, 270
Semiconductor
    25, 51, 68, 77, 110, 123, 214, 217, 219, 391-409
    paramagnetic, 399, 402
Semimetal, 76-80, 189, 347-355
Selenium-nitrogen contacts, 58, 354-359

Se_4N_4, 190, 253
S_4N_4, 190
$(SeN)_x$, 190
Singlet superconductivity, 3, 7, 238-291
Sinusoidal modulation, 329
Site energy, 392
Sliding conductivity, 267
$(SN)_x$, 16, 88, 167-195, 253, 303, 342-344
   films of, 185
   imperfections in, 168
   hydrogen impurity in, 186
   β-monoclinic phase, 170, 171
   β-phase, 170
   orthorhombic phase, 170
   superconducting properties, 168
S-N separations, interchain, 57, 58
Softening of phonon modes, 21, 187, 361
Solid state
   syntheses, 297-314
   transition, 234, 299
Solitons, 10-14
Space filling structure, 395
Specific heat, 115, 182-188
   anomaly, 115, 187, 326
Spin
   correlation function, 160, 356, 357
   degeneracy, 139
   density wave (SDW), 3-15, 49, 87, 158, 269, 270
   distribution, 25, 26, 143
   excitation, 139-160
   flip scattering, 10, 63
   lattice relaxation, 64, 356
Spinless fermions, 152
Spin-orbit coupling, 61, 357
Spin-Peierls anomaly, 155-159
Spin-Peierls transition, 149, 150, 159, 236
Spin relaxation time, 61, 63
Spin susceptibility, 64
Spin wave excitation, 8, 148
Spin waves, 49, 137-161, 202, 331-336
Spin wave phonon interaction, 98

Spin wave scattering, 149, 150
Square planar d^8 metal complexes, 205-207, 215-231, 393-409
Stacks, 1, 143, 234-238
   mixed, 25, 248, 399, 401-408
   molecular, 317-335
Stereoalloys, 240
Strong coupling limit, 7-13, 97, 98, 331, 336
Superconductivity, 3, 7, 10, 11, 16, 127, 167-193, 257-259, 267-271, 275, 279-296, 388
Superconducting fibers, 187
Superconducting fluctuations, 157, 161
Superexchange, 397, 402, 403
Superfluidity, 3, 7, 295
Superlattice
   71, 88, 95, 199, 226, 326-329
   distortion, 20, 21
   incommensurate, 88, 95, 98, 111, 152, 328, 382
Susceptibility
   8, 61-65, 99-115, 147, 233-240, 270, 347-349, 359, 375
   Bonner-Fisher, 146
   Pauli, 8, 64, 100, 103, 104, 146, 348, 349
Tantalum
   sulfide, 377
   selenide, 377-390
TCNQ-salts, 25-166, 233-256, 315-339, 403-406
Temperature, characteristic, 234
Ternary compounds, 199
Tetracyanoplatinates
   cation deficient, 226, 229, 406
   optical absorption spectra, 210
   structure, 225-231
Thermal expansion, 187
Thermoelectric power, 76, 180
Thiographite $(C_4S)_x$, 252, 259
Three-dimensional order, 204, 323, 334, 344, 360
Tight-binding-approximation, 53, 95, 137, 172, 179, 284, 356

TMMC, 203, 204, 401
TMTSF-TCNQ, 25-45, 52-82
TMTTF-salts, 27, 33, 55-80, 118, 120, 240, 360
TNAP, 27, 80, 244
Topochemical polymerization, 297-314
Transfer integral, 137, 172, 198, 351, 356, 369
Transition
    7, 71, 234, 333, 343, 376, 381, 387
    allowed, 211
    antiferromagnetic, 17
    band-band, 207
    d-d, 208, 213, 394, 399, 401, 403, 406
    density, 284
    first order, 73, 198, 372, 376, 381
    metal-insulator, 3, 28, 64, 73, 80, 88, 236, 315, 319, 341, 344
    metal-semiconductor, 346, 348
    structure sensitive, 399
Transition metal
    complexes, 197-223, 225-231, 236, 391-412
    dichalcogenides, 370
Transition temperature, 3, 16, 19, 80, 112, 119, 167, 185, 240, 268-275, 279, 283, 341, 344
Transport
    176, 189, 234
    nonlinear, 121, 122
    properties, 189, 234, 383
Triplet
    exciton systems, 398, 401, 406
    pairing, 270, 277
    state, 3, 10, 271, 277, 303
    superconductivity, 3, 8, 288-295
TSeF-TCNQ, 27, 30, 38, 60, 66, 76, 105, 118-120, 161, 323, 342-344, 354
TSeT, 244
TTF
    alloys, 105, 120, 236, 258
    salts, 25-166, 233-252, 315-367

TTF
    salts, methylated, 33, 240
    stacks, 111, 113-115, 234-237, 248
    metal complexes, 236, 238
    halides, 234, 239, 323
    TCNQ, 2-21, 25-168, 176, 234, 252, 315-367
TTF(D_4)-TCNQ, 359
TTF-TCNQ(D_4), 357-359
TTT, 244, 255
TTN-salts, 244, 255
Two dimensional metals, 369-390
Two dimensional systems, 342, 369-390
Two particle excitation, 203
Ultraviolett photoemission experiments, 184
Umklapp
    induced correlations, 20
    process, 268
    scattering, 10, 11, 20, 177
Valence
    delocalization, 216
    disproportionation, 381
Van der Waal's interaction, 55-58, 258, 297
Vertex corrections, 230, 290, 293
Vibrational modes, 219
Vibronic mixing, 206
Viscosity measurements, 299
Vanadium(IV)oxide, 198, 200
Vanadium(V)oxide, 341
Voltage
    autocorrelation function, 71
    thermoelectric, 356
    power, 105, 177
Voltammetry cyclic, 30
Wave packet, 261
Wigner crystal, 142, 154, 160
Wigner crystal limit, 138, 140
Wigner wave antiferromagnetic, 158, 159
Wolffram's Red Salt, 214-216, 404

X-ray investigations, 7, 17, 29,
    60, 95-98, 152, 157, 168, 170,
    225-231, 315-339, 344, 351,376
X-ray photoemission, 134
Zero gap, 354
    semiconductor, 53
"Zero-sound" channel, 269
Zig-zag chains, 226
Zwitterionic structure, 248